Silicon Nanophotonics
Basic Principles, Present Status and Perspectives

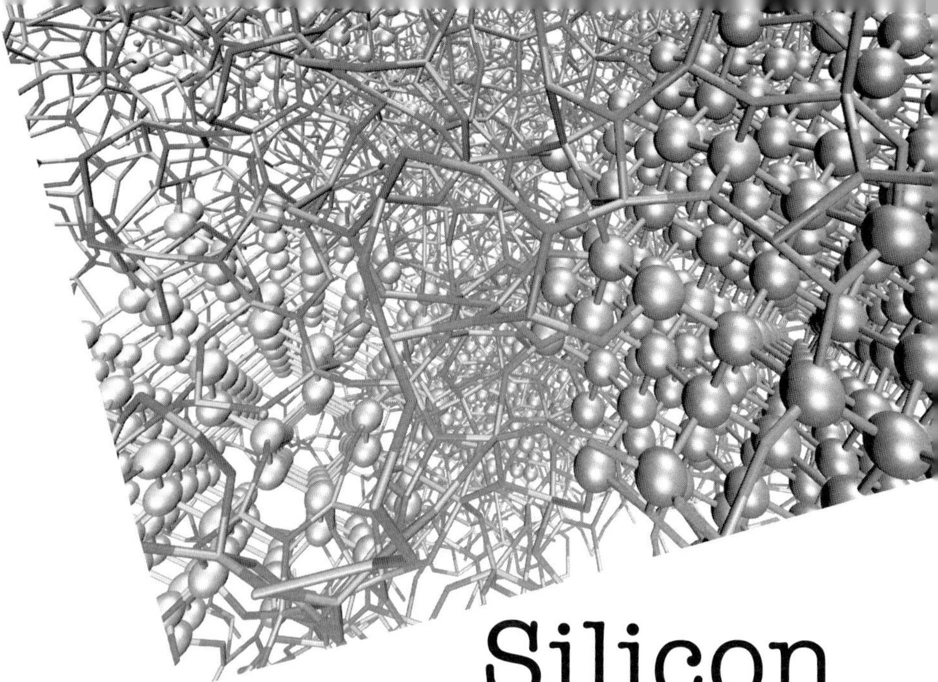

Silicon Nanophotonics

Basic Principles, Present Status and Perspectives

Editor

Leonid Khriachtchev

University of Helsinki, Finland

Pan Stanford Publishing

Published by

World Scientific Publishing Co. Pte. Ltd.

5 Toh Tuck Link, Singapore 596224

USA office: 27 Warren Street, Suite 401-402, Hackensack, NJ 07601

UK office: 57 Shelton Street, Covent Garden, London WC2H 9HE

British Library Cataloguing-in-Publication Data
A catalogue record for this book is available from the British Library.

SILICON NANOPHOTONICS
Basic Principles, Present Status and Perspectives

ISBN-13 978-981-4241-11-3
ISBN-10 981-4241-11-3

Printed in Singapore by B & JO Enterprise

To my daughters

Vera and Ksenia, with love

Leonid

PREFACE

Nanoscience is a rapidly developing area of research which promises a lot in physics, chemistry, and medicine, and some of the ideas have been already realized. Nanoscale materials are particularly interesting for photonics, which can be defined as the science and technology of light. Photonics supplements electronics in the form of optoelectronics, and it is considered as one of the key technology areas of the 21st century. Silicon is the leading material for electronics; hence integration of all optical functions into silicon technology is practically very important and widely recognized as a great challenge. This book combines the concepts of nanoscience, photonics, and silicon technology. A lot of research activity has been carried out in these fields, and it is impossible to cover all aspects. Our book presents a special viewpoint of Silicon Nanophotonics, and the content is mainly limited by photonic properties of silicon nanocrystals and by closely related topics. We believe that silicon nanocrystals offer a promising practical perspective for photonics and the related materials are exciting also from the fundamental and educational points of view.

Research on silicon nanocrystals was strongly activated by Leigh T. Canham who discovered in 1991 bright visible emission from porous silicon. Very many studies have been devoted to understanding of the light-emission mechanisms and a number of models have been suggested. An important opinion was published by Philippe Fauchet and co-workers in 1999 when they provided strong arguments in favor of surface origin of the light emission from oxidized porous silicon. Light amplification (optical gain) in silicon nanocrystals in silica was reported first by an Italian research group led by Lorenzo Pavesi. Indeed, generation of light in silicon is a challenging perspective; however, the issue of a laser and other light emitting devices limits neither the activity in the field nor the contents of this book. The studies cover light modulators, optical waveguides and interconnectors, optical amplifiers, detectors, memory elements, *etc*.

The present book collects recent results of a number of groups worldwide. The contributors of our book work in United States, Japan, and eight European countries. The book contents include: (i) Basic principles of the most important photonic elements based on silicon nanocrystals; (ii) Theoretical analysis of optical properties, light emission and optical gain of silicon nanocrystals; (iii) Experimental studies of the most important phenomena and optoelectronic properties of silicon nanocrystals such as light emission, optical gain and lasing, structure, optical properties, optical waveguiding, optical and electrical memory. The experimental results are illustrated by simple modeling; (iv) Experimental methods (transmission electron microscopy, Raman spectroscopy, *etc*.), preparation technique (molecular beam deposition, laser ablation, ion implantation, *etc*.), and sample architecture (silicon-rich silicon oxide films, Si/SiO_2 superlattices, free-standing films) described in appropriate places; (v) Silicon-based material with additional doping (Er-doped silicon nanocrystals and SiN materials) and single silicon dots; (vi) Perspective applications and some related topics. The authors present rich bibliography helping further reading. Some overlap between the chapters is inevitable; however, this allows the chapters to be understood independently. Some differences in opinions and interpretations between the authors can be found, which is also understandable for this hot and quickly developing field. In any case, we have tried to indicate in our book where the field is now and where it is going. We hope that this information will be useful for a broad readership including young researchers coming to the field of nanoscience and nanotechnology.

The Editor thanks all contributors for accepting his invitation to participate in the book and writing exciting stories.

Leonid Khriachtchev
Editor
January 21, 2008

CONTENTS

CHAPTER 1

SILICON NANOCRYSTALS ENABLING SILICON PHOTONICS

Nicola Daldosso and Lorenzo Pavesi

Nanoscience Laboratory, Physics Department, University of Trento, via Sommarive 14, Povo 38050, Trento, Italy

Silicon Photonics is an emerging field of research and technology, where nano-silicon can play a fundamental role. In this chapter, the main building blocks of Silicon Photonics (waveguides, modulators, sources and detectors) are reviewed and compared to their counterparts made by Si nanocrystals. In addition, non-linear optical effects in Si nanocrystals which will enable fast all-optical switches are presented as well as our recent research efforts to obtain optical amplification at 1550 nm by using Er ions and the sensitizing properties of Si nanocrystals.

1. The Need of a Silicon Photonics

Optical communications, optical storage, imaging, lighting, optical sensors or security are just a few examples of the increasing pervasion of Photonics into the day life. The world market for Photonics is larger than the one of semiconductors and of the automotive industry. Photonics is getting also more and more importance in electronics since it can take pace with both the "more-Moore" and "beyond-Moore" evolution trends of microelectronics. These are dictated by the requests posed by speed, signal delay, packaging, fanout, and power dissipation of nowadays multiprocessors and memories where ever-increasing chip sizes, decreasing feature sizes and increasing clock frequencies beat the physical limitations of electrical signaling. By manipulating photons instead of electrons, some of the limitations placed on electronic devices may be overcome. Integrated optics is capable of signal splitting and combining, switching and amplification; the last function being a key component in compensating transmission, insertion and distribution losses. Even if photonics could bring new functionalities to electronic components as low propagation losses, high bandwidth, wavelength multiplexing and immunity to electromagnetic noise, the high cost of photonic components and their assembly is a major obstacle to their deployment in most of application fields.

Silicon photonics (or better CMOS Photonics) is a viable way to tackle the problem by developing a small number of integration technologies with a high level of functionality that can address a broad range of applications.

1.1. *Silicon Photonics*

The basics of Silicon Photonics have been pioneered by Soref[1,2] across 1980s - 1990s, but only in these last years a consistent number of breakthroughs have been achieved.[3,4,5,6]

In the world of Silicon Photonics, different approaches of integration have been developed during the years. These differentiate by the integration degree. The first, where Si is only used to channel the light signal, was pioneered by Bookham Technology.[7] It is comparable to the silica on silicon technology, where waveguides have large cross sections. This technology, with waveguide dimensions typically in the μm range, is actually used by Kotura[8] for their products. A few components developed by INTEL were also based on this technology.[9] A second approach is based on a hybrid technology where silicon, germanium and III-V semiconductors are integrated together. Based on this, a device has been recently released by Luxtera Inc.: a monolithic optoelectronic Optical Active Cable assembly containing four complete fiber optic transceivers per end, each operating at data rates from 1 to 10.5 Gbps and supporting a reach up to 300 meters.

If we move to academy research, it has to be remarked that from 2000, most of the work has been devoted to Silicon Photonics components and systems where waveguide dimension are in the sub-micrometer range. In the following, we will briefly review the main building blocks of Silicon Photonics with the only exception of an all-silicon injection laser, since it has still to be demonstrated.

1.1.1. *Waveguides*

Optical waveguides (WG) are fundamental components of integrated optical circuits because they provide the connections among the various devices. Bending radius and device size scale down with the refractive index contrast (Δn), while scattering losses increase proportionally to Δn^2.[10] The optical absorption is mainly characterized, at least in semiconductors materials, by the inter-band transitions and by free-carrier absorption. For glass or dielectric waveguides, absorption losses are due to molecular bonds, usually associated to hydrogen content in the core layer. Therefore the choice of the waveguide material determines the wavelength of the signal, the integration density and the minimum intrinsic losses. A natural choice is to look for dielectrics and/or semiconductors already used in microelectronics: Si oxynitride (SiON) and Si nitride (Si_3N_4), Si on insulator (SOI) and Si nanocrystals in Si oxide (SiO_2+Si-nc). The appeal of Si oxynitride WGs stems from the tunability of their refractive index contrast and their transparency over a wide wavelength range, including the visible. Propagation losses in the visible range as low as 0.1-0.2 dB/cm have been found in silicon nitride waveguides,[11,12,13] while losses in the NIR are limited essentially by the residual stress which limits the growth of thick core layer (in order to get large confinement factor) by LPCVD and by molecular absorption (mainly OH) for PECVD grown layer due to the gas precursor. Strain release and control is possible by using a multilayer structure where alternating Si_3N_4 and SiO_2 layers allows thick cores, as shown by Melchiorri *et al.*[14] In these structures the propagation losses were about 1.5 dB/cm at 1544 nm thanks to a large optical mode confinement factor and to the good quality of the interfaces.

As for the loss figure, SOI waveguides have the best performances in the near infrared (NIR) range due to a low optical absorption. Propagation losses as low as 0.4 dB/cm at 1523 nm have been reported.[15] Scattering losses can be made negligible by improving the waveguide processing, while free carrier and defect absorption related losses are intrinsic. Free carrier absorption related optical losses have a limit of about 0.33 dB/cm at 1523 nm in large mode WG. One way to reduce it is to decrease the free carrier lifetime by reducing the mode size of the WG and, hence, reducing the lifetime by surface recombination. In addition, since the refractive index contrast is very large in a SOI WG (at 1550 nm, the Si refractive index is about 3.5, against 1.45 of Si oxide) small mode size WGs keep a large optical confinement factor. Losses in the range of 0.1–3 dB/cm depending on the dimensions and processing conditions have been obtained. Extremely small mode size SOI waveguides are usually called Si wires. Si wires as small as 0.1 μm^2 have been fabricated at IMEC and at IBM with losses lower than 3 dB/cm opening the possibility of realizing photonics structures on the same scale of CMOS VLSI.[16,17]

1.1.2. *Modulators*

Silicon is a centro-symmetric material and, hence, has no electro-optic effect. The only way to achieve a modulator is to use the free-carrier effect where the free carrier concentration is controlled in a *pn* junction by injection, accumulation or depletion.[18] In 2005, University of Surrey proposed a four terminal p^+pnn^+ vertical modulator integrated into a SOI rib waveguide, based on carrier depletion in a *pn* junction formed in one arm of a Mach-Zehnder interferometer.[19] In 2007, based on a similar design, INTEL developed a high speed and high scalable optical modulator based on depletion of carrier in a vertical *pn* diode showing data transmission up to of 30 Gbit/s at 1.55 μm.[20] Recently Luxtera Inc.[21] and D. Marris-Morini *et al.*[22] realized a lateral modulator either with *pn* or *pipin* structure both achieving about 10 GHz roll-off frequency and insertion losses of 3 and 5 dB, respectively. Due to the low refractive effects, these modulators have to be mm-long. In order to reduce the devices dimensions optical resonators can be used with the caveat that the wavelength range is reduced with respect to Mach-Zehnder modulators.[23,24] As an example, Lipson *et al.*[25] reported compact device using a ring resonator (10 μm diameter) with comparable performances.

1.1.3. *Sources*

At the present, the only viable technology for an on-chip light source is a hybrid technology where III-V semiconductors are used. Some convergence is appearing towards the use of InP-based materials. All the work done in the nineties on the heterogrowth of III-V on silicon proved to be unsuccessful and, nowadays, integration is done by bonding the III-V layer on top of the silicon layer. Two main approaches are followed. The first, pioneered by the work of the PICMOS consortium, aims at integrating InP μlasers on top of a silicon lightwave circuitry. Here the active layer is bonded to silicon and then it is processed to a laser. The other approach uses the concept of evanescently coupling an active layer to a silicon optical cavity.[26] The laser is thus self-aligned to the silicon lightwave circuitry.

In addition to these two successful technologies, other efforts are directed to an all Si-based light source, where the extensive experience in Si fabrication and processing could be put to best use.[3,27] Many think that it will be the light source that will make Silicon Photonics even more appealing than what it is now. The main limitation to the use of silicon as a light source is related to its indirect band-gap, which implies very long radiative lifetimes (ms range). Long radiative lifetimes mean that most of the excited carriers recombine non-radiatively. Moreover, when population inversion is looked for to achieve lasing, high excitation is needed. Under this condition, fast non-radiative processes turn on such as Auger recombination (three-particles non radiative processes) or free carrier absorption, which depletes the population inversion and provides a further loss mechanism. Despite this, many different strategies have been employed to turn silicon into a light emitting material. Some rely on band structure engineering, such the use of SiGe quantum well or Si/Ge superlattices, while others rely on quantum confinement effects in low dimensional silicon. Still another approach is impurity-mediated luminescence from, for example, isoelectronic impurities or rare earth ions. In Table 1, a summary of the different approaches towards a Si-based light source is reported together with their main characteristics.

Table 1. Summary of the different approaches to a Si-based light source.

System	λ (μm)	Results	Ref
High quality bulk Si in a forward biased solar cell	1.1	LED with a power efficiency >1% at 200 K	[28]
Small junctions in a p-n diode	1.2	Stimulated emission observed	[29]
Stimulated Raman scattering in silicon wires	1.6	CW optically pumped Raman laser	[30]
Nanopatterned silicon	1.28	Optically pumped stimulated emission at cryogenic temperature	[31]
Dislocation loops formed by ion implantation in a silicon *pn* junction	1.1	LED with a power efficiency <1%	[32]
Silicon nanocrystals in a dielectric	0.75	High optical gain at room temperature, efficient field-effect LED demonstrated	[33,34]
Er coupled to silicon nanocrystals in a dielectric	1.53	Internal gain demonstrated in waveguides	[35]
Strained Ge on Si	1.55	Theory predicts high gain	[36]

In early 2000 a series of papers paved the way towards a silicon laser.[31,32,33,37] In October 2004, the first report on a silicon Raman laser appeared,[38] while in January 2005 the first all-continuous wave CW silicon Raman laser was reported.[30]

1.1.4. *Detectors*

Photodetectors below 1000 nm are generally made of silicon but for long waveleghts application silicon is transparent. Among CMOS compatible materials, bulk Ge can absorb infrared light over distances of a few microns thanks to its smaller bandgap. High bandwidth and high responsivity Ge photodetector integrated into SOI waveguide have been reported.[39,40] Germanium processing is compatible with complementary-metal-oxide-semiconductor (CMOS) technology.

With bonding technologies either molecular or polymer, InGaAs photodetector coupled to a silicon waveguide have been demonstrated both by IMEC and Technical University of Eindhoven. The measured responsivity was 1 A/W at 1.55 μm with nA range dark current.

2. Nanosilicon for Photonics

The possibility of low dimensional silicon to tune on one side its electronic properties and on the other side its dielectric properties allows for new phenomena and device concepts.[41,42] In this section, the exploitation of low dimensional silicon (i.e. Si-nc) to demonstrate various optical components for an all Si nanophotonics is reviewed. It is worth to note that the main advantage of using Si-nc is to integrate light sources and/or amplifiers within CMOS photonics platform. This is the biggest challenge facing Si nanophotonics.

Various techniques are used to form Si-nc, whose size can be tailored to a few nanometers. Bottom-up approaches rely on the direct chemical synthesis of Si-nc by chemical reactions of suitable precursors. Since the precursors are usually in a liquid phase these methods are mostly suitable for bio-applications. On the contrary, other methods are based on a thermodynamically induced self-aggregation of Si-nc in non stoichiometric dielectrics. Starting from a Si-rich oxide, which can be formed by deposition, sputtering, ion implantation, cluster evaporation, *etc.*, partial phase separation into a stoichimetric oxide and silicon is induced by thermal annealing. The duration of the thermal treatment, the annealing temperature, and the starting excess Si content all determine the final silicon cluster sizes, their size dispersion, and their crystalline nature. Recently, thermal anneal of amorphous SiO/SiO_2 superlattices has been proposed to better control the size distribution: almost monodispersed size distribution has been demonstrated.

Basics of Si-nc, how they are fabricated and their fundamental properties are discussed in the following chapters. Hereafter, we emphasize how Si-nc can serve Silicon Photonics by reviewing performances and possibilities of low dimensional silicon in guiding, modulating and, above all, generating and/or amplifying the light.

2.1. *Si-nc waveguides*

As the Si-nc rich region has an effective Si content larger than SiO_2, its refractive index is larger than that of silica (1.45). Refractive indices ranging between 1.45 and 2.2 at 780 nm have been reported depending on the Si excess content. Hence, Si-nc rich SiO_2 can be used to form the core region of a planar waveguide where the cladding is made by SiO_2. 2-dim waveguides (channel, rib or stripe-loaded waveguides) can be easily fabricated by using standard optical lithography and reactive ion etching. Signal light which travels trough the waveguide can loose power due to various mechanisms: direct absorption in the Si-nc, scattering due to roughness at the core boundaries (both between the core/cladding interface and at the stripe/rib edges), radiation into the substrate, scattering due to the composite nature of the core layer.

Optical losses of 120-160 dB/cm have been reported for Si-nc in the visible range.[43,44] Lower values (about 10 dB/cm) have been reported for thick slab waveguides at 780 nm and about 3.5 dB/cm at 1000 nm, where Rayleigh scattering is decreased according to the well-known $1/\lambda^4$ law.[45] Pellegrino *et al.*[46] found at 780 nm in ion-implanted samples, losses as low as 10 dB/cm (about 2 cm^{-1}), mainly due to scattering and absorption. It is clear that the assessment of the losses strongly depends on the density and the size of the nanostructures. A detailed study as a function of the probe wavelength has been recently performed by us.[47] We studied Si-rich silicon oxide and SiO_2 multilayers samples grown by reactive magnetron sputtering and then annealed at high temperature to induce the formation of Si-nc with mean size of 3-4 nm and density of about 3.5×10^{18} cm^{-3}. Propagation losses decrease with increasing the wavelength from about 73 dB/cm (at 785 nm) to 2 dB/cm (at 1630 nm). An analysis of the different contributions to the optical losses such as Mie scattering and scattering due to waveguide roughness has been done, allowing to isolate the contribution due to the absorption losses and thus to extract the absorption cross section at different wavelengths (see Figure 1). Values of about 3.5×10^{-18} cm^2 have been found at 830 nm, increasing with decreasing the wavelength.

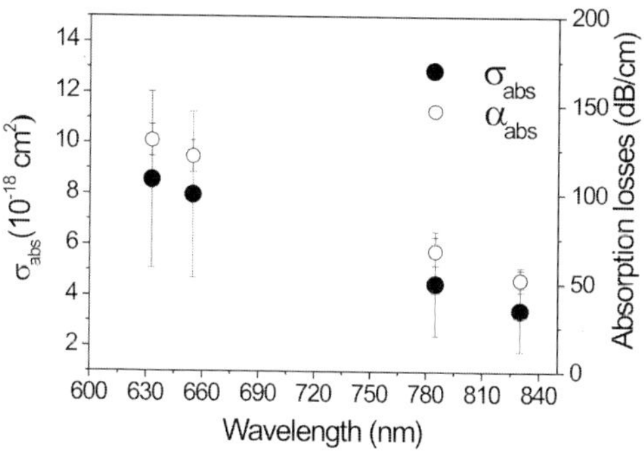

Fig. 1. Absorption losses α_{abs} (empty symbols) and absorption cross sections σ_{abs} (full symbols) as in Ref. 47.

As for passive component the performance of Si-nc waveguides are far from the SOI ones due to the intrinsic composite nature of the system.

2.2. *Non-linear effects: fast optical switches*

Nonlinear photonic materials are widely used in many key-devices for the telecom industry such as switches, routers, wavelength converters. As an example, optical logic gates realized with nonlinear MZI offer a very attractive feature for mass-manufacturing such as scalability and flexibility.

Silicon nanocrystals have been already demonstrated as a promising material for nonlinear applications both in the form of porous silicon and Si-nc embedded in SiO_2.[48,49] At 800 nm, the Kerr coefficient has been shown to be two and four orders of magnitude higher than the one of bulk Si and SiO_2, respectively and the nonlinear properties can be tuned depending on Si-nc size and density. One of the main advantages of using Si-nc in SiO_2 as nonlinear optical material relies on their full process compatibility with mainstream CMOS technology, thus allowing the realization of practical, low cost, compact and low-switching power all-silicon devices. Furthermore, the Si/SiO_2 system provides excellent dielectric properties and excellent stability (chemical, thermal, mechanical). It is worth to note that with respect to SOI

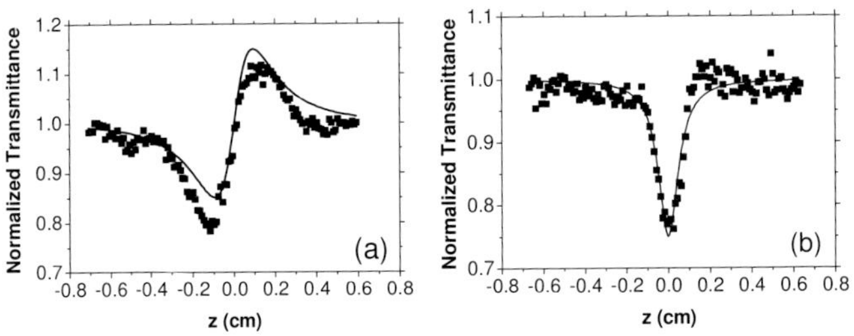

Fig. 2. z-scan traces with closed (a) and open (b) aperture for a sample with 1 nm large Si-nc. The theoretical fitting is also included as a solid curve. The laser had a λ = 1550 nm, 1 kHz repetition rate with 100 fs pulse duration and a peak intensity of 6×10^{11} W/cm^2.

modulators, all optical switching can be possible without the need of carrier injection. It is clear the advantage of this approach in terms of size (no need of metal contacts), device architecture (only short MZI is needed), power consumption and inherent heating.

Non-linear optical properties at 1550 nm of Si-nc deposited by PECVD and LPCVD technique have been recently investigated in details by using z-scan technique as a function of Si-nc size within the PHOLOGIC European Project.[50]

Representative experimental data are shown in Fig. 2, where positive z-scan traces at 1550 nm have been found.[51] Free carrier and bound electronic nonlinear response were investigated. Measurements at different power excitation permitted the separation and the evaluation of these two contributions. In particular, it was found that the nonlinear response arising from bound electron is of the order of ~10^{-13} cm^2/W. In addition, a nonlinear absorption is observed, whose coefficient β is in the range of (10^{-9}-10^{-8}) cm/W. It has been found that the non-linear refractive index n_2 is higher for higher linear refractive index n_0, which can be related to the Si-nc size and density. This effect is in contrast with expectations dictated by quantum confinement effects, where it is predicted that the nonlinearities increase as the nanocrystal size decreases. A possible explanation is that the nonlinearities are affected by the dielectric mismatch between the Si-nc and the oxide. The local electric field experienced by the nanocrystals is enhanced compared to the incident field because of the dielectric mismatch. The relative importance of the two effects (dielectric mismatch and quantum confinement) is weighted by the energy at which the nonlinearities are measured. For infrared light, the nonlinearities seem to be largely influenced by the dielectric mismatch.

2.3. *Light emission and optical gain in Si nanocrystals*

Si-nc have been initially investigated because of the improvement in their emission performances due to quantum confinement. Indeed, quantum confinement:

- maximizes carrier confinement in small spatial region reducing the encounter probability with non-radiative recombination centers;
- increases the radiative probability by delocalizing the wavefunction in the momentum space and, thus, increasing the electron-hole wavefunction overlaps;
- shifts the emission wavelength to the visible and controls it by silicon nanocrystal size,
- decreases the confined carrier absorption due to the increased emission wavelength,
- increases the light extraction efficiency by reducing the dielectric mismatch between the materials and the air.

Porous silicon was initially studied because of the easy of its fabrication,[52,53,54] and indeed it was found that once silicon is reduced to nanometric dimensions (as a porous quantum sponge of nanoparticles and/or nanowires) bright red luminescence at room temperature occurs. Unless microelectronics compatibility of porous silicon has been demonstrated and integration of driving circuits with light emitting elements has been performed, the disordered distribution of nanocrystal sizes and interconnectivities, and the porous silicon surface reactivity to chemical agents hamper a real engineering of porous silicon properties. The enormous and active inner surface causes time dependent properties, ageing effects and uncontrolled deterioration of device performances.

In comparison with porous silicon, Si nanocrystals (Si-nc) embedded in amorphous silica are better candidates for photonics, because of their robustness and stability and their full compatibility with the mainstream CMOS technology. The generation of visible light from Si-nc embedded in *a*-SiO_2 matrix has been extensively studied to obtain optically tunable quantum systems by modifying the dimensions of the nanoparticles.[55]

The Si-nc system is very promising to achieve a laser and many breakthroughs have been recently demonstrated in this field.[27,41] Among the others, we have shown amplified spontaneous emission (ASE) from Si-nc grown by a wealth of different techniques (PECVD, superlattices, magnetron sputtering) by means of VSL (variable stripe length) technique in the CW (continuous wavelength) and TR (time-resolved) regime.[44,56]

The recent achievements on this hot topic are critically reviewed in Chapter 4 of this book.

2.4. *Si nanocrystals LEDs*

One of the more appealing properties of Si-nc is the possibility to get light by current injection into the Si-nc and thus enabling LEDs.[57] The main difficulty in obtaining efficient Si nanocrystal LEDs is to get efficient carrier injection.[41] Interesting results have been obtained by the Stockholm group[58] in ion implantated samples, showing maximum external quantum efficiency of about 3×10^{-5}. Similar data have been obtained in PECVD Si-nc.[59] Field effect induced luminescence has been achieved by alternate tunnel injection of electrons and holes into Si nanocrystals with external quantum efficiencies of 0.03%.[60]

Materials with higher bandgap and higher optical transparency than SiO_2, such as Si nitride, have also been used as host matrices.[61,62] LEDs based on Si/SiO_2 superlattices were fabricated either by MBE[63] or LPCVD:[64,65] both PL and EL were observed. Life tests of several LEDs showed stable continuous operation for over one year. Vertical optical micro-cavities based on a Fabry-Perot structure with mirrors constituted by Distributed Bragg Reflectors (DBR) and where the central layer is formed by Si-nc dispersed in SiO_2 have been fabricated by Iacona *et al.*[69] The presence of thick SiO_2 layer needed to form the DBR can be a problem for electrical injection when current has to flow through the DBR. Lateral injection schemes can avoid these problems.

On the other hand the electrical injection into the Si-nc is a delicate task by itself. Indeed in most of the reported devices the electroluminescence is produced either by black-body radiation (the electrical power is converted into heat which raises the sample temperature which, in turn, radiates) or by impact excitation of electron-hole pairs in the Si-nc by energetic electrons which tunnel through the dielectric by a Fowler-Nordheim process (see Figure 3). Electron-hole pairs excited in this way recombine radiatively with an emission spectrum which is very similar to that obtained by photo-luminescence. The problem with impact excitation is its inefficiency (maximum quantum efficiency of 0.1%) and the damage it induces in the oxide due to the energetic electron flow. To get high electroluminescence efficiency one should try to get bipolar injection. However, bipolar injection is extremely difficult to achieve. In fact the effective barrier for tunnelling of electrons is much smaller than the one for holes (see Figure 3).

Despite some claims, most of the reported Si-nc LEDs are impact ionisation devices: electron-hole pairs are generated by impact ionisation by the energetic free-carriers injected through the electrode. Another recent work reports on a FET structure where the gate dielectric is a thin oxide with a layer of Si-nc.[34] In this way, by changing the sign of the gate bias, separate injection of electrons and holes in the Si-nc is achieved. Luminescence is observed only when both electrons and holes are injected into the Si-nc. By using this pulsing bias technique, alternate charge carrier injection is achieved which lead to high efficiency in the emission of the LED. Electrical charge injection and charge trapping effects in Si-nc based LEDs prepared by PECVD have been examined in details by I-V, C-V, and impedance measurements.[66]

A detailed description of the more recent achievements in this field is presented in Chapter 5.

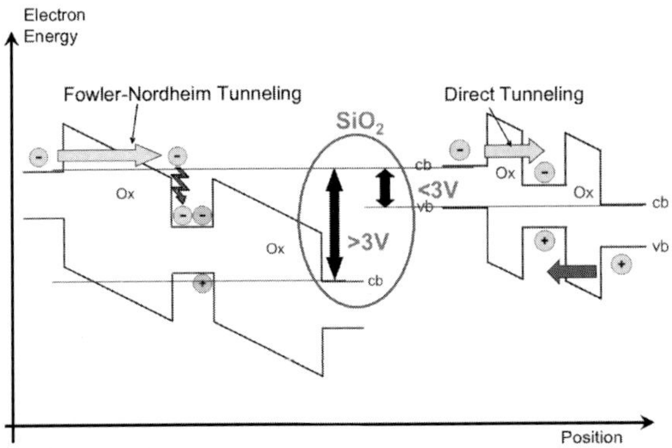

Fig. 3. Schematic view of the process of generation of electron-hole pairs in silicon nanocrystals by impact excitation or direct tunneling: cb or vb refer to the conduction or valence band-edges, while Ox refers to the silicon oxide barrier.

2.5. *Er coupled to nano-Si for optical amplifiers*

A major limitation of present-day optical networks is the difficulty of exploiting the enormous bandwidth available in local area networks. While erbium doped fiber amplifiers (EDFA) are well established in long-haul transmission, reducing their size and cost for widespread integration presents major difficulties: ion pair interactions, combined with the small excitation cross-section of the Er^{3+} ion, necessitate the use of long lightly doped fibers. Moreover, high power (and therefore expensive) laser diodes tuned to specific electronic transitions are required as pump sources. Clearly, a breakthrough would be a new gain medium that enables broad-band optical or electrical excitation of rare-earth ions, with a potential hundred-fold reduction in pump costs, and that provides order-of-magnitude enhancements in effective absorption cross-section, with corresponding reductions of amplifier length. The use of broadband sensitizers relaxes the stringent conditions for the pump source and raises the efficiency of the optical amplifier.[67,68] A good sensitizer has to have a high absorption cross section and has to transfer efficiently energy to Er^{3+}.

Si-nc have absorption spectra that depend on the average size of the Si-nc but that usually start to be appreciable near 600 nm and

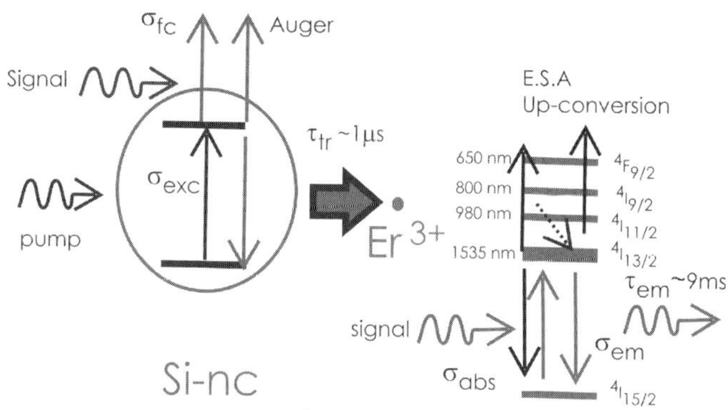

Fig. 4. Diagram of the excitation process of Er^{3+} ions via a Si-nc, with the main related cross sections.

grow towards shorter wavelengths. The absorption cross sections are of the order of 10^{-16} cm^2 around the 488 nm region that is five orders of magnitude higher than the direct absorption cross section of Er^{3+} in silica. In addition, it has been demonstrated that Er^{3+} doped silica containing Si-nc produced by co-sputtering, plasma-enhanced chemical vapor deposition (PECVD) or ion implantation exhibits a strong energy coupling between Si-nc and Er^{3+}. Quantum efficiencies greater than 60% and Si-nc to Er^{3+} transfer rates higher than 1 μs^{-1} by pumping at 488 nm have been also measured. In addition to the increase of effective excitation cross-section (σ_{exc}), Si-nc increase the average refractive index of the dielectrics, allowing good light confinement, and conduct electrical current,[60,69] which opens the route to electrically pumped optical amplifiers.

It has been also seen that the shape of the Er^{3+} photoluminescence spectra at low pumping powers when placed in a SiO_2 matrix is almost independent of the presence of the Si-nc in the matrix, which indicates that Er is surrounded by oxide. However, many aspects of the exact nature of the interaction between Si-nc and Er are still controversial. In particular, to engineer the system, with the aim of achieving a net optical gain in the amplifiers, the role of detrimental processes is to be figured out. Figure 4 summarizes various mechanisms and defines the related cross-sections for this system. Excitation of Er^{3+} occurs via an energy

Table 2. Summary of the various cross sections at 1.535 μm of Er^{3+} in various hosts.

	Er in SiO_2 (cm^2)	Er in Si (cm^2)	Er in Si-nc (cm^2)
Effective excitation cross section of luminescence by pumping at 488 nm	$1\text{-}8\times10^{-21}$	3×10^{-15}	$1.1\text{-}0.7\times10^{-16}$ [73,74]
Effective excitation cross section of electroluminescence		4×10^{-14}	1×10^{-14} [69]
Emission cross section	6×10^{-21}		2×10^{-19} [75]
Absorption cross section	4×10^{-21}	2×10^{-20}	5×10^{-21} [79]

transfer from photoexcited e-h pairs which are excited in the Si-nc: the overall efficiency of light generation at 1.535 μm through direct absorption in the Si-nc is described by an effective Er^{3+} excitation cross section σ_{exc}. On the other hand, the direct absorption of the Er^{3+} ions, without the mediation of the Si-nc, and the emission from the Er ions are described by absorption (σ_{abs}) and emission (σ_{em}) cross sections, respectively. The typical radiative lifetime is of 6–9 ms, which is similar to the one of Er^{3+} in pure SiO_2. Several authors have suggested different channels for the quenching of the Er emission such as cooperative up-conversion,[70] excited state absorption[71] and Augèr de-excitation.[72]

Table 2 summarizes the various cross sections reported in the literature. It is important to notice the five order of magnitude increase in σ_{exc} and the fact that this value is conserved also when electrical injection is used to excite the Si-nc. In addition, despite erroneous literature reports[76] on an enhanced σ_{abs}, more reliable data shows that its value is almost the same as that of Er in silica.[77]

If one places the Er^{3+} ions in a Si-nc ridge waveguide (see inset of Figure 5) one can perform experiments on signal amplification at 1.535 μm with the aim to demonstrate an Er doped waveguide amplifier (EDWA). Few groups have performed such an experiment. Han *et al.*[75] reported an internal gain of 7 dB/cm in samples with a very low Si-nc concentration. A successful experiment of pumping the EDWA with a LED battery was also reported.[78] In other experiments, with larger Si-nc concentration, no or weak signal enhancement has been observed.[35,79] The reason is attributed to the presence of a strong confined carrier absorption[79,80] which introduces a loss mechanism at the signal

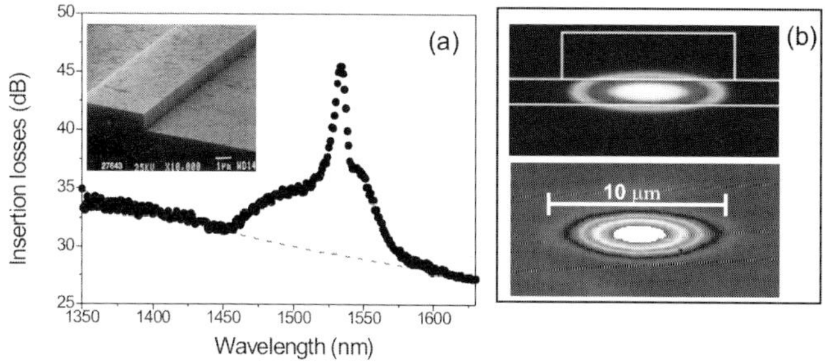

Fig. 5. (a) Transmission spectrum of a rib waveguide (SEM picture in the inset) and (b) output optical mode measured (bottom) and simulated (top) as in Ref. 77.

wavelength and prevents the sensitizing action of the Si-nc (see next section). Moreover, a limited number of Er ions is found to be efficiently coupled to the Si-nc: only those Er in close proximity to the Si-nc can be excited and the amount of excitable Er depends strongly on the Si-nc density. For low Si-nc density the distance dependent interaction model predicts an Er excitable fraction in the range of few %.[81] This model is discussed in more details in Chapter 5 of this book together with recent pump and probe experiments.

2.6. *Carrier absorption within Si nanocrystals waveguides*

Considering that both all-optical logical gates based on the positive nonlinear refractive index of Si-nc (discussed in section 2.2) and electrical light amplifiers based on Er-coupled Si-nc (discussed in section 2.5) are based on excitation of Si-nc in a waveguides, it is fundamental to characterize not only the linear losses (as discussed in previous sections) but even the non-linear optical losses at the wavelength of interest. One of the main deleterious loss mechanisms is due to the excited carrier absorption. This process generates additional losses at signal wavelength and is particularly detrimental for devices where excitons are created within the Si-nc due to one photon or two photons absorption, such as in the amplifier or logic gate, respectively. Free carrier absorption has been extensively studied in bulk silicon,[82] and it appears as responsible of the poor performance of many Si based photonics devices, such as IR detectors, Raman amplifiers, etc. A positive application has been, however, found in realizing fast electro-absorption modulators of SOI waveguides as previously presented in section 1.1.2.

Concerning Si-nc, only few works has studied in details this effect. Elliman *et al.*[83] performed optical pump-probe measurements by prism-coupling a probe beam into the waveguide (defined by the Si nanocrystal distribution) and exposing a portion of the guide to a UV pump beam. Induced absorption has been attributed to carrier absorption within the Si-nc. Similar results have been recently reported in Er doped Si-nc waveguides.[79] An extensive study of the carrier absorption (CA) mechanism in multilayered Si-nc rib waveguides has been performed at our laboratory. The rib-waveguides are based on an active layer (waveguide core layer) formed by Si-rich silicon oxide or SiO_2 multilayers, grown by reactive magnetron sputtering and annealed at high temperatures.[84] Si-nc size of 3–4 nm and a density of $(3.5 \pm 0.9)\times 10^{18}$ cm^{-3} have been determined by HRTEM analysis.[47] A pump (532 nm) and probe (1535 nm) technique is used to assess the loss mechanisms due to optical excitation of the system. When the waveguide is optically pumped, carriers are excited within the Si-nc and contribute to an additional loss term which is proportional to the number of excited carriers (N_{carr}) and to their absorption cross section at the signal wavelength (σ_{CA}). Thus, CA loss coefficient can be written as a function of SE, the ratio between the transmitted signal when the waveguide is pumped to the transmitted signal when the waveguide is not pumped (usually named signal enhancement), in the following way:

$$\sigma_{CA} N_{Carr} = -\frac{\ln(SE)}{\Gamma L_{pump}} \tag{1}$$

where Γ is the optical mode confinement factor and L_{pump} is the length of the waveguide which is actually excited by the pump. In Fig. 6a, the transmitted signal is shown when the pump is switched on. A rapid decrease in the transmission is observed. The dynamics of the decrease is characterized by two time scales, one fast (order of μs) and one slow

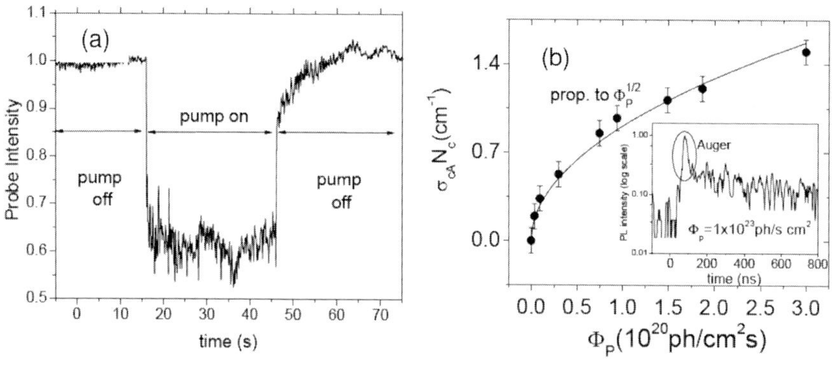

Fig. 6. Direct measurement of the intensity of 1535 nm probe signal in presence of the pump at 532 nm: (a) full temporal dynamics; (b) carrier absorption losses as a function of the photon flux. A square root fit to the experimental data is also shown (solid line). Inset: time resolved PL intensity.

(order of s). The slow one is due to thermal effects while the other is due to excited CA. Fig. 6b shows the maximum of the excited CA losses as a function of the pump photon flux Φ_p. A square root Φ_p dependence of $\sigma_{CA}N_{carr}$ is observed. CA losses increase with pumping flux up to 6 dB/cm for 3x10^{20} ph/cm^{2}s. Since σ_{CA} is independent of Φ_p, $N_{carr} \sim \Phi_p$½. This is an indication of Auger dominated recombination processes in the Si-nc, possibly between close Si-nc due to their particular close distribution in the multilayered samples. Time resolved luminescence measurements performed on the same active core material support the presence of a strong Auger recombination mechanism at high pump photon fluxes (see inset of Fig. 6b).

As a consequence of the presence of the pumping light, two independent loss mechanisms appear that behave with very different temporal dynamics: one slow (seconds) that is related with thermal effects and affects the waveguiding properties and a second fast (microseconds) that is associated to the excited carrier absorption. The excited carrier absorption has the same characteristic dynamics of the recombination of exciton luminescence in large Si-nc. This indicates that the way to reduce the excited carrier absorption is to decrease the Si-nc size in the waveguide.

3. Conclusions

In this Chapter, we have briefly introduced various concepts about silicon nanophotonics. In particular, we have underlined how quantum size effects which occur in low dimensional silicon together with the delicate interplay between the silicon and the embedding matrix allow to underpin new phenomena which in turn enable new devices. Silicon has still a lot to say, especially when its surface and dimension are tailored.

Acknowledgments

This work is supported by EC through the LANCER (FP6-033574), PHOLOGIC (FP6-017158), POLYCERNET (MCRTN-019601) projects and by a grant from INTEL. We acknowledge the help of many co-workers both from national and from international collaboration. They can be recognized in the cited literature. In particular, we would like to recognize the hard work of many present and past collaborators of the Nanoscience Laboratory without whom the research here reported would not have been performed.

References

1. R. Soref and J. Lorenzo, *IEEE J. Quantum Electron.* **22**, 873 (1986).
2. R. Soref, J. Schmidtchen and K. Petermann, *IEEE J. Quantum Electron.* **27**, 1971 (1991).
3. *Silicon Photonics*, edited by L. Pavesi and D. Lockwood, *Topics in Applied Physics* **94** (Springer-Verlag, Berlin 2004).
4. M. Lipson, *J. Lightwave Technol.* **23**, 4222 (2005).
5. B. Jalali and S. Fathpour, *J. Lightwave Technol.* **24**, 4600 (2006).
6. R. Soref, *IEEE J. Selected Topics Quantum Electronics* **12**, 1678 (2006).
7. www.bookham.com
8. www.kotura.com
9. M. Salib, L. Liao, R. Jones, M. Morse, A. Liu, D- Samara-Rubio, D. Alduino and M. Paniccia, *Intel Technology Journal* **8**, 143 (2004).
10. F. P. Payne and J. P. R. Lacey, *Optical and Quantum Electronics* **26**, 977 (1994).
11. R. M. Ridder, K. Wörhoff, A. Driessien, P. V. Lambeck and H. Albers, *IEEE J. Selected Topics Quantum Electronics* **4**, 930 (1998).
12. C. Netti, M.D. B. Charlton, G. J. Parker and J. J. Baumberg, *Appl. Phys. Lett.* **76**, 991 (2000).

13. N. Daldosso, M. Melchiorri, F. Riboli, M. Girardini, G. Pucker, M. Crivellari, P. Bellutti, A. Lui and L. Pavesi, *J. Lightwave Technol.* **22**, 1734 (2004).
14. M. Melchiorri, N. Daldosso, F. Sbrana and L. Pavesi, G. Pucker, C. Kompocholis, P. Bellutti and A. Lui, *Appl. Phys. Lett.* **86**, 121111 (2005).
15. G. Rickman and G. T. Reed, *IEEE Proc.-Optoelectron.* **141**, 391 (1994).
16. P. Dumon, W. Bogaerts, V. Wiaux, J. Wouters, S. Beckx, J. V. Campenhout, D. Taillaert, B. Luyssaert, P. Bienstman, D. V. Thourhout and R. Baets, *IEEE Photon. Technol. Lett.* **16**, 1328 (2004).
17. Y. Vlasov and S. McNab, *Opt. Express* **12**, 1622 (2004).
18. R. Soref and B. Bennett, *IEEE J. Quantum Electron.* **23**, 123 (1987).
19. F. Gardes, G. Reed, N. Emerson and C. Png, *Optics Express* **13** (22), 8845 (2005).
20. A. Liu, L. Liao, D. Rubin, H. Nguyen, B. Ciftcioglu, Y. Chetrit, N. Izhaky and M. Paniccia, *Optics Express* **15** (2), 660 (2007).
21. T. Pinguet, V. Sadagopan, A. Mekis, B. Analui, D. Kucharski and S. Gloeckner, *Proceedings of Group IV Photonics 2007*, **1-3**, 186 (2007).
22. D. Marris-Morini, L. Vivien, S. Maine, E. Cassan, S. Laval, P. Lyan and J-M. Fédéli, *Proceedings of Group IV Photonics 2007*, **1-3**, 195 (2007).
23. C. E. Png, G. T. Reed, W. R. Headley, K. P. Homewood, A. Liub, M. Paniccia, R. M. H. Atta, G. Ensell, A. G. R. Evans, D. Hak and O. Cohen, *Proc. SPIE* **5356**, 44 (2004).
24. G. Cocorullo, A. Cutolo, F. G. Della Corte and I. Rendina, *Opt. Commun.* **86**, 228 (1991).
25. Q. Xu, B. Shmidt, S. Pradhan and M. Lipson, *Nature* **435**, 325 (2005).
26. A. W. Fang, H. Park, Y.-h. Kuo, R. Jones, O. Cohen, D. Liang, O. Raday, M. N. Sysak, M. J. Paniccia and John E. Bowers, *Materials Today* **10**, 28 (2007).
27. L. Pavesi, *Materials Today* **8**, 23 (2005).
28. M. A. Green, J. Zhao, A. Wang, P. J. Reece and M. Gal, *Nature* **412**, 805 (2001).
29. M. J. Chen, J. L. Yen, J. Y. Li, J. F. Chang, S. C. Tsai and C. S. Tsai, *Appl. Phys. Lett.* **84**, 2163 (2004).
30. H. Rong, J. Liu, M. Jones, O. Cohen, D. Hak, R. Nicolaescu, A. Fang and M. Paniccia, *Nature* **433**, 292 (2005).
31. S. G. Cloutier, P. A. Kossyrev and J. Xu, *Nature Materials* **4**, 887 (2005).
32. W. L. Ng, M. A. Lourenço, R. M. Gwilliam, S. Ledain, G. Shao and K. P. Homewood, *Nature* **410**, 192 (2001).
33. L. Pavesi, L. Dal Negro, C. Mazzoleni, G. Franzò and F. Priolo, *Nature* **408**, 440 (2000).
34. R. J. Walters, R. I. Bourianof and H. Atwater, *Nature Materials* **4**, 143 (2005).
35. N. Daldosso, D. Navarro-Urrios, M. Melchiorri, C. Garcia, P. Pellegrino, B. Garrido, C. Sada, G. Battaglin, F. Gourbilleau, R. Rizk and L. Pavesi, *IEEE Journal of Selected Topics in Quantum Electronics* **12**, 1607 (2006).

36. J. Liu, X. Sun, D. Pan, X. Wang, L. C. Kimerling, T. L. Koch and J. Michel *Optics Express* **15**, 11272 (2007).
37. G. Dehlinger, L. Diehl, U. Gennser, H. Sigg, J. Faist, K. Ensslin and D. Grützmacher, *Science* **290**, 2277 (2000).
38. O. Boyraz and B. Jalali, *Optics Express* **12**, 5269 (2004).
39. D. Ahn, C-yin Hong, J. Liu, W. Giziewicz, M. Beals, L. C. Kimerling, J. Michel, J. Chen and F. X. Kärtner, *Optics Express* **15**, 3916 (2007).
40. L. Vivien, M. Rouvière, J.-M. Fédéli, D. Marris-Morini, J. F. Damlencourt, J. Mangeney, P. Crozat, L. El Melhaoui, E. Cassan, X. Le Roux, D. Pascal and S. Laval, *Optics Express* **15**, 9843 (2007).
41. *Nanostructured Silicon for Photonics – from materials to devices* – Z. Gaburro, P. Bettotti, N. Daldosso, M. Ghulinyan, D. Navarro, M. Melchiorri, F. Riboli, M. Saiani, F. Sbrana and L. Pavesi, *Materials Science Foundation* **27-28**, TransTecH Publications LTD (2006).
42. *Nanosilicon* edited by Vijay Kumar (Elsevier Ltd, 2007) Hardbound, Sept. 2007.
43. J. Ruan, P. M. Fauchet, L. Dal Negro, M. Cazzanelli and L. Pavesi, *J. Appl. Phys.* **83**, 5479 (2003).
44. L. Dal Negro, M. Cazzanelli, B. Danese, L. Pavesi, F. Iacona, G. Franzò and F. Priolo, *J. Appl. Phys.* **96**, 5747 (2004).
45. L. Khriachtchev, M. Räsänen and S. Novikov, *Appl. Phys. Lett.* **83**, 3018 (2003).
46. P. Pellegrino, B. Garrido, C. Garcia, J. Arbiol, J. R. Morante, M. Melchiorri, N. Daldosso, L. Pavesi, E. Schedi and G. Sarrabayrouse, *J. Appl. Phys.* **97**, 074312 (2005).
47. N. Daldosso, M. Melchiorri, L. Pavesi, G. Pucker, F. Gourbilleau, S. Chausserie, Ali Belarouci, X. Portier and C. Dufour, *J. Luminescence* **121**, 344 (2006).
48. G. Vijaya Prakash, M. Cazzanelli, Z. Gaburro, L. Pavesi, F. Iacona, G. Franzò and F. Priolo, *J. Mod. Opt.* **49**, 719 (2002).
49. G. Vijaya Prakash, M. Cazzanelli, Z. Gaburro, L. Pavesi, F. Iacona, G. Franzó and F. Priolo, *J. Appl. Phys.* **91**, 4607 (2002).
50. A. Martínez, S. Hernández, P. Pellegrino, Y. Lebour, G. Carles, S. Marco, B. Garrido, R. Spano, M. Cazzanelli, N. Daldosso, L. Pavesi, E. Jordana and J. M. Fedeli, *Proceedings of SPIE* **6591**, 65910A (2007).
51. R. Spano, M. Cazzanelli, N. Daldosso, Z. Gaburro, L. Ferraioli, L. Tartara, J. Yu, V. Degiorgio, S. Hernandez, Y. Lebour, P. Pellegrino, B. Garrido, E. Jordana, J. M. Fedeli and L. Pavesi, *MRS Proceedings* **958**, L08-06 (2007).
52. L. T. Canham, *Appl. Phys. Lett.* **57**, 1046 (1990).
53. O. Bisi, S. Ossicini and L. Pavesi, *Surf. Sci. Rep.* **38**, 1 (2000).
54. *Porous Silicon*, Z. Gaburro, N. Daldosso, L. Pavesi in *Encyclopedia of Condensed Matter Physics* edited by Franco Bassani, Jerry Liedl and Peter Wyder (Elsevier Ltd) – 2005.

55. S. Ossicini, L. Pavesi and F. Priolo, *Light Emitting Silicon for Microphotonics*, Springer Tracts in Modern Physics **194** (Springer-Verlag, Berlin 2003).
56. L. Dal Negro, M. Cazzanelli, N. Daldosso, Z. Gaburro, L. Pavesi, F. Priolo, D. Pacifici, G. Franzò and F. Iacona, *Physica E* **16**, 297 (2003).
57. L. Brus in *Light Emission in Silicon. From Physics to Devices*, Ed. by D. Lockwood, Semiconductors and Semimetals **49**, Academic Press, 303 (1998).
58. N. Lalic and J. Linnros, *J. Luminescence* **80**, 263 (1999).
59. N.-M. Park, T.-S. Kim and S.-J. Park, *Appl. Phys. Lett.* **78**, 2575 (2001).
60. M. Peralvarez, C. Garcia, M. Lopez, B. Garrido, J. Barreto, C. Dominguez and J. A. Rodriguez, *Appl. Phys. Lett.* **89**, 051112 (2006).
61. Z. H. Lu, D. Lockwood and J.-M. Baribeau, *Nature* **378**, 258 (1996).
62. K. S. Cho, N.-M. Park, T.-Y. Kim, K.-H. Kim, G. Y. Sung and Jung H. Shin, *Appl. Phys. Lett.* **86**, 0713909 (2005).
63. R. Tsu, A. Filios, C. Lofgren, K. Dovidenko and C. G. Wang, *Electrochem. Solid State Lett.* **1**, 80 (1998).
64. L. Heikkila, T. Kuusela and H. P. Hedman, *J. Appl. Phys.* **89**, 2179 (2001).
65. Z. Gaburro, L. Pavesi, G. Pucker and P. Bellutti, *MRS. Proc.* **638**, F18.5.1 (2001).
66. A. Anopchenko, S. Prezioso, Z. Gaburro, L. Ferraioli, G. Pucker, P. Bellutti and L. Pavesi, *Proceedings of Group IV Photonics 2007*, **1-3**, 70 (2007).
67. A. J. Kenyon, *Progress in Quantum Electronics* **26**, 223 (2002).
68. A. Polman and F. C. J. M. van Veggel, *J. Opt. Soc. Am.* **B21**, 871 (2004).
69. F. Iacona, D. Pacifici, A. Irrera, M. Miritello, G. Franzò and F. Priolo, *Appl. Phys. Lett.* **81**, 3242 (2002).
70. D. Pacifici, G. Franzò, F. Priolo, F. Iacona and L. Dal Negro, *Phys. Rev.* **B67**, 245301 (2003).
71. C. J. Oton, W. H. Loh and A. J. Kenyon, *Appl. Phys. Lett.* **89**, 031116 (2006).
72. I. Izeddin, T. Gregorkiewicz and M. Fujii, *Physica E* **38**, 144 (2007).
73. F. Priolo, G. Franzò, D. Pacifici, V. Vinciguerra, F. Iacona and A. Irrera, *J. Appl. Phys.* **89**, 264 (2001).
74. A. J. Kenyon, C. E. Chryssou, C. W. Pitt, T. Shimizu-Iwayama, D. E. Hole, N. Sharma and C. J. Humphreys, *J. Appl. Phys.* **91**, 367 (2002).
75. H.-S. Han, S.-Y. Seo, Jung H. Shin and N. Park, *Appl. Phys. Lett.* **81**, 3720 (2002).
76. P. G. Kik and A. Polman, *J. Appl. Phys.* **91,** 534 (2002).
77. N. Daldosso, D. Navarro-Urrios, M. Melchiorri, L. Pavesi, C. Sada, F. Gourbilleau and R. Rizk, *Appl. Phys. Lett.* **88**, 161901 (2006).
78. I. Lee, Jung H. Shin and N. Park, *J. Lightwave Technol.* **23**, 19 (2005).
79. N. Daldosso, D. Navarro-Urrios, M. Melchiorri, L. Pavesi, F. Gourbilleau, M. Carrada, R. Rizk, C. Garcia, P. Pellegrino, B. Garrido and L. Cognolato, *Appl. Phys. Lett.* **86**, 261103 (2005).
80. R.G. Elliman, M. Forcales, A. R. Wilkinson and N. J. Smith, *Nucl. Instr. and Meth.* **B257**, 11 (2007).

81. B. Garrido, C. Garcia, S.-Y. Seo, P. Pellegrino, D. Navarro-Urrios, N. Daldosso, L. Pavesi, F. Gourbilleau and R. Rizk, *Phys. Rev.* **B76**, 245308 (2007).
82. W. Spitzer and H. Y. Fan, *Phys. Rev.* **108**, 268 (1957).
83. R. G. Elliman, M. J. Lederer, N. Smith and B. Luther-Davies, *Nucl. Instr. and Meth.* **B206**, 427 (2003).
84. C. Ternon, F. Gourbilleau, X. Portier, P. Voivenel and R. Rizk, *Thin Solids Films* **419**, 5 (2002).

CHAPTER 2

THEORETICAL STUDIES OF ABSORPTION, EMISSION AND GAIN IN SILICON NANOSTRUCTURES

Elena Degoli[1], Roberto Guerra[2], Federico Iori[2], Rita Magri[2], Ivan Marri[2] and Stefano Ossicini[1*]

[1]*Dipartimento di Scienze e Metodi dell'Ingegneria*
Universitá di Modena e Reggio Emilia
via Amendola 2, 42100 Reggio Emilia, Italy
[*]*stefano.ossicini@unimore.it*
[2]*Dipartimento di Fisica*
Universitá di Modena e Reggio Emilia
via Campi 213/A, 41100 Modena, Italy

Density-functional and many body perturbation theory calculations have been carried out in order to study the structural, electronic, and optical properties both in ground and excited state configuration, of silicon nanocrystals in different conditions of surface passivation and doping. Starting from hydrogenated clusters, we have considered different Si/O bonding geometries at the interface. We provide strong evidences that not only the quantum confinement effect but also the chemistry at the interface has to be taken into account in order to understand the physical properties of these systems. In particular we show that only the presence of a surface Si-O-Si bridge bond induce an excitonic peak in the emission-related spectra, redshifted with respect to the absorption onset, able to provide an explanation for both the observed Stokes shift and the near-visible photoluminescence (PL) experimentally observed in Si-nc. For the silicon nanocrystals embedded in a SiO_2 matrix, the electronic and optical properties are discussed in detail. The strong interplay between the nanocrystal and the surrounding host environment and the active role of the interface region between them is pointed out, in very good agreement with the experimental results. Finally, concerning doping, we will show that, thanks to electronic transitions between donor and acceptor states present at the band edges and considering also the effect of quantum confinement it is possible to engineer the absorption and emission spectra of Si nanocrystals. For each considered system optical gain calculations have been carried out giving some insights on the system characteristics necessary to optimize the gain performance of Si-nc.

1. Introduction

Several strategies have been researched over the last years for light generation and amplification in silicon. One of the most promising is based on silicon nanocrystals (Si-nc) with the idea of taking advantage of the reduced dimensionality of the nanocrystalline phase (1-5 nm in size) where quantum confinement, band folding and surface effects play a crucial role.[1,2] Indeed, it has been found that Si nanocrystals band-gap increases with decreasing size and a photoluminescence external efficiency in excess of 23% has been obtained.[3] Silicon nanocrystals based LED with high efficiency have been obtained by using Si nanocrystals active layers[4] and achieving separate injection of electrons and holes.[5] Moreover, optical gain under optical pumping has been already demonstrated in a large variety of experimental conditions.[6–11] All these facts show that the expectations of achieving an injection laser in silicon are well founded. The possibility of a silicon based laser suggested by the optical gain measurements on Si-nc formed by ion implantation[6] and by PECVD[8] in SiO_2 has attracted a lot of interest on this type of system and produced a lot of debates especially on the gain mechanisms.[12] Interface states localized at the surface of the light emitting silicon nanocrystals have been proposed as good candidates for explaining the peculiar optical behavior observed.[6,8] Thus the nature and the properties of the interface between the Si-nc and the SiO_2 host matrix has become crucial in order to confirm or not this model. A model based instead on an effective rate equation for a four level system was[8,13,14] proposed in order to interpret and explain the observed optical gain. The physical origin of the four level is nevertheless not yet individuated. A first principle approach being capable of providing a very detailed description of the system atomistic structure could help greatly in shading some light on this problem. In this work we present the first investigation of the optical gain in Si-nc performed including the effects of spatial confinement, of equilibrium geometries, of interface properties and of doping. This ab-initio study of optical gain has the final goal of investigate the possible mechanisms and determine the best conditions for the electronic population inversion and for lasing in Si-nc trying to create a reasonable model, coherent with experimental data, in terms of nanocrystals and interface energy levels.

A preliminar work in order to interpret the optical gain calculation has required the study of the structural, electronic and optical properties of simplified models. We have first of all considered pure Si-nc fully passivated by hydrogen in order to determine the properties strictly due to cluster itself. We have then considered the role played by oxidation substituting an H interface atom with an O in different bond configurations. Actually, Wolkin et al.[15] observed that oxidation introduces states in the gap, which pin the transition energies. They and others[16,17] suggest that the formation of a Si=O double bond is responsible of the red shift of the optical absorption edge upon oxidation. On the contrary, Vasiliev et al.[18] have shown that similar absorption gaps can be obtained also for O connecting two Si atoms (bridge bond) at the Si-nc surface. Recently Gatti and Onida[19] considered six small different prototypical oxidized Si clusters and found results similar to those of Refs. 16 and 17, i.e. that the red shift of the absorption edge is much more pronounced in the case of the double Si=O bond than for bridge bonds. Ramos et al.[20] have found a blueshift of the absorption onset for strongly oxidized Si-nc. Actually, heavy oxidation (the formation of complete oxide shells) originates a reduction of the effective size of the Si-nc, i.e. an increase of quantum confinement. All these calculations address only the problem of absorption, theoretically very few work has been done concerning the excited-state configurations, which are mainly responsible for emission energies. Nevertheless from an experimental point of view large interest has been given to optical properties such as photoluminescence or gain that play a fundamental role in a wide range of applications. Tuning the optical response of such systems cannot prevent from a complete understanding of phenomena such as Stokes shift (SS) (difference between emission and absorption energies) or optical gain (obtained from silicon quantum dots in a SiO_2 matrix[6,8]): the understanding of the physical mechanisms at the origin of this phenomena implies the analysis of both the ground and excited state of the optically excited nanostructures, and that is our interest in this paper. Finally we have analized the case of a small Si-nc in a SiO_2 matrix in order to better simulate the experiments. Contemporary to this we have also studied the role of doping in Si-nc. Actually, Si-nc still have a memory of the indirect band gap of the bulk phase and this is evidenced by the clearly observed structures related to momentum-conserving phonons.[1,21,22] One way to overcome this problem, i.e. the low radiative recombination efficiency with respect to non-radiative recombination processes typical of bulk silicon but also of nanocrystalline Si, is the introduction of isoelectronic impurities or the simultaneous doping with $n-$ and $p-$type impurities. In a series of recent very interesting papers, Fujii et al.[23–25] have proven the possibility of a detailed control of the Si-nc photoluminescence by this last technique showing that a (B and P) codoped Si-nc always has a higher photoluminescence intensity than that of both a single (B or P) doped nanocrystal and an undoped nanocrystal and that the quasidirect optical transitions are predominant. Starting from the already described hydrogenated Si-nc and following the work of Fujii et al.,[23] we have doped the nanocrystals with B and P impurities in substitutional positions just below the nanocrystal surface and we have then studied their structural, electronic and optical properties.

The paper is organized as follows. A complete description of the theoretical methods used is given in Sec. 2 considering, in the different subsections, the Density Functional Theory (DFT) (section 2.1), the Δ-self-consistent (Δ-SCF) approach (section 2.1.1), and the Many Body Perturbation Theory (MBPT) through the GW (section 2.2.1) and the Bethe-Salpeter (BS) (section 2.2.2) methods. The study of the physical systems is presented in Sec. 3. Starting from the analisys of hydrogenated Si-nc (section 3.1) we then consider the effect of oxidation (section 3.2), of doping (section 3.3) and finally of an embedding matrix (section 3.4). Conclusions are presented in section 4.

2. Theoretical Methods

Electronic and optical properties of complex systems are nowaday accessible thanks to the impressive development of theoretical approaches and of computer power. Surfaces, nanostructures, and even biological systems can now be studied within ab-initio methods.[26,27] In principle within the Born-Oppenheimer approximation to decouple the ionic and electronic dinamic, the equation that governs the physics of all those systems is the Many-Body equation:

$$(\sum_i -\frac{1}{2}\nabla^2{}_{\mathbf{r}_j} + V_{ext} + \frac{1}{2}\sum_{i\neq j}\frac{1}{|\mathbf{r}_i - \mathbf{r}_j|})\Psi(r_1, r_2, ..r_N) = E\Psi(r_1, r_2, ..r_N) \quad (1)$$

with V_{ext} ionic potential. Unfortunately a direct solution of equation (1) is a formidable task not only for realistic systems but already for N≥2. It is hence necessary to resort to approximations, or to different approaches. The static (Section 2.1) Density Functional Theory (DFT) avoid dealing directly with the Many-Body equation by mapping the interacting system into a fictitious non-interacting system (which is then described by single particle equations); the Green's function approach (Section 2.2.1), instead, maps the Many-Body electronic problem to a system of quasi-particles, which describe the excitations in terms of a particle of finite life time, that represent the extra electron (and/or the extra hole added to the system) plus its screened interaction with the electrons of the system. In next sections, we review the theoretical approaches used to calculate ground state properties, band structures, and optical spectra. We show in Fig. 1 the three schemes used: DFT for ground state properties, GW for band structure calculations (charged excitations), and the Bethe-Salpeter approach for optical spectra (neutral excitations).

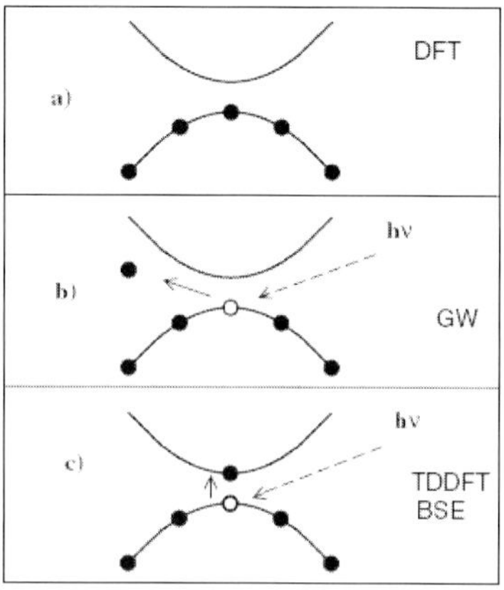

Fig. 1. Scheme of the 3 physical processes we describe here: a) ground state, b) photoemission experiment (charged excitation), c) absorption experiment (neutral excitation); a) can be treated with DFT, b) within GW, c) within the Bethe-Salpeter approach and/or the Time Dependent DFT approach.

2.1. *The Density Functional Theory*

Density Functional Theory (DFT) treats the case of an external, time independent potential, and is based on the Hohenberg and Kohn theorem[28] which proves that all the ground state properties of an interacting electronic system, including, in principle, the many-body wave function, could be expressed as unique functionals of the electronic density alone. In particular this assertion is valid also for the total energy E of the system. For the total energy functional E[n], the variational principle ensures that for a given density n(r), $E[n(\mathbf{r})] \geq E_{GS}$ where E_{GS} is the ground state energy of the system. The energy functional finds its minimum at the ground state density n_{GS}, for which $E[n_{GS}] = E_{GS}$. This theorem leads, in principle, to a straightforward method for computing ground state properties, but the explicit form of the functional in terms of the density is not known and such an explicit functional may not exist. Once given an approximation for the functional dependence of E, the ground state density can be obtained by a minimization procedure of the energy functional. Kohn and Sham,[29] introduced an ad hoc separation of the terms contributing to the total energy E: the electron-electron interaction and the kinetic energy T of the interacting N-electron system. The advantages of this separation are: 1) provide a single particle scheme to obtain the ground state density and total energy, and 2) have an expression in which the approximations to the unknown part would be, in many cases, not as relevant as before. In the KS scheme the total energy functional of the interacting system is written as:

$$E[n] = T_0[n] + E_H[n] + \int d\mathbf{r} n(\mathbf{r}) v_{ext}(\mathbf{r}) + E_{xc}[n] \tag{2}$$

T_0 is the kinetic energy of a non-interacting system with density n, E_H is the Hartree contribution to the total energy, and E_{xc} is the remaining part of the total energy which contains exchange-correlation contributions plus the difference between the kinetic energy of the interacting N-electron system T and the kinetic energy T_0 of the non-interacting system. The main idea underlying this approach is to map the study of the (complicated) interacting system into the study of a (simpler) non interacting fictitious system whose Hamiltonian is written as:

$$-\frac{1}{2}\nabla^2 + v_{ext} + v_H + V_{xc} = \varepsilon_i \phi_i(\mathbf{r}) \tag{3}$$

and which has by construction the same ground state density of the interacting system:

$$n(r) = \sum_i f_i |\phi_i(\mathbf{r})|^2 \tag{4}$$

withf_i being the occupation number of the state i. In Eq. (3) $V_{xc} = \delta E_{xc}[n]/\delta n$ is the exchange and correlation potential of the interacting system. It is now possible, given an approximation for $E_{xc}[n]$, to solve the Kohn-Sham Eqs. (3), (4) self consistently and calculate the density of the interacting (real) system via Eq. (2). Once the density is known, is possible to calculate the energy of the ground state of the interacting system and hence, by proper minimization, to find its ground state geometry. The simplest and most common approximation to E_{xc} is the Local Density Approximation (LDA) given by:[30]

$$E_{xc}^{LDA} = \int d\mathbf{r} n(\mathbf{r}) \varepsilon_{xc}^{heg}(n) \tag{5}$$

where $\varepsilon_{xc}^{heg}(n)$ is the exchange-correlation energy per electron of a homogeneous electron gas of density n. It is worth noting that the Kohn-Sham equations, representing a fictitious auxiliary system, have in principle no physical meaning. Nevertheless, their eigenvalues are often interpreted as one electron excitation energies corresponding to the excitation spectra of the system upon removal or addition of an electron, and DFT is in this way used to calculate band structures. The qualitative agreement with experiments is often remarkable, but not quantitative: the electronic gaps of semiconductors are, as a matter of fact, always systematically underestimated within DFT.

2.1.1. *The Δ-self-consistent approach: Absorption, emission and gain*

What happens when an e-h pair is excited in the system is that the charge density distribution changes, thus perturbing the forces acting on the atoms. The ions follow the excitation relaxing to a new equilibrium geometry, provided that the time taken by this process is shorter than the e-h pair recombination time. Therefore, the excitation and de-excitation processes involve two atomic configurations, that we can call ground-state (GS) and excited-state (ES) geometries. This gives rise to four possible electronic/atomic configurations with their corresponding total energies, as follows:

(i) ground state as the starting point, whose total energy is E_1,
(ii) one e-h pair induced by the excitation, described by the excited-state level occupation but in the ground-state geometry, whose total energy is E_2,
(iii) cluster relaxation in the presence of the e-h pair, toward the excited-state minimum energy geometry, whose total energy is E_3,
(iv) e-h pair recombination, ground-state occupation, excited-state geometry, total energy E_4.

After the recombination the system can eventually relax toward the starting, ground-state geometry. As we will see the excitation can induce significant changes on the electronic and structural properties of the system and, mostly, there is a strong interplay between the electronic levels occupation and the cluster geometry. For systems with a spatial confinement, i.e., with free-exciton radii larger than, or, of the order of their spatial extent, the lowest pair-excitation energies ϵ_{ex} can be rigorously expressed in terms of DFT-LDA total energies. Usually one defines $\epsilon_{ex} = E(N+1)+E(N-1)-2E(N)$

using the ground-state energies E of the $N+1$, $N-1$, and N electron system. However, in this definition the effect of the attractive screened Coulomb interaction of the excited electron and hole pair is missing. In order to account for the excitonic effect, we use the Δ-SCF approach that is based on the study of two different electronic configurations for the system:

- initial state: Ground State

$$\vec{\gamma}_{gs} = \begin{cases} \gamma_i = 1 \; i = 1, ..., N \\ \gamma_i = 0 \; i > N \end{cases}$$

- final state: Neutral Excited State

$$\vec{\gamma}_{exc} = \begin{cases} \gamma_i = 1 \; i = 1, ..., (h-1), (h+1), ..., N \quad \text{and} \quad i = e \; (e > N) \\ \gamma_i = 0 \; i = h \quad \text{and} \quad i > N, i \neq e \end{cases}$$

In the last configuration a particle-hole pair is considered in the system promoting an electron from the valence band (i=h) to a conduction band (i=e). For this reason the method is also called "constrained DFT". The excitation energy of the many-electron system is the difference in total energy, between two self-consistent calculation with the occupations described above, i.e.:

$$E_{exc} = E_{\vec{\gamma}_{exc}} - E_{\vec{\gamma}_{gs}} \tag{6}$$

and should give the optical gap (since the initial and final states are both neutral). In our particular calculation the total energy $E_{\vec{\gamma}_{exc}}$ is calculated with the constraint that the HOMO of the ground-state system contains a hole h, placing the corresponding electron e into the LUMO of the ground-state system. When the atomic geometry of the nanocrystal is optimised in the ground-state configuration, the corresponding calculated pair-excitation energies $\epsilon^A_{ex} = E_2 - E_1$ define the optical absorption (A) edges. We also relax the nanostructure with an excited electron-hole pair. At the resulting geometry, the lowest emission (E) energy $\epsilon^E_{ex} = E_3 - E_4$ is calculated. The difference of the two energies ($\epsilon^A_{ex} - \epsilon^E_{ex}$) defines the Stokes or Frank-Condon shift which is due to the lattice relaxation induced by the electronic excitation. Through this method one can extract the absorption and emission energies that are different with respect to the HOMO-LUMO gaps obtained for both the ground and excited state configurations; the latter are actually affected by the well known problem of underestimation of the gap of all DFT calculations. Δ-SCF can yield good results when it is possible to simulate the excitations occupying just one-particle orbital; this immediately excludes the possibility to describe those excitations that are not easily described in terms of isolated single particle transitions. Furthermore, this method works in finite systems, but not in infinite ones. This is because the main contribution in Δ-SCF is the Hartree relaxation, that for extended system, is negligible leaving the system energy dominated by other contributions. Through the Δ-SCF method it is not possible to obtain the lineshape of the optical spectra that allow a direct comparison with experimental data, anyway, using the ground and excited state optimized geometry calculated within this scheme we can calculate the absorption and emission spectra evaluating $Im(\epsilon_{NS}(\omega))$ (the imaginary part of the nanostructures dielectric function) through first-principle calculations also beyond the one-particle approach. The emission spectrum is calculated, in a first approximation, as the time reversal of the absorption.[31] Strictly speaking, $Im(\epsilon_{NS}(\omega))$ corresponds to an absorption spectrum in a new structural geometry, the ES geometry, with the electronic configuration of the GS. For the first time, we consider the electron-hole interaction also in the emission geometry. This different approach where the many-body effects are combined with the study of the structural bond distortion at the Si-nc surface in the ES, accounts both for the absorption and PL spectra.

For some of the considered systems we will also show the first optical gain calculations. Optical gain corresponds to a light-matter interaction regime for which generation of photons by stimulated emission dominates over photon absorption. As in other lasing media, optical gain in nanocrystals requires population inversion that is, the situation in which the population of the excited states is larger than that of the ground state. Within a single particle approach, using electric dipole approximation, according to the Ehrenreich-Cohen formula, we will calculate the dielectric function

$$\varepsilon_2(\omega) = \frac{8e^2\pi}{m^2V\omega^2}\sum_{v,c,k}|P_{v,c}(k)|^2\delta(\varepsilon_{v,c}-\hbar\omega)[f_v - f_c] \quad (7)$$

where $|P_{v,c}(k)|$ is the dipole matrix element between the v and c states, respectively, and f_v and f_c define the quasi Fermi level occupation probability. Obviously, in the ground state, f_v=1 and f_c=0. As previously noted optical gain occurs only when some higher lying level have an higher occupation probability f_c than some lower lying level, i.e. when a population inversion is present. For the wave intensity inside a medium we have $I(z) = I_0e^{\gamma(\omega)z}$. Therefore $\gamma(\omega) < 0$ means absorption, $\gamma(\omega) > 0$ means gain. The absorption coefficient $\gamma(\omega)$ is proportional to the imaginary part of the dielectric function through $\gamma(\omega) = -[\omega/n(\omega)c]\varepsilon_2$, where n is the refractive index, thus it is possible to calculate the coefficient $\gamma(\omega)$ through $\varepsilon_2(\omega)$. When the condition of population inversion is verified, $\varepsilon_2(\omega)$ is negative and $\gamma(\omega)$ assume a positive value. For bulk systems (as for instance for the Si-nc embedded in SiO_2) population inversion will be simulated introducing quasi-Fermi levels F_n and F_p and thus filling states above the Fermi level and depleting a corresponding number of states below the Fermi level. We have:

$$f_v = \frac{1}{1+e^{\frac{E_v(k)-F_p}{KT}}} \qquad f_c = \frac{1}{1+e^{\frac{E_c(k)-F_n}{KT}}} \tag{8}$$

where K is the Boltzmann coefficient and T is the temperature. The number of electrons/holes created in the system can be determined from the calculated density of states $\rho(E)$, through the equation $N = \int \rho(E) f_m(E, F_m) dE$. Concerning the results that we will show in the rest of this paper they have been obtained fixing the temperature at 10 K and the electron-hole pair density for bulk structures to $2x10^{19}\ cm^{-3}$.

2.2. *The many body perturbation theory*

2.2.1. *The GW approach*

Underestimation of the electronic gaps in DFT calculations finds its origin not in a deficiency of the theory, but in the uncorrect use we make of the Kohn-Sham equations. In fact, even if the Kohn-Sham equations have the form of single-particle Schrödinger-like equations, their eigenvalues are not the excited state energies. DFT is an exact theory for ground state properties, but there is no strict theoretical justification to use it to obtain excitation energies. Looking at case (b) in Fig. 1 we can realize that, in the determination of the electronic band structure, we are dealing with energy differences between the system with N-electrons and the system with N-1 (direct photoemission spectroscopy) or N + 1 (inverse photoemission spectroscopy) electrons. It is hence natural to change over to the Green's function formalism.[32] The one-electron Green's function is defined as the probability amplitude that an electron or a hole, created at time t_1 and point r_1 will be detected at time t_2 and point r_2, that is, the probability amplitude of the electron (hole) propagation from r_1 to r_2 in the time interval $(t_2 - t_1)$. Because of this interpretation, the Green's function is also called one-electron propagator. After a Fourier transform into the frequency domain, the Green's function G can be formally written as

$$G(\mathbf{r}_1, \mathbf{r}_2, \omega) = \sum_s \frac{\Psi_{0s}(\mathbf{r}_1)\Psi^+_{s0}(\mathbf{r}_2)}{\omega - \varepsilon_s - \mu + i\delta} + \sum_{s'} \frac{\Psi^+_{0s'}(\mathbf{r}_2)\Psi_{s'0}(\mathbf{r}_1)}{\omega + \varepsilon_{s'} - \mu - i\delta} \tag{9}$$

where s and s' run over the N+1 and N-1 electrons excited states, $\Psi_{0s}(\mathbf{r})$ ($\Psi_{0s'}(\mathbf{r})$) is the expectation value of the creation operator of an electron at position $\mathbf{r}$ between the N particle ground state and the N+1 (N-1) particle excited state labeled by $s(s')$, and $\varepsilon_s(\varepsilon_{s'})$ is the energy of the N+1 (N-1) excited states; the infinitesimally small imaginary term $i\delta$ is needed for the convergence of the Fourier transform over the time variable. In expression (8) all the main ingredients of photoemission experiments appear, i.e. the photon energy ω and the energy differences between the N-electrons system and the N$\pm$1 electrons excited system. This explains why Green's functions are the main tools for band structure calculations: the excitation energies of a system are obtained by the determination of the poles of G. The Green's function cannot be calculated exactly for realistic systems. It obeys a Dyson equation of the form

$$G(1,2) = G_0(1,2) + \int d(34)G_0(1,3)\Sigma(3,4)G(4,2), \tag{10}$$

where G_0 is the non-interacting Green's function and Σ is the self-energy, a non-hermitian, non-local and energy dependent operator. In this framework, by introducing the equation of motion for G, we can obtain the so called 'quasiparticle equation':

$$\left[-\frac{\nabla^2}{2} + V_{ext}(\mathbf{r}_1) + V_H(\mathbf{r}_1)\right]\phi_s(\mathbf{r}_1,\omega) + \int d\mathbf{r}_2\Sigma(\mathbf{r}_1,\mathbf{r}_2,\omega)\phi_s(\mathbf{r}_2,\omega) = \varepsilon_s(\omega)(\mathbf{r}_1,\omega). \tag{11}$$

The energies ε are the poles of the Green's function, so that the solution of Eq. (11) gives directly the electronic band structure of the system. It is worth noticing that this equation reduces to the Hartree equations when $\Sigma = 0$, to the Hartree-Fock ones when $\Sigma = iGV$, and to the Kohn-Sham equations when, instead of Σ, a local, hermitian and energy independent operator is taken: $\Sigma \rightarrow V^{KS}_{xc}(r)$. This observation gives a *a−posteriori* justification of the use of the Kohn-Sham equations to calculate the excitation energies; the qualitative agreement between DFT and experimental band structure comes from the fact that, somehow, the Kohn-Sham equations already carry a lot of physics, containing, at least approximately, exchange and correlation effects. In other words, $V^{KS}_{xc}(r)$ is often a quite good approximation for the *true* self-energy Σ. In the same way in which in DFT we have to use an approximation for V_{xc}^{KS}, we have now to face the problem of finding a suitable expression for Σ. But at odds with the DFT case, within the Many-Body theory we have a closed set of equations, the Hedin Eq.,[33] which together with Eq. (10) define implicitly the self-energy Σ:

$$\Sigma(12) = i \int G(13)\Gamma(324)W(41)d(34); \tag{12}$$
$$W(12) = \nu(12) + \int \nu(13)P(34)W(42)d(34);$$
$$P(12) = -i \int G(13)G(41)\Gamma(342)d(34);$$
$$\Gamma(123) = \delta(12)\delta(13) + \int \frac{\delta\Sigma(12)}{\delta G(45)} G(46)G(75)\Gamma(673)d(4567),$$

where Γ is called vertex function, P is the polarization operator and W is the dynamically screened potential, $W = \varepsilon^{-1}\nu$.

In principle, the Hedin equations could be solved self-consistently, with an iterative procedure; in practice full self-consistency is never achieved for realistic systems. As a first step we can neglect the vertex corrections by assuming $\Gamma(123) = \delta(12)\delta(13)$; in this way the self-energy operator takes the form $\Sigma = iGW$, which is the famous 'GW' approximation.[34,35] Computing quasiparticle energies within the GW approximation, nowadays represents the most refined method for band structure calculation. The GW method can be applied to the calculation of the band structures of solids and surfaces and to the determination of the energy levels of molecules and of atoms. In order to describe optical spectra, we have to take into account excited states of a completely different nature with respect to the ones involved in photoemission spectroscopy. In the latter the final state of the system is charged since one electron has been removed or added to it; in the first, instead, the system exchanging energy with the probe, is left in what is called a neutral excited state (see Fig. 1c). Excitons are examples of neutral excited states that are commonly detected by optical spectroscopy; in a very schematic way excitons can be thought as bounded electron-hole states. The physical quantity which is directly connected to this spectrum is the macroscopic dielectric function ε_M that, following the derivation of Adler and Wiser,[36] is given, in terms of the microscopic dielectric function,$\varepsilon_{G,G'}(\mathbf{q},\omega)$ by:

$$\varepsilon_M(\omega) = \lim_{\mathbf{q}\to 0} \frac{1}{(\varepsilon(q,\omega))^{-1}_{\mathbf{G}=0,\mathbf{G}'=0}} \tag{13}$$

The absorption spectrum is proportional to the imaginary part of the macroscopic dielectric function. Adopting the same level of approximation that we have introduced to obtain GW quasiparticle energies, i.e. neglecting the vertex correction by putting $\Gamma = \delta\delta$, we get the so called random phase approximation (RPA) for the dielectric matrix. Within this approximation, neglecting local field effects, the response to a longitudinal field, for $q \rightarrow 0$, is:

$$\varepsilon_M(\omega) = 1 - \lim_{\mathbf{q}\rightarrow 0} \nu(q) \sum_{\nu c} \frac{|\langle \nu | e^{-iqr} | c \rangle|^2}{\omega - (\varepsilon_c - \varepsilon_\nu) - i\eta} \tag{14}$$

$|\nu\rangle(|c\rangle)$ represent valence (conduction) single particle state and $\varepsilon_\nu(\varepsilon_c)$ is its corresponding energy. From Eq. (14) we see that the sum of all possible independent transitions gives rise to the spectra, its structures are located at quasiparticle energy differences and the line shape is determined by the matrix elements of e^{iqr} between a valence state and a conduction state independently for each transition. In fact, in a non interacting quasiparticle scheme, the neutral excited state can be seen to be a sum of independent excitations of one electron from the valence to the conduction band. The spectrum may also be strongly dependent on local field effects, which must be included for highly anisotropic systems such as surfaces, nanowires and clusters. Even including local field effects the optical spectra of solids obtained within RPA are often not even in qualitative agreement with experiments. Due to some fortuitous cancellation of errors, DFT-RPA can sometimes yield very good agreement with experiment, but it should be stressed, however, that such good quantitative agreement is not generally found.

2.2.2. *The Bethe-Salpeter equation*

The vertex corrections cannot be completely neglected if we want to obtain optical spectra in agreement with experiments. Neglecting vertex corrections means to neglect the interaction between the hole and the electron that are formed within the polarization processes of the system. In fact putting $\Gamma = \delta\delta$ in the Hedin equation for P, leads to an expression for it, namely P = -iGG, in which the hole and the electron propagators are independent of one another. In the description of absorption experiments, in which the excited electron remains in the sample, it is evident that the effects of the electron-hole interaction, i.e. the excitonic effects, cannot be neglected. The effects of the electron-hole interaction in the spectra can be included in MBPT through an effective two-body Hamiltonian, the so called excitonic Hamiltonian which is the projection of the Bethe-Salpeter equation for the four-point modified polarizability ${}^4\overline{P}$

$$ {}^4\overline{P} = {}^4 P_{IQP} + {}^4 P_{IQP} K {}^4\overline{P} \tag{15} $$

in the transition space, i.e. into a basis made of couples of single quasiparticle states, usually labeled by band and wave vector indices. The kernel K of this integral equation contains an electron-hole exchange $\overline{\nu}$ and an electron-hole attraction $-W$ term. The excitonic Hamiltonian is not necessarily hermitian, however, in calculations just its hermitian part is usually taken into account, which is:

$$ H^{2p,exc}_{(\nu ck)(\nu' c' k')} = (E_{ck} - E_{vk'})\delta_{\nu\nu'}\delta_{cc'}\delta_{kk'} + 2\nu^{\nu' c' k'}_{\nu ck} - W^{\nu' c' k'}_{\nu ck} \tag{16} $$

Here $E_{ck}, (E_{vk'})$ are the quasiparticle energies, calculated within the GW approximation, of the states (ck) and (vk'). In terms of the eigenvalues and eigenvectors of the excitonic Hamiltonian, namely:

$$ H^{2p,exc}_{(n_1 n_2)(n_3 n_4)} A^{(n_3 n_4)}_{\lambda} = E^{exc}_{\lambda} A^{(n_1 n_2)}_{\lambda} \tag{17} $$

the macroscopic dielectric function is:

$$ \varepsilon_M(\omega) = 1 - \lim_{\mathbf{q}\to 0} \nu(q) \sum_{\lambda} \frac{\left| \sum_{(n_1 n_2)} \langle n_1 | e^{-iqr} | n_2 \rangle A^{(n_1 n_2)}_{\lambda} \right|^2}{E^{exc}_{\lambda} - \omega - i\eta} \tag{18} $$

where each label n_i represents a couple of band and wave vector indices. Comparing this formula with the RPA dielectric function (Eq. (14)), we can see that in this case the dielectric function has poles at the eigenvalues of the excitonic Hamiltonian which are not anymore differences of occupied and unoccupied quasiparticle state energies. This fact usually moves the position of the structures in the spectra to lower energies. Moreover, as we can see from Eq. (17), each eigenstate of the excitonic Hamiltonian (17) is made by a sum of independent quasiparticle transitions; each transition contributes to it with a weight equal to $A^{n_1 n_2}_{\lambda}$. This mixing of independent transitions is present in the numerator of the dielectric function Eq. (18) and its effect is to strongly modify the lineshape of the spectra. From Eq. (18), we can see that to obtain the macroscopic dielectric function we need to diagonalize the excitonic Hamiltonian and to know the quasiparticle wavefunctions (in order to compute the dipole matrix elements present in the numerator). Usually in most $ab-initio$ calculation, it is assumed that Kohn-Sham wave functions are a very good approximation to the quasiparticle ones, the calculation of W is limited to a static RPA screening, and the quasiparticle energies are computed within the GW approximation of the self energy. To summarize, in order to compute the excitonic spectra we need a well converged structure calculation, Kohn-Sham wave functions, quasiparticle energies within GW and the (statically) screened coulomb interaction present in the excitonic Hamiltonian.

3. Physical Systems

3.1. *Hydrogenated silicon nanocrystals*

In this section we present a comprehensive analysis of the structural, electronic and optical properties of hydrogenated silicon nanocrystals (H-Si-nc) as a function of size and symmetry and in particular we will point out the main changes induced by the nanocrystal excitation. The study of H-Si-nc has been done within DFT, using a pseudopotential, plane-wave approach. All the calculations have been performed with the ABINIT code.[37] Norm-conserving, non-local Hamann-type pseudopotentials have been used. The Kohn-Sham wave functions have been expanded within a plane-wave basis set, choosing an energy cutoff of 32 Ry. The calculations performed are not spin-polarised. Each H-Si-nc has been embedded within a large cubic supercell, containing vacuum in order to make nanocrystal-nanocrystal interactions negligible. The calculations for each cluster have been performed both in ground and excited state as described in section 2.1.1. The starting configuration for each cluster has been fixed with all Si atoms occupying the same position as in the bulk crystal, and passivating the surface with H atoms placed along the bulk crystal directions, at a distance determined by studying the SiH_4 molecule. It is worth pointing out that the starting H-Si-nc has T_d symmetry, which is kept during relaxation in the ground state configuration. Nevertheless for excited state configurations such symmetry is generally lost, due to the occupation of excited energy levels. We have first of all investigated the structural distortions caused by the relaxation of these structures in different electronic configurations. To qualitatively appreciate the structural changes we have plotted in Fig. 2 the relaxed structures of some of the considered clusters, both in ground- and excited-state configurations. The analysis of the structural properties reveals that the average Si-Si bond approaches the bulk bond length as the cluster dimension increases. In particular, moving from the center of the cluster toward the surface, a contraction of the outer Si shells is observed. The presence of a electron-hole pair in the clusters causes a strong deformation

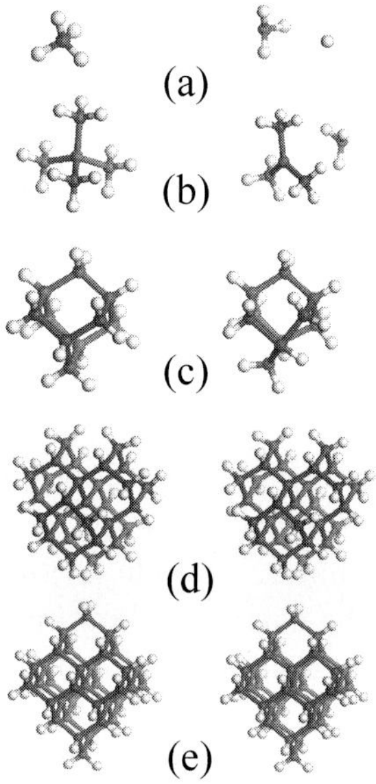

Fig. 2. Structural models for the a) Si_1H_4, b) Si_5H_{12}, c) $Si_{10}H_{16}$, d) $Si_{29}H_{36}$ and e) $Si_{35}H_{36}$ clusters at relaxed geometry in the ground- (left panels) and excited-state (right panels) configuration.

of the structures with respect to their ground-state configuration, and this is more evident for smaller systems and at the surface of the H-Si-nc. This is what we expect, since for large clusters the charge density perturbation is distributed throughout all the structure, and the effect it locally induces becomes less evident (small adjustments of bonds and angles occur with respect to the ground state). Baierle et al.[38] and G. Allan et al.[39] stressed the importance of bond distortion at the Si-nc surface in the excited state (ES) in creating an intrinsic localized state responsible of the PL emission. The structural analysis is immediately reflected into the electronic structure. In Fig. 3 we report the Kohn-Sham energy levels for the Si_5H_{12}, $Si_{10}H_{16}$, $Si_{29}H_{36}$ and $Si_{35}H_{36}$ clusters in both the (a) ground- (on the left) and (b) excited-state (on the right) configurations. It can be noted the expected decrease of the energy gap on increasing the cluster dimension and also that the excitation of the electron-hole pair causes a reduction of the energy gap as much significant as smaller is the cluster. For small excited clusters the HOMO (Highest Occupied Molecular Orbital) and LUMO (Lowest Unoccupied Molecular Orbital) become strongly localised in

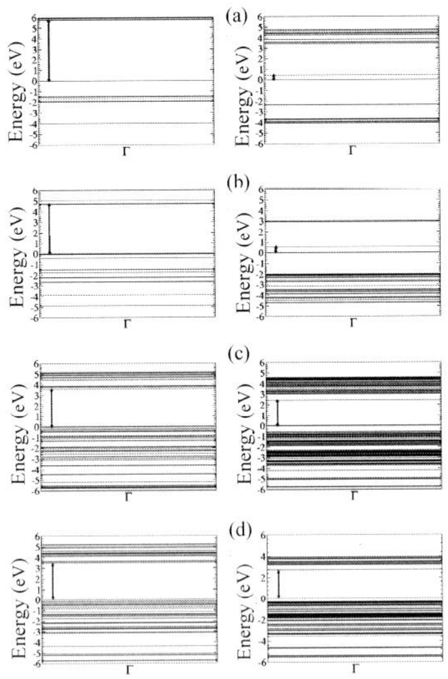

Fig. 3. Calculated energy levels at the Γ point for the a) Si_5H_{12}, b) $Si_{10}H_{16}$, c) $Si_{29}H_{36}$ and d) $Si_{35}H_{36}$ clusters in ground (left) and excited (right) state configuration. The energies are referred to the higher valence level.

correspondence of the distortion, giving rise to defect-like states which reduce the gap. The distortion induced by the nanocluster excitation can give a possible explanation of the observed Stokes shift in these systems. The radiation absorption of the cluster in its ground state configuration induces a transition between the HOMO and LUMO levels, which for all these clusters is optically allowed. Such a transition is followed by a cluster relaxation in the excited state configuration giving rise to distorted geometries (as previously shown) and to new LUMO and HOMO, whose energy difference is smaller than that in the ground-state geometry (see Fig. 3). It is between these two last states that emission occurs, thus explaining the Stokes shift. It is also worth pointing out how such a shift changes as a function of the dimension. Being the distortion smaller for larger clusters, it is expected that the Stokes shift decreases on increasing the dimension. This is shown in table 1 where the absorption and emission energies together with the Stokes Shift calculated as described in section 2.1.1 are reported. Gain calculations have also been performed preliminary on the $Si_{10}H_{16}$ cluster through the evaluation of the oscillator strengths between different energy levels at the band edges and the calculation of the imaginary part of the dielectric function in the cases in which population inversion was induced between

Table 1. Absorption and emission energy gaps and Stokes shift calculated within the Δ-SCF approach for the considered H-Si-nc. All values are in eV.

	Absorption	Emission	Stokes shift
$\mathbf{Si_1H_4}$	8.76	0.38	8.38
$\mathbf{Si_5H_{12}}$	6.09	0.42	5.67
$\mathbf{Si_{10}H_{16}}$	4.81	0.41	4.40
$\mathbf{Si_{29}H_{36}}$	3.65	2.29	1.35
$\mathbf{Si_{35}H_{36}}$	3.56	2.64	0.92

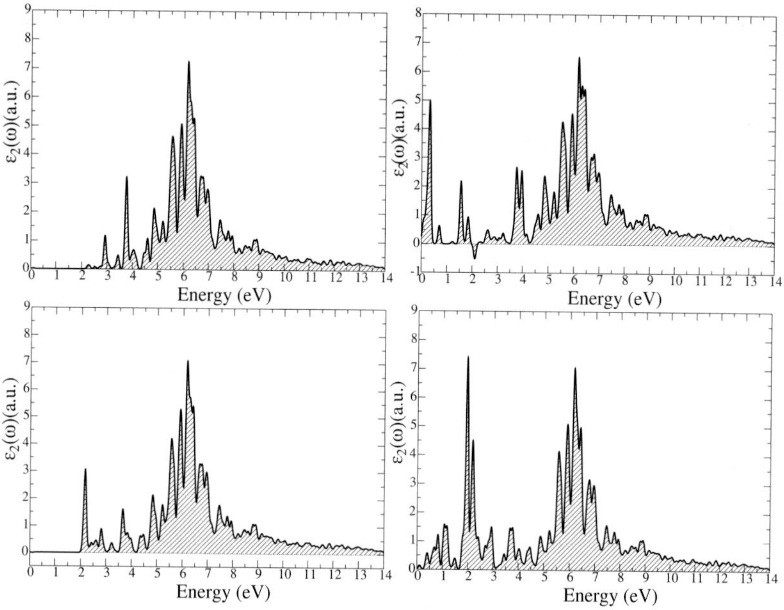

Fig. 4. Imaginary part of the dielectric function for the $Si_{10}H_{16}$ cluster in the excited state geometry with four different electronic configurations. Top panel: ground state configuration on the left, population inversion between the (HOMO-1) and LUMO states on the right. Bottom panel: population inversion between the HOMO and LUMO states on the left and between the HOMO and (LUMO+1) on the right.

these states. The spectra obtained are shown in Fig. 4 for the ground state configuration, i.e. absorption spectra (top panel on the left) and for three different models of population inversion. In the first case (see top panel on the right of Fig. 4) we have depleted the (HOMO-1) state and filled the LUMO state, in the second one the population inversion has been induced between the HOMO and LUMO states (bottom panel on the left) and finally in the last case the HOMO has been depleted and the (LUMO+1) filled (bottom panel on the right). As one can see the absorption spectra in the three cases of population inversion present new optical features in the low energy range due to new inter- and intra-band transitions induced by the considered population inversion, but negative features, i.e. gain, are present only in the first and second population model. In the first case the negative peak is due to the (HOMO-1)-HOMO intraband transition, while in the second case a really negligible gain peak is present near the zero energy due to the HOMO-LUMO transition (in this case the energy gap is of 0.11 eV and the oscillator strength between HOMO and LUMO is really small. No gain features instead are present for the third model due to the forbidden nature of the HOMO-LUMO+1 transition.

3.2. *Oxidized silicon nanocrystals*

Recent experimental data have shown a strong evidence that the surface changes of silicon nanocrystals exposed to oxygen produce substantial impact on their optoelectronics properties, thus oxidation at the surface has to be taken into account. In this section we will analize the structural, electronic and optical properties of oxidized Si-nc first of all, using the Δ-SCF method (see section 2.1.1).[40–43] Two classes of systems have been studied, based on the Si_{10} and the Si_{29} core based nanoclusters, and considering three type of Si/O bonds: the Si-O back bond, the Si$>O$ bridge and the Si=O double bond. Through formation energy calculation we have found that the configuration with the back-bonded oxygen is not favored with respect to the other two and moreover the bridge bonded configuration has been demonstrated to lead to the stablest isomer configuration by Gatti and Onida,[19,44] too. The geometries for the $Si_{10}H_{14}$ >O (bridge bond) and the $Si_{10}H_{14}$=O (double bond) are shown in Fig. 5. Full relaxation with respect to the atomic positions is performed within DFT limit for all systems both in the ground and excited configurations using norm conserving LDA pseudopotential with an energy cutoff of 60 Ry.[45] The ionic relaxation has produced structural changes with respect to the initial geometry which strongly depend on the type of surface termination. In the case of $Si_{10}H_{14}$=O, the changes are mainly localized near the O atom, in particular the angle between the double bonded O and its linked Si atom is modified (see Fig. 5). In the bridge structure, instead, the deformation is localized around the Si-O-Si bond determining a considerable strain in the Si-Si dimer distances.[44] Similar results are obtained for the larger Si_{29} based clusters. The only difference is that now the distortion induced by the promotion of an electron is smaller, as expected, since for larger clusters the charge density perturbation is distributed throughout the structure, and the effect locally induced becomes less evident. These structural changes are reflected in the electronic and optical properties.

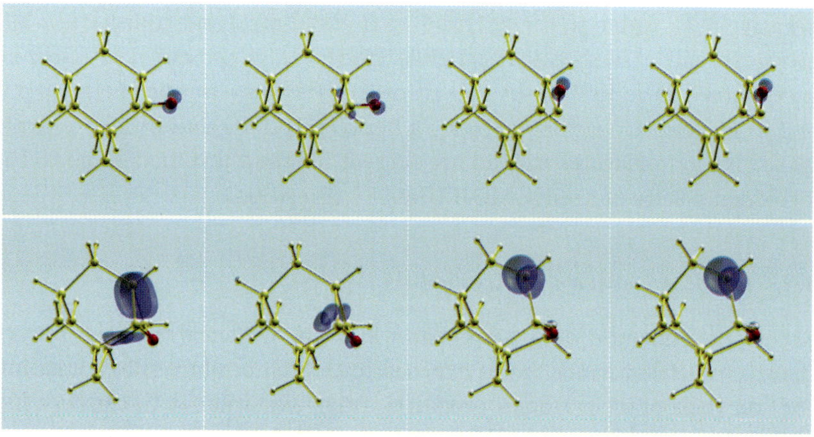

Fig. 5. Calculated HOMO(first and third column) and LUMO (second and fourth column) charge density for $Si_{10}H_{14}O$-double (top panel), $Si_{10}H_{14}O$-bridge (bottom panel) for the ground (first and second column) and excited (third and fourth column) states; the structures shown are the relaxed ones.

In Fig. 5 we also show the HOMO and LUMO charge density for the systems studied. It is evident that in all cases these states are mainly localized on the distorted region of the cluster.

In Table 2 absorption and emission gaps are reported: the red shift of the emission gap with respect to the absorption is less evident for the case of the cluster with the double-bonded oxygen (see the Stokes shift values); the same can be observed for the double-bonded $Si_{29}H_{34}O$. The

Table 2. Absorption and emission energy gaps and Stokes shift calculated within the Δ-SCF approach. All values are in eV.

	Absorption	**Emission**	**Stokes shift**
$\mathbf{Si_{10}H_{14}{=}O}$	2.79	1.09	1.70
$\mathbf{Si_{10}H_{14} > O}$	4.03	0.13	3.90
$\mathbf{Si_{29}H_{34}{=}O}$	2.82	1.17	1.65
$\mathbf{Si_{29}H_{34} > O}$	3.29	3.01	0.28

Table 3. Absorption and emission gaps calculated as HOMO-LUMO differences within DFT-LDA, GW, and with the inclusion of excitonic and local field effects (BS-LF). In parenthesis also the lowest dark transitions (when present) are given. All values are in eV.

	Absorption LDA GW BS-LF	**Emission** LDA GW BS-LF	**Stokes shift**
$\mathbf{Si_{10}H_{14}=O}$	3.3 (2.5), 7.3 (6.5), 3.7 (2.7)	0.8, 4.6, 1.0	2.7
$\mathbf{Si_{10}H_{14}>O}$	3.4, 7.6, 4.0	0.1, 3.5, 1.5	2.5
$\mathbf{Si_{29}H_{34}=O}$	2.5, 6.0, 3.7 (3.1)	0.9, 4.1, 1.2	2.5
$\mathbf{Si_{29}H_{34}>O}$	2.3, 4.8, 2.3	0.4, 3.0, 2.2 (0.3)	0.1

oxygen double-bonded seems hence almost size independent: actually, the presence of this kind of bond creates localized states within the gap that are not affected by quantum confinement as previously predicted by Luppi and Ossicini.[17]

Our aim is to allow a direct comparison between experimental data and theoretical results, thus we have calculate not only the transition energies within the Δ-SCF approach but also directly the absorption and emission optical spectra. Actually, for both the calculated GS and ES optimized geometry, we have evaluated the optical response $Im(\varepsilon_{NC}(\omega))$ (the imaginary part of the nanocrystal dielectric function) through first-principles calculations also beyond the one-particle approach. We have considered the self-energy corrections by means of the GW method and the excitonic effects through the solution of the Bethe-Salpeter equation. The effect of the local fields (LF) is also included, to take into account the inhomogeneity of the systems. In Table 3, the calculated gaps (as energy differences between LUMO and HOMO) at different levels of approximation are reported for both the Si_{10} and Si_{29} based nanocrystals. The main result common to absorption and emission is the opening of the LDA band-gap with the GW corrections by amounts weakly dependent on the surface termination but much larger than the corresponding 0.6 eV of the Si bulk case. Looking at the BS-LF calculations, we note a sort of compensation (more evident in the GS than in the ES) of the self-energy and excitonic contributions: the BS-LF values return similar to the LDA ones. The only exception are the BS-LF calculations for the excited state geometries of the clusters with Si-O-Si bridge bonds at the surface. Concerning the differences between the values of the Stokes shifts calculated through the Δ-SCF approach in Table 2 or through the MBPT in Table 3 are essentially due to the ability or not of the two methods of distinguish dark transitions. In the MBPT the oscillator strengths of each transition are known while the Δ-SCF approach only gives the possibility to find the energy of the first excitation: if this transition is dark (and the Δ-SCF approach do not give this information) the associated energy is not the real optical gap. A clearer insight on the MBPT results is offered by Fig. 6 (left panel), where the calculated absorption and emission spectra for all the oxidized Si_{10} based clusters are depicted and compared with the fully hydrogenated cluster. Self-energy, local-field and excitonic effects are fully taken into account. Concerning the absorption spectra (Fig. 6, dashed lines), all three cases

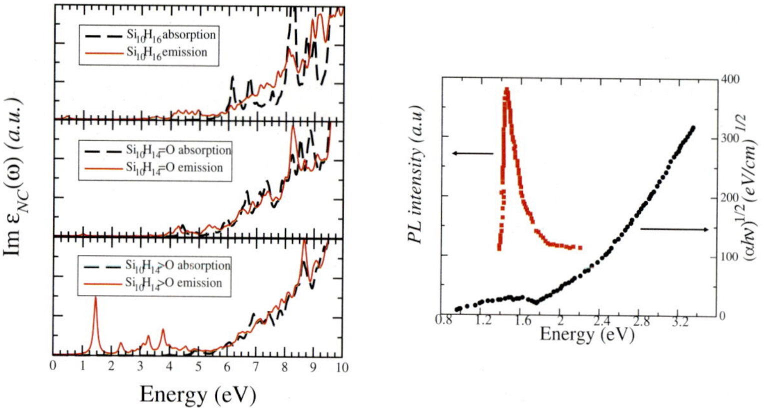

Fig. 6. Emission (solid line) and absorption (dashed line) spectra: imaginary part of the dielectric function for the three considered Si-nc. On the left: $Si_{10}H_{16}$ (top panel), $Si_{10}H_{14}{=}O$ (central panel) and $Si_{10}H_{14} > O$ (bottom panel). On the right: experimental results for emission (red curve on the left) and absorption (on the right) by Ma et al.[46]

show a similar smooth increase in the absorption features. Different is the situation for the emission related spectra (Fig. 6, solid lines). Here, whereas the situation remain similar for the fully hydrogenated $Si_{10}H_{16}$ (top panel) cluster and for the $Si_{10}H_{14}{=}O$ (central panel) cluster, in the case of a Si-O-Si bridge bond (Fig. 6 (bottom panel)) an important excitonic peak, separated from the rest of the spectrum, is evident at 1.5 eV. Actually bound excitons are present also in the fully hydrogenated (at 0.4 eV) and in the $Si_{10}H_{14}{=}O$ (at 1.0 eV) clusters, nevertheless, the related transitions are almost dark and the emission intensity is very low. Only in the case of the Si-O-Si bridge bond a clear PL peak appears thanks to the strong oscillator strength of the related transition. The right panel of Fig. 6 shows the experimental absorption and emission spectra measured by Ma et al.[46] for Si-nanodots embedded in SiO_2 matrix. A strong photoluminescence peak appears around 1.5 eV. Comparison of the experimental spectra with our results suggest that the presence of a Si-O-Si bridge bond at the surface of Si-nc and the relative deformation localized around the Si-O-Si bond can explain the nature of luminescence in Si nanocrystallites: only in this case the presence of an excitonic peak in the emission related spectra, red shifted with respect to the absorption onset, provides an explanation for both the observed SS and the near-visible PL in Si-nc. Similar results have been obtained in the case of Si_{29}-based clusters. Only in the case of O in bridge position there is a cage distortion at the interface that allows the presence of significant emission features in the optical region.

Optical gain calculations have been performed for the Si_{10} clusters with the oxygen atom double bonded (see Fig. 7) and bridge bonded (see Fig. 8) at the cluster surface (we remember that gain calculation have been performed within a single particle approach and this means that absorption spectra of Fig. 6 cannot be compared with that of Figs. 7 and 8). Excited state geometry has been considered in order to take into account the structural relaxation due to population inversion between band edges states. Four different electronic configurations have been considered in both cases: i) the ground state configuration (top panel on the left), the population inversion between ii) the (HOMO-1) and LUMO states (top panel on the right) iii) the HOMO and LUMO states (bottom panel on the left) and iv) the HOMO and (LUMO+1) states (bottom panel on the right). In each condition of population inversion new features, with respect to the absorption peaks in the ground state configuration, are present in the optical spectra in the low energy range due to both intra- and inter-band transitions. Anyway, optical gain, i.e. negative absorption peaks, can be observed only for population inversion between HOMO-LUMO and (HOMO-1)-LUMO states in both type of oxygen bond. On the contrary, the HOMO-(LUMO+1) transition is dark and the population inversion between these two states do not gives rise to gain peaks.

In the case of $Si_{10}H_{14}{=}O$ whose energy gap in the ground state configuration in 0.74 eV the negative peaks are due to the HOMO-LUMO transition in the case of HOMO-LUMO population inversion and to the (HOMO-1)-LUMO transition when the inversion is betweem (HOMO-1) and LUMO states. For the $Si_{10}H_{14}>O$ instead the gain in the HOMO-LUMO population inversion case is again due to the HOMO-LUMO transition while is due to the (HOMO-1)-HOMO intraband transition when the (HOMO-1) is depleted and the LUMO filled.

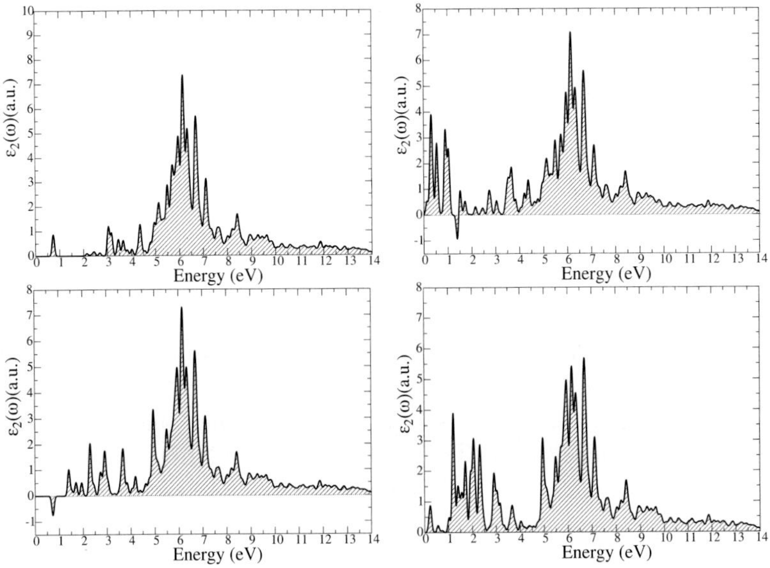

Fig. 7. Imaginary part of the dielectric function for the $Si_{10}H_{14}{=}O$ cluster in the ES geometry with four different electronic configurations. Top panel: ground state configuration on the left, population inversion between the (HOMO-1) and LUMO states on the right. Bottom panel: population inversion between the HOMO and LUMO states on the left and between the HOMO and (LUMO+1) on the right.

3.3. *Doped silicon nanocrystals*

As already said simultaneous doping with $n-$ and $p-$type impurities represent a way to overcome the low radiative recombination efficiency in our systems so, starting from the already described hydrogenated Si-nc and following the work of Fujii et al.,[23] we have doped the $Si_{35}H_{36}$ cluster locating the B and P impurities in substitutional positions just below the nanocrystal surface. It is worth mentioning that this arrangement represents the most stable configuration, as confirmed by theoretical and experimental works.[47–49] Full relaxation with respect to the atomic positions has been allowed and electronic properties have been computed through DFT calculations. The choice of studing the small $Si_{33}BPH_{36}$ cluster (see Fig. 10) (diameter around 1nm) is due to the fact that the GW-BSE calculation,[50] necessary to obtain the optical spectra, are very computing demanding. The energy levels of this system at the Γ point calculated at the optimized geometries are shown in Fig. 9 where only the levels corresponding to the HOMO, LUMO, HOMO-1 and LUMO+1 states are depicted. Calculated

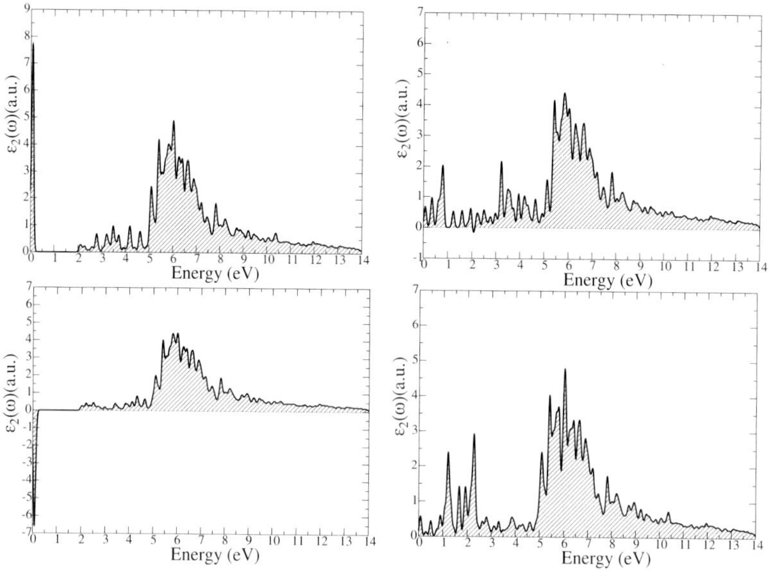

Fig. 8. Imaginary part of the dielectric function for the $Si_{10}H_{14}$ >O cluster in the ES geometry with four different electronic configurations. Top panel: ground state configuration on the left, population inversion between the (HOMO-1) and LUMO states on the right. Bottom panel: population inversion between the HOMO and LUMO states on the left and between the HOMO and (LUMO+1) on the right.

square modulus contour plots related to HOMO and LUMO states, have shown their localization within the Si-nc, in particular the HOMO state is localized on the B impurity while the LUMO is localized on the P one. The presence of these donor and acceptor states lowers the energy gap from 3.51 eV for the pure cluster to 2.86 eV for the doped one. In principle, starting with a bigger cluster, for which the energy gap is smaller than in this case, it is possible through codoping to tune the gap also below the bulk Si band gap as experimentally observed by Fuji and coworkers.[25] In order to give a complete description, within the many-body framework, of the codoped Si-nc response to an optical excitation, we consider both the self-energy corrections by means of the GW method[51] to obtain the quasiparticle energies and the excitonic effects through the solution of the Bethe-Salpeter equation. The effect of local fields is also included, to take into account the inhomogeneity of the systems.

To carry out emission spectra calculations, we have used the excited state geometry and the ground state electronic configuration as already

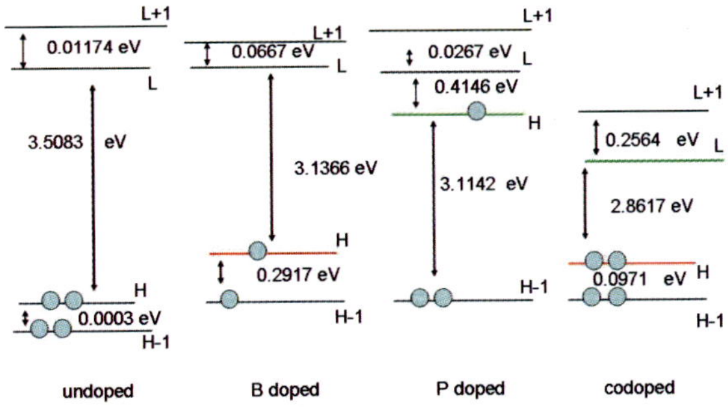

Fig. 9. Calculated energy levels at Γ point for the $Si_{33}BPH_{36}$-nc. Alignment has been performed locating at the same energy the fully occupied levels with the same type of localization.

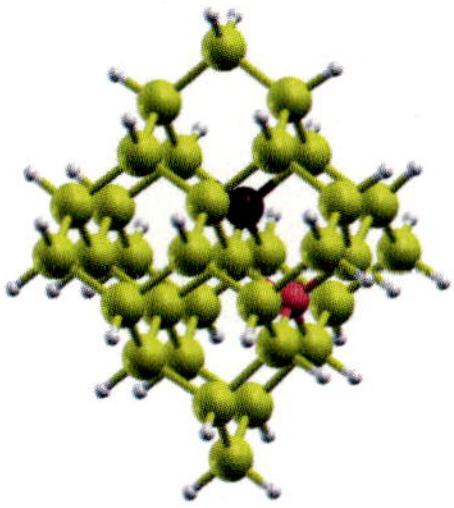

Fig. 10. Relaxed structure of the $Si_{33}BPH_{36}$ codoped nanocrystal (diameter=1.10 nm). Gray balls represent Si atoms, while the light gray balls are the hydrogens used to saturate the dangling bonds. B (dark gray) and P (black) impurities have been located at subsurface position in substitutional sites on opposite sides of the nanocrystals. The relaxed impurity distance is DBP=3.64 Å.

described in section 2.1.1. Thus, the electron-hole interaction is here considered also in the emission geometry. Figure 11 (right panel) shows the calculated absorption and emission spectra fully including the many-body effects. The electron-hole interaction yields significant variations with respect to the single-particle spectra (shown in the left panel), with an important transfer of the oscillator strength to the low energy side. Moreover, in the emission spectrum the rich structure of states characterized, in the low energy side, by the presence of excitons with largely different oscillator strengths, determines excitonic gaps well below the optical absorption onset. Thus the calculated emission spectrum results to be red shifted to lower energy with respect to the absorption one. This energy difference between emission and absorption, the Stokes shift, can be lead back to the relaxation

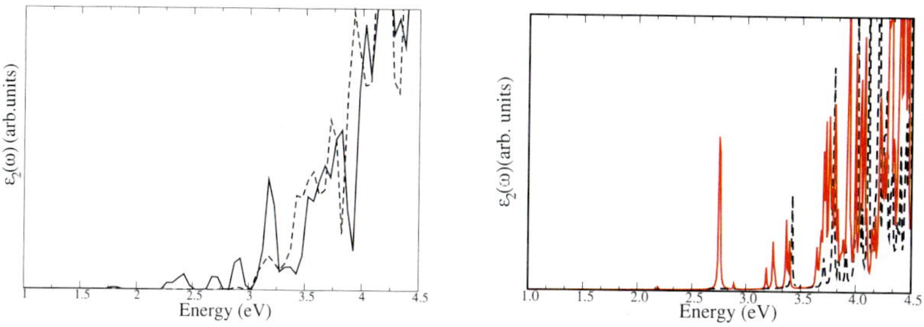

Fig. 11. Left panel: Single-particle imaginary part of the dielectric function for the codoped $Si_{33}BPH_{36}$ nanocrystal in the ground (dashed line) and excited (solid line) geometries. Right panel: absorption (dashed line) and emission (solid line) many-body spectra of $Si_{33}BPH_{36}$.

of the Si-nc after the excitation process. The new important features that appear in the emission many-body spectra are related to the presence of both B and P impurities as showed by Fig. 12, which gives the real-space probability distribution $|\psi_{exc}(r_e, r_h)|^2$ for the bound exciton as a function of the electron position r_e when the hole is fixed in a given r_h position. In this case the hole is fixed on the boron atom and we see that the bound exciton is mainly localized around the phosphorus atom. From Table 4, it can be seen that the single-particle DFT results strongly underestimate the absorption and emission edge with respect to the GW+BSE calculation, in which the excitonic effect are taken exactly into account. This means that, in this case, the cancellation between GW gap opening (which gives the electronic gap) and BSE gap shrinking (which originates the excitonic gap) is only partial.[52] The difference between the GW electronic gap and the GW+BSE optical excitonic gap gives the exciton binding energy E_b. We note the presence of exciton binding energies as big as 2.2 eV, which are very large if compared with bulk Si ($\sim$ 15 meV) or with carbon nanotubes[53,54] where $E_b \sim 1$ eV, but similar to those calculated for undoped Si-nc[55] of similar size and for Si and Ge small nanowires.[56,57] It is interesting to note that the HOMO-LUMO transition in the emission spectrum at 2.20 eV is almost dark while an important excitonic peak is evident at about 2.75 eV (see Fig. 11), red-shifted with respect to the first absorption peak. A preliminary optical gain calculation within single particle approach is shown in Fig. 13 for the considered codoped cluster. The same conditions of temperature and population inversion as for the other systems are considered. In this case no gain features can be observed except for an very negligible peak at around 1.75 eV in the case of HOMO-LUMO population inversion

Fig. 12. Excitonic wave function of $Si_{33}BPH_{36}$ (atom colors as in Fig. 10). The gray isosurface represents the probability distribution of the electron, with the hole fixed on the B impurity.

Table 4. Absorption and Emission energies calculated as HOMO-LUMO energy difference within the single-particle DFT, the many-body GW and the GW+BSE approaches.

$Si_{33}BPH_{36}$	DFT	GW	GW+BSE
Abs. (eV)	2.80	5.52	3.35
Ems. (eV)	1.79	4.37	2.20 (almost dark)
Stokes shift (eV)	1.01	1.15	1.15

(bottom panel on the left). This peak is due to the HOMO-LUMO transition but the oscillator strength between these two states is really small.

3.4. *Silicon nanocrystals embedded in a SiO_2 matrix*

In this section our goal is to build up a simple model to study the properties of Si nanocrystals embedded in SiO_2 matrix from a theoretical point of view.[58] We wanted, at the same time, two fundamental qualities: a silicon skeleton with a crystalline behavior for simulating the Si-nc and the simplest Si-SiO_2 interface, with the minimum number of dangling bonds or defects. For these reasons we have started with a cubic cell ($l = 14.32$ Å) of SiO_2 beta-cristobalite (BC) which is well known to have one of the simplest Si/SiO_2 interface because of its diamond-like structure.[59] We get the cell repeating twice along each cartesian axe the unit cell of SiO_2 BC. Then we obtained a small cluster simply deleting some oxygen atoms of the SiO_2 matrix and linking together the silicon atoms left with dangling bonds, as shown in Fig. 14. In this way we have built an initial supercell of 64 Si and 116 O atoms with 10 Si bonded together to form a small crystalline skeleton (T_d interstitial symmetry) with a very highly strained bond length

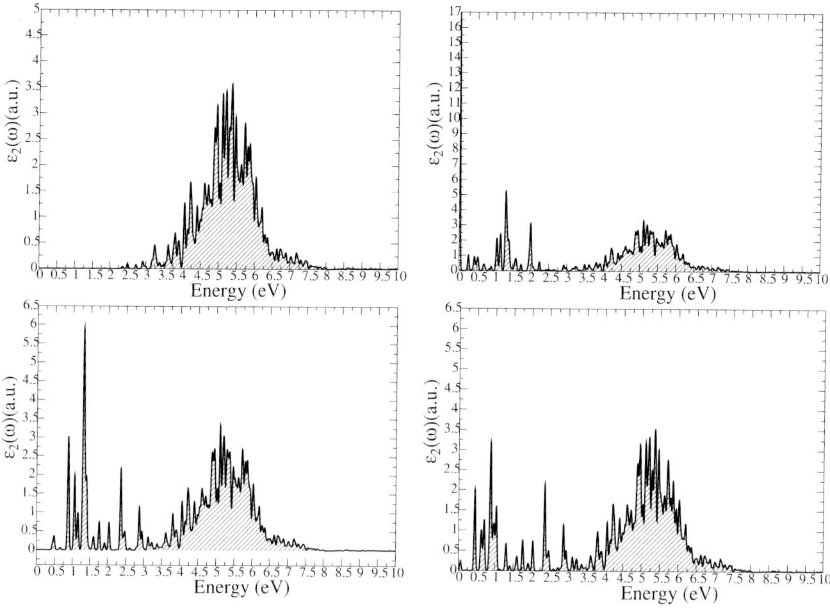

Fig. 13. Imaginary part of the dielectric function for the $Si_{33}BPH_{36}$ cluster with four different electronic configurations in the geometry of the excited state. Top panel: ground state configuration on the left, population inversion between the (HOMO-1) and LUMO states on the right. Bottom panel: population inversion between the HOMO and LUMO states on the left and between the HOMO and (LUMO+1) on the right.

with respect to the bulk case: 3.1 Å. We have performed a molecular dimanics simulation through a Car-Parrinello approach to fully relax the system. Figure 14 shows the final relaxed supercell structure after the geometry optimization for the Si-nc (Si_{10}) in SiO_2. Looking at the cluster behavior we find that the skeleton is still crystalline-like (diamond) with a Si-Si bond length of 2.67 Å, that means a strain of 14% respect to the bulk case. This rearrangement causes a complex deformation of the SiO_2 matrix around the cluster both in bond lengths and angles. Nevertheless the deformation doesn't affect the all SiO_2 matrix. It is actually possible to find still a good BC crystalline structure, in terms of angles and bond-length, at a distance from the cluster's atoms of 0.8-0.9 nm. This means that the cluster is surrounded by a cap-shell of stressed SiO_2 BC with a thickness of about 1 nm which progressively goes towards a pure crystalline BC. We have in summary a three-region picture: i) a strained Si-nc, ii) a cap shell, with a thickness of 8-9 Å, of distorted SiO_2, and iii) an unstressed crystalline

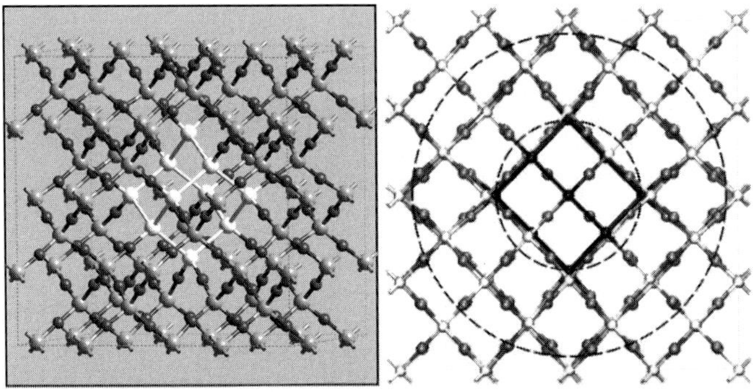

Fig. 14. Stick and ball pictures of the final optimized structure of Si_{10} in SiO_2. The dark gray spheres represent O atoms, light gray Si and white the Si atoms of the nanocrystal.

SiO_2. These three regions are highlighted in Fig. 14 (right panel). Despite the simplicity of the model this picture is in close agreement with what emerges from energy-filtered transmission electron microscopy (EFTMEM) and x-ray measurements on Si nanocrystals in SiO_2 samples produced by PECVD.[60,61] Our result is very important because it provides theoretical information on the structural properties of these types of materials, suggesting the existence of a not-sharp interface between Si-nc and the host SiO_2 matrix in which they are immersed.

This view is also supported by the analysis of the electronic properties reported in Fig. 15. The calculated E_g for the cluster in the matrix is 1.48 eV, that must be compared with the value of 4.58 eV for the E_g of BC SiO_2 (bulk).[62]

The strong reduction is originated by the presence, at the valence and conduction band edges, of confined, flat, states completely related to the Si_{10}-SiO_2 interface. Actually nor the isolated, H-passivated cluster (see

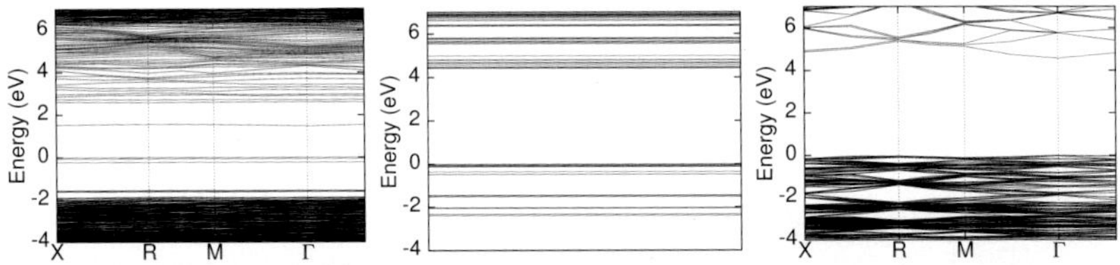

Fig. 15. Band structure along high symmetry points of the BZ for the Si_{10} cluster in SiO_2 (left panel) compared with the band structure of beta-cristobalite bulk (right panel) and the energy levels at the Γ point for the isolated $Si_{10}H_{16}$ cluster (central panel).

Fig. 15, central panel), neither the pure SiO_2 matrix see Fig. 15, right panel) show these states, whereas deep inside the valence and conduction bands the more k-dispersed states related to the SiO_2 matrix are still present. In Fig. 16 the HOMO and LUMO isosurfaces at the fixed value of 10% are reported; we clearly see that the distribution is totally confined in the Si-nc region with some weight on the interface O atoms. These cluster-related states originate strong absorption features in the optical region as witnessed from Fig. 17 where absorption spectra of the isolated Si_{10} cluster, of the pure matrix and of the composed system are reported. These features are entirely new, they do not exist for the isolated cluster or for the pure matrix, and so have to be due to the nature of the interface region. The

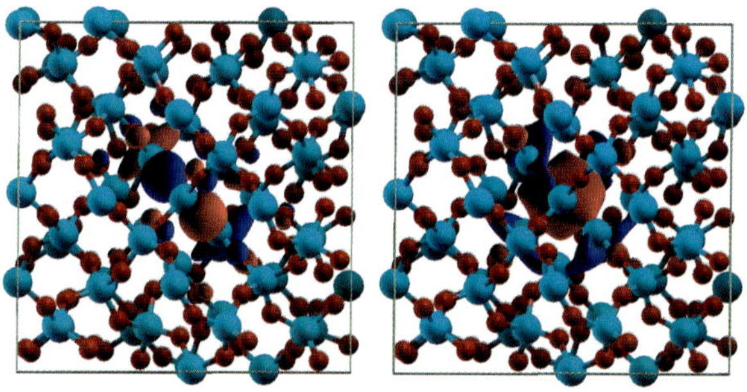

Fig. 16. Isosurfaces at fixed value (10% of max. amplitude) of the square modulus of highest occupied (HOMO) and lowest unoccupied (LUMO) Kohn-Sham orbitals for the Si_{10} cluster in the SiO_2 matrix.

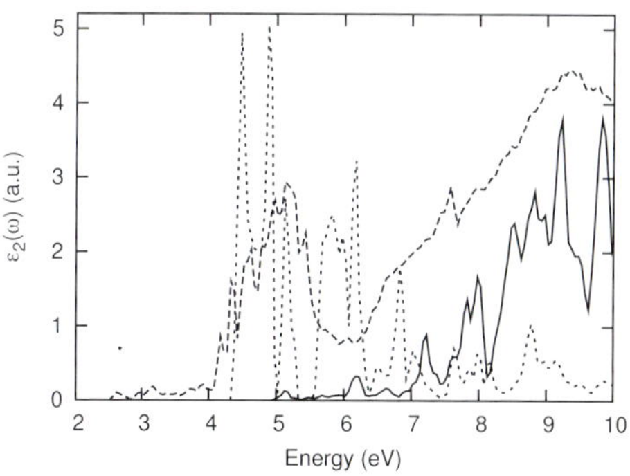

Fig. 17. Imaginary part of the dielectric function for the Si_{10} cluster in BC matrix (dashed line) compared with that of isolated, hydrogenated Si_{10} cluster (dotted line) and of the beta-cristobalite bulk (solid line).

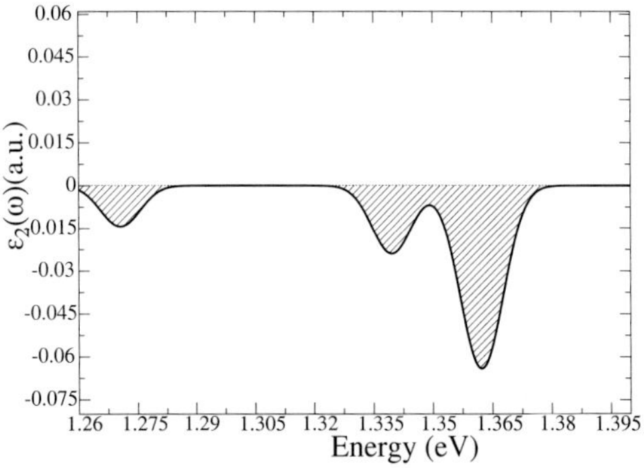

Fig. 18. The optical gain spectrum for the Si_{10} nanocluster embedded in a SiO_2 matrix in the low energy region.

origin of the PL observed in the red optical region for Si-nc immersed in a SiO_2 matrix can than be found in the properties of this interface. Our result concerning the role of both Si-nc and the interface Si-O region with respect to the absorption process is in close agreement with the X-ray absorption fine structure measurements[60] that indicate the presence of an intermediate region between the Si-nc and the SiO_2 matrix about 1 nm thick that participates in the light emission process. Very recently we have also calculated, for the first time, the gain spectra (as described in section 2.1.1) for this small Si-nc in the beta-cristobalite matrix, finding the presence of negative features in the low energy region of the spectrum for transitions related to electronic states centered on the nanocluster. Actually, the three negative peaks shown in Fig. 18 are really near one to the other (the region that include the three peaks is about 0.10 eV wide) and correspond to the HOMO-LUMO transition at different k-points in the reciprocal space.

4. Conclusions

In conclusion, we have performed a theoretical study of the properties of Si nanocrystals. We have demonstrated, by first-principle calculations also beyond the single particle approach, that the structural, electronic, and optical properties of Si nanocrystals strongly depend on the different passivation regimes and on the doping. Starting from hydrogen-covered Si nanoparticles we have shown that:

i) the creation of an electron-hole pair in the system with a change in the charge density induces significant distortion in particular on the smaller

clusters. Such distortions modify the energy gaps and induce energy levels arrangement that can explain the observed Stokes shift.
ii) Structural, electronic and optical properties calculations concerning the role of oxidation performed fully including excitonic effects suggests, if compared with experimental data, that only the presence of a Si-O-Si bridge bond at the surface of Si-nc can explain the nature of luminescence in these systems: only in this case,in the many body calculations, the presence of an excitonic peak in the emission-related spectra, redshifted with respect to the absorption onset, provide an explanation for both the observed SS and the near-visible PL in Si-nc.
iii) In the case of BP codoped Si-nc the study of the electronic properties shows that both HOMO and LUMO are localized around the impurity sites thus strongly lowering the enegy gap with respect to that of the pure Si-nc. Moreover, the emission spectra show a Stokes shift with respect to the absorption, which is due to the structural relaxation after the creation of the electron-hole pair. These facts allows electronic transitions between donor and acceptor states, making it possible to engineer the absorption and emission spectra of Si nanocrystals.
iv) When Si nanocrystals are embedded in a SiO_2 matrix, our results show that the SiO_2 cage is slightly deformed by the presence of the nanocrystals, that new electronic states are originated within the silica band gap and that both the Si atoms in the nanocrystals and the O atoms at the interface play a role in the optical properties. This theoretical analysis clearly point out the important role played not only by the Si nanocrystals but also by a modified silica capping region in determining the optoelectronic properties of the system. Our results help to clarify the experimental outcomes regarding the structural, electronic, and optical properties of Si-nc dispersed in SiO_2.[60,63]

For each considered system optical gain calculations have been performed within a single particle scheme. The obtained results could give some insight concerning the best nanoclusters characteristics and the energy levels important in order to observe optical gain.

Acknowledgments

We would like to acknowledge Dr Olivia Pulci from the University of Rome 2, Tor Vergata, Prof. Domenico Ninno and Dr Giovanni Cantele from the University of Naples “Federico II”, Dr Eleonora Luppi and Dr Marcello Luppi, for their contribution to the realization of this paper. All the calculations have been performed at CINECA-Bologna (we acknowledge the "Iniziativa calcolo parallelo" of CNR-INFM) and CICAIA-Modena advanced computing facilities.

References

1. S. Ossicini, L. Pavesi and F. Priolo, "Light Emitting Silicon for Microphotonics", *Springer Tracts on Modern Physics* **194** (Springer-Verlag, Berlin, 2003).
2. O. Bisi, S. Ossicini and L. Pavesi, *Surf. Sci. Rep.* **38**, 5 (2000).
3. B. Gelloz, A. Kojima and N. Koshida, *Appl. Phys. Lett.* **87**, 031107 (2005).
4. G. Franzó, A. Irrera, E. C. Moreira, M. Miritello, F. Iacona, D. Sanfilippo, G. Di Stefano, P. G. Falica and F. Priolo, *Appl. Phys.* **A77**, 57 (2003).
5. R. J. Walters, G. I. Bourianoff and H. A. Atwater, *Nature Materials* **4**, 143 (2005).
6. L. Pavesi, L. Dal Negro, C. Mazzoleni, G. Franzó and F. Priolo, *Nature* (London) **408**, 440 (2000).
7. L. Khriachtchev, M. Rasanen, S. Novikov and J. Sinkkonen, *Appl. Phys. Lett.* **79**, 1249 (2001).
8. L. Dal Negro, M. Cazzanelli, L. Pavesi, S. Ossicini, D. Pacifici, G. Franzó, F. Priolo and F. Iacona, *Appl. Phys. Lett.* **82**, 4636 (2003).
9. J. Ruan, P. M. Fauchet, L. Dal Negro, M. Cazzanelli and L. Pavesi, *Appl. Phys. Lett.* **83**, 5479 (2003).
10. M. Cazzanelli, D. Kovalev, L. Dal Negro, Z. Gaburro and L. Pavesi, *Phys. Rev. Lett.* **93**, 207042 (2004).
11. K. Luterova, K. Dohnalova, V. Servcek, I. Pelant, J.-P. Likforman, O. Crgut, P. Gilliot and B. Hönerlage, *Appl. Phys. Lett.* **87**, 3280 (2004).
12. *Towards the First Silicon Laser*, Nato Science Series No. 93, edited by L. D. Negro, S. Gaponenko and L. Pavesi (Kluwer Academic Publishers, Dordrecht, 2003).
13. S. Ossicini, C. Arcangeli, O. Bisi, E. Degoli, M. Luppi, R. Magri, L. Dal Negro and L. Pavesi "Gain theory and models in silicon nanostructures" in *Towards the First Silicon Laser*, ed. by L. Dal Negro, S. Gaponenko and L. Pavesi, Nato Science Series vol. 93, pp. 271-290 (Kluwer Academic Publisher, Dordrecht, 2003).
14. L. Dal Negro, M. Cazzanelli, Z. Gaburro, P. Bettotti, L. Pavesi, F. Priolo, G. Franzó, D. Pacifici and F. Iacona, "Stimulated emission in silicon nanocrystals: Gain measurement and rate equation modeling", in *Towards the First Silicon Laser*, edited by L. Pavesi, S. Gaponenko, L. Dal Negro, NATO Science Series, Vol. 93, pp. 145-164 (Kluwer Academic Publishers, Dordrecht 2003).
15. M. V. Wolkin, J. Jorne, P. M. Fauchet, G. Allan and C. Delerue, *Phys. Rev. Lett.* **82**, 197 (1999).
16. A. Puzder, A. J. Williamson, J. C. Grossman and G. Galli, *Phys. Rev. Lett.* **88**, 097401 (2002).

17. M. Luppi and S. Ossicini, *J. Appl. Phys.* **94**, 2130 (2003).
18. I. Vasiliev, J. R. Chelikowsky and R. M. Martin, *Phys. Rev.* **B65**, 121302(R) (2002).
19. M. Gatti and G. Onida, *Phys. Rev.* **B72**, 045442 (2005).
20. L. Ramos, J. Furthmüller and F. Bechstedt, *Appl. Phys. Lett.* **87**, 143113 (2005).
21. *Properties of Porous Silicon*, edited by L. T. Canham (INSPEC, London, 1997).
22. C. Delerue, G. Allan and M. Lannoo, *Phys. Rev.* **B64**, 193402 (2001).
23. M. Fujii, Y. Yamaguchi, Y. Takase, K. Ninomiya and S. Hayashi, *Appl. Phys. Lett.* **87**, 211919 (2005).
24. M. Fujii, K. Toshikiyo, Y. Takase, Y. Yamaguchi and S. Hayashi, *J. Appl. Phys.* **94**, 1990 (2003).
25. M. Fujii, Y. Yamaguchi, Y. Takase, K. Ninomiya and S. Hayashi, *Appl. Phys. Lett.* **85**, 1158 (2004).
26. O. Pulci, M. Marsili, E. Luppi, C. Hogan, V. Garbuio, F. Sottile, R. Magri and R. Del Sole, *Phys. Stat. Sol. (b)* **242**, No. 13, 2737-2750 (2005).
27. M. Palummo, M. Bruno, O. Pulci, E. Luppi, E. Degoli, S. Ossicini and R. Del Sole, *Surf. Sci.* **601**, Issue 13, 2696 (2007).
28. P. Hohenberg and W. Kohn, *Phys. Rev.* **B136**, 864 (1964).
29. W. Kohn and L. J. Sham, *Phys. Rev.* **A140**, 1113 (1965).
30. D. M. Ceperley and B. J. Alder, *Phys. Rev. Lett.* **45**, 566 (1980). J. P. Perdew and A. Zunger, *Phys. Rev.* **B23**, 5048 (1981).
31. F. Bassani and G. Pastori Parravicini, *Electronic States and Optical Transitions in Solids* (Pergamon Press, New York, 1975).
32. L. Fetter and J. D. Walecka, *Quantum Theory of Many Body Systems* (McGraw-Hill, New York, 1981).
33. L. Hedin, *Phys. Rev.* **A139**, 796 (1965). R. D. Mattuck, *A Guide to Feynman Diagrams in the Many Body Problem* (McGraw-Hill, New York, 1976).
34. L. Hedin and B. J. Lundquist, in: *Solid State Physics, Vol. 23*, edited by H. Ehrenreich, F. Seitz and D. Turnbull (Academic Press, New York, 1969), p. 1.
35. F. Aryasetiawan and O. Gunnarsson, *Rep. Prog. Phys.* **61**, 237 (1998). Aulbur W. G. Johsson and J. W. Wilkins, *Solid State Phys.* **54**, 1-218 (2000).
36. S. L. Adler, *Phys. Rev.* **126**, 413 (1962). N. Wiser, *Phys. Rev.* **129**, 62 (1963).
37. First-Principles computation of material properties: The ABINIT software project (URL http://www.abinit.org).
38. R. J. Baierle, M. J. Caldas, E. Molinari and S. Ossicini, *Solid State Comm.* **102**, 545 (1997).
39. G. Allan, C. Delerue and M. Lannoo, *Phys. Rev. Lett.* **76**, 2961 (1996).
40. E. Luppi, E. Degoli, G. Cantele, S. Ossicini, R. Magri, D. Ninno, O. Bisi, O. Pulci, G. Onida, M. Gatti, A. Incze and R. Del Sole, *Opt. Mater.* **27**, 1008 (2005).
41. A. Puzder, A. J. Williamson, J. C. Grossman and G. Galli, *J. Am. Chem. Soc.* **125**, 2786 (2003).
42. A. Franceschetti and S. T. Pantelides, *Phys. Rev.* **B68**, 033313 (2003).

43. E. Degoli, G. Cantele, E. Luppi, R. Magri, D. Ninno, O. Bisi and S. Ossicini, *Phys. Rev.* **B69**, 155411 (2004).
44. The $Si_{10}H_{14}$ >O cluster we considered here corresponds to the $Si_{10}H_{14}O$-sym of Ref. 19, where O make a bridge between "second neighbors" Si atoms. We have obtained similar results for the $Si_{10}H_{14}O$-asym case, where O is in between two "first neighbors" Si atoms.
45. The DFT calculations for oxidized clusters have been performed using the ESPRESSO package: S. Baroni, A. Dal Corso, S. de Gironcoli, P. Giannozzi, C. Cavazzoni, G. Ballabio, S. Scandolo, G. Chiarotti, P. Focher, A. Pasquarello, K. Laasonen, A. Trave, R. Car, N. Marzari and A. Kokalj, http://www.pwscf.org/.
46. Z. Ma, X. Liao, G. Kong and J. Chu, *Appl. Phys. Lett.* **75**, 1857 (1999).
47. S. Ossicini, E. Degoli, F. Iori, E. Luppi, R. Magri, G. Cantele, F. Trani and D. Ninno, *Appl. Phys. Lett.* **87**, 173120 (2005).
48. L. C. Ciacchi and M. C. Payne, *Phys. Rev. Lett.* **95**, 196101 (2005).
49. E. Garrone, F. Geobaldo, P. Rivolo, G. Amato, L. Boarino, M. Chiesa, E. Giamello, R. Gobetto, P. Ugliengo and A. Vitale, *Adv. Mater.* (Weinheim, Ger.) **17**, 528 (2005).
50. EXC Code, V. Olevano, http://www.bethe-salpeter.org.
51. We have used the non selfconsistent G_0W_0 approach within the RPA plasmon pole approximation. We use a planewave-frequency space code.
52. C. Delerue, M. Lannoo and G. Allan, *Phys. Rev. Lett.* **84**, 2457 (2000).
53. C. D. Spataru, S. Ismail-Beigi, L. X. Benedict and S. G. Louie, *Phys. Rev. Lett.* **92**, 077402 (2004).
54. E. Chang, G. Bussi, A. Ruini and E. Molinari, *Phys. Rev. Lett.* **92**, 196401 (2004).
55. E. Luppi, F. Iori, R. Magri, O. Pulci, E. Degoli, S. Ossicini and V. Olevano, *Phys. Rev.* **B75**, 033303 (2007).
56. M. Bruno, M. Palummo, A. Marini, R. Del Sole, V. Olevano, A. N. Kholod and S. Ossicini, *Phys. Rev.* **B72**, 153310 (2005).
57. M. Bruno, M. Palummo, A. Marini, R. Del Sole and S. Ossicini, *Phys. Rev. Lett.* **98**, 036807 (2007).
58. M. Luppi and S. Ossicini, *Phys. Rev.* **B71**, 035340 (2005).
59. H. Kageshima and K. Shiraishi, in: M. Scheffler, R. Zimmermann (Eds.), *Proc. 23rd Int. Conf. Phys. Semicon.*, World Scientific, Singapore, p. 903, (1996).
60. N. Daldosso, M. Luppi, S. Ossicini, E. Degoli, R. Magri, G. Dalba, P. Fornasini, R. Grisenti, F. Rocca, L. Pavesi, S. Boninelli, F. Priolo, C. Spinella and F. Iacona, *Phys. Rev.* **B68**, 085327 (2003).
61. T. Takagahara and K. Takeda, *Phys. Rev.* **B46**, 15 578 (1992).
62. The differences with respect to Ref. 58 are due to the different computational code used.
63. N. Daldosso, G. Dalba, R. Grisenti, L. D. Negro, L. Pavesi, F. Rocca, F. Priolo, D. Pacifici, G. Franzò and F. Iacona, *Physica E* (Amsterdam) **16**, 429 (2003).

CHAPTER 3

COMPUTATIONAL STUDIES OF FREE-STANDING SILICON NANOCLUSTERS

Olli Lehtonen and Dage Sundholm

Department of Chemistry,
P.O.Box 55, A.I. Virtasen aukio 1,
FIN-00014 University of Helsinki, Finland
olehtone@chem.helsinki.fi, sundholm@chem.helsinki.fi

In this review, we describe computational studies of silicon nanoclusters and compare the results with reported experiments. *Ab initio* and density functional theory methods for studies of optical properties of silicon nanoclusters are briefly discussed. The accuracy of optical gaps and oscillator strengths obtained using different computational methods is addressed. Various types of silicon structures, including hydrogen-capped, oxidized, silane-capped, and alkyl-capped nanoclusters have been investigated. Short oligosilanes have been studied in order to benchmark density functional theory with more accurate coupled-cluster methods. In the basis-set limit, density functional theory tends to underestimate optical gaps, whereas coupled-cluster calculations at the approximate singles and doubles level seem to yield accurate excitation energies for oligosilanes provided that large basis sets are employed. The oscillator strengths for the hydrogen-capped and oxidized silicon nanoclusters are significantly lower than observed experimentally, suggesting that functional groups of some sort on the cluster surface may be responsible for the observed bright light emission. Computations on silicon nanoclusters capped with linear silane chains show that they have very strong light absorption, the intensity of which is tuned by the length of the silane chain, whereas the absorption wave length depends mainly on the size of the nanocluster itself.

1. Introduction

Crystalline silicon is known to be a poor light emitter due to its indirect optical band gap.[1] The research activities in semiconductor optoelectronics have therefore mainly been focussed on compound semiconductors, like

GaAs[1] which has a direct band gap enabling efficient light emission. The discovery that porous silicon and nano-sized silicon clusters can be strongly luminescent initiated intense research activities. Porous silicon was observed to be photoluminescent at cryogenic temperatures already in 1984,[2] but with the discovery that silicon structures can be photoluminescent at room temperature[3] it became apparent that silicon can after all be a useful optical material for technical applications and devices.[4]

The intense visible photoluminescence (PL) from porous silicon occurs in the energy range of red to yellow, while the blue emission is rather weak.[5] Freestanding silicon nanoclusters are light emitters with strong luminescence in the energy range from red to blue depending on their size.[6–9] Many different mechanisms for the light emission from porous silicon have been proposed. These comprise quantum confinement effects, nanocrystal surface states, defects, molecular functional groups, and structural disorder.[10] The quantum confinement model has proven to be very successful to predicting PL properties that are in good agreement with observations, at least as far as the energy of the emitted light is concerned. However, the confinement model provides limited information about the emission mechanism at the atomic level. Actually, hitherto little is known about the luminescence mechanism and the microscopic nature of the structure at the light-emission centers. Besides the cluster size, there are other variables that affect the optical properties of the silicon nanoclusters. The shape and crystal structure might be important, but the surface chemistry including active functional groups certainly influences the optical properties of silicon nanoclusters. In preparative chemistry, new routes to synthesize silicon nanoclusters have been developed and also new approaches to tailor the surface chemistry are being evolved.[8,11] Jurbergs *et al.*[12] obtained strongly luminescent silicon nanoclusters with an optical yield exceeding 60%. To achieve such bright silicon nanoclusters they modified the surface properties by attaching octadecacene groups to the surface. In technical applications and for the microelectronic industry an interesting research direction is the design and construction of strongly luminescent silicon nanoclusters embedded in amorphous SiO_2 or in other transparent and robust matrices protecting the delicate light emitters.[13] Polysilylenes also show promise as materials for strong luminescence and electroluminescence with applications in light-emitting diodes and flat-panel screens.[11,14,15]

The origin of the PL is by no means been understood and no consensus prevails. A large number of luminescence mechanisms have been proposed.[9,10,16] It has even been proposed that the observed light would not be PL, but emitted by silicon nanoparticles due to thermal radiation.[17] The most popular explanation of the bright luminescence is quantum confinement. The exciton generated by the applied radiation is confined inside a nanocrystal in porous silicon and the emitted light is created through radiative recombination processes of the electron-hole pair. In crystalline silicon and other semiconductors with indirect band gaps, the optical recombination transition is forbidden, unless lattice vibrations (phonons) are activated in order to conserve the crystal momentum. The spatial confinement of electrons and holes inside nanocrystals increases the uncertainty of the crystal momentum and optical transitions become allowed.[18] According to the quantum confinement model the recombination rate of electrons and holes increases with decreasing size of the cluster. The present computational studies of absorption and emission spectra of freestanding silicon nanoclusters show that the confinement model alone cannot explain the observed strong PL of silicon nanoclusters because the calculated oscillator strengths for the hydrogen capped silicon nanoclusters[19–22] are several orders of magnitude smaller than observed in PL the experiments.[23]

2. Computational Methods

The two computational methods employed in these studies are density functional theory (DFT) and low-order coupled-cluster (CC) theory. Both methods are suitable for studies of excited electronic states of large molecules. In DFT calculations, the electron-correlation contributions are accounted for by an approximate exchange-correlation potential. DFT is thus computationally feasible for large systems and it provides accurate results already with moderate basis-set sizes. On the other hand, the CC approaches provide a rigorous way to obtain accurate results in the basis-set limit. In contrast to DFT, the accuracy of the CC calculations can at least in principle be increased by applying higher-order CC methods in combination with large basis sets. Unfortunately, the high computational costs render higher-order CC calculations on nanoclusters still unattainable; only the lowest order CC methods can be applied. The calculations presented in this chapter have been performed using the TURBOMOLE program package.[24] In the following sections, we provide a short description of the DFT and CC methods employed in our studies of silicon nanoclusters.

2.1. *Time-dependent density functional theory*

The density functional theory (DFT) model was originally developed for computations of ground-state properties,[25,26] but it was later extended to treat also time-dependent phenomena.[27] DFT is widely used by the chemistry and physics communities in both molecular and solid state applications. The fundamental variable in DFT is the electron density and the basic premise of DFT is that the knowledge of the electron density of the system is enough to compute other properties as well. Thus, no information about the actual wave function of the system is needed. The first approximation one has to introduce when performing computations is the choice of the approximate density functional.[28] The DFT equations include in principle all many-body effects through the exchange-correlation potential whose exact form is though unknown. Approximate functionals of various kinds yield slightly different results. Some functionals are suited for energy calculations and others provide more reliable properties. Currently, no approximate functional providing superior results exists. The main advantage of DFT is its relatively low computational cost compared to other methods of similar accuracy.

The time-dependent density functional theory (TDDFT) approach employed here to study excited-states properties is derived by transforming the time-dependent Kohn-Sham equations into the frequency domain and treating the external (electric) field as a perturbation. The first-order change of the electron density due to the perturbation is determined by using response theory.[29–32] Finally, this problem reduces into solving a matrix eigenvalue problem which determines the excitation energies and transition moments. Expressions for calculations of excited state properties including molecular gradients for excited state structure optimizations are most conveniently derived through a Lagrangian formalism.[33–36] Similar to ground-state DFT, TDDFT is also an approximate method and it has its own shortcomings as well.[32,37,38] Most TDDFT computations are performed within adiabatic approximation, which means that the exchange-correlation kernel is assumed to be frequency independent and one uses standard ground-state density functionals to obtain an approximation for the kernel.

The employed approximate functionals have proven to yield too small optical gaps for silicon nanoclusters. Hybrid functionals which include certain fraction of the Hartree-Fock exchange compensate partly the errors present in non-hybrid functionals resulting in more accurate excitation energies than obtained with functionals based purely on the local density approximation (LDA) or on the generalized gradient approximation (GGA). However, calculations employing hybrid functionals are computationally 10–100 times more expensive than those using the LDA and GGA functionals because of the nonlocal Hartree-Fock exchange terms. For nonhybrid functionals, the computation of the Coulomb energy and its derivatives can be accelerated by expanding the density in an auxiliary basis using the Coulomb norm (RI-J) without sacrificing the accuracy of the excited state properties. The computational savings due to the RI-J approximation render excited state optimizations for nano-sized molecules at the DFT level feasible.[39]

2.2. *Coupled-cluster methods*

Ab initio or wave function based methods for electronic structure calculations are computationally more expensive than DFT calculations. The main advantage with the *ab initio* approaches is that their accuracy can in principle be systematically improved to approach the solution of the Schrödinger equation (within the given finite basis set). In the *ab initio* quantum chemistry community, one of the aims of the method development is to find efficient algorithms and rational approximations that yield accurate molecular properties using a reasonable amount of computational resources.[40] The hierarchy of coupled-cluster (CC) models has proven to be a very successful approach. Lowest-order methods in the CC hierarchy can be applied to rather large molecules, whereas the accuracy and computational costs increase as high-order excitation amplitudes are considered in the calculation. The wave function of the lowest order method in the CC hierarchy of models is parameterized by the amplitudes of the single excitations from the occupied space to the virtual orbitals (CCS). The CCS and configuration interaction singles (CIS) calculations considering only singles excitation parameters yield identical excitation energies but slightly different wave functions.[41] The accuracy of the CIS and CCS calculations is usually not enough for studies of excited states for silicon nanoclusters as they neglect most of the electron correlation effects. The coupled-cluster approximate singles and doubles (CC2) model, which is an approximation to the coupled-cluster singles and doubles (CCSD) method, comes next in the CC series. The CC2 model in combination with the resolution of the identity (RI) approximation has proven to be a reliable and still feasible *ab initio* approach for studies of excited states of large molecules.[42–45] The computational costs for the CC2 calculations are significantly larger than for DFT, although only the single excitation amplitudes of the wave function need to be stored. Thus, the same number of parameters describes the left and right eigenvectors of the CC2 wave function as needed for parameterization of the corresponding TDDFT state.

The expressions for calculating the CC excited state energy and the energy derivatives can be derived using a Lagrange function for the CC energy of the excited state.[42,46] The excitation energies and transition moments are obtained from the poles and the residues of the linear response function by diagonalizing the nonsymmetric CC Jacobian. The excitation energies are thus obtained as the eigenvalues of the Jacobian; the matrix of the first derivative of the CC equations with respect to the cluster amplitudes.

3. Accuracy of TDDFT and CC2 Calculations

In *ab initio* and DFT calculations of excitation energies, the two largest sources of uncertainties are the level of correlation treatment and the size of the one-particle basis sets. The reliability of the employed computational level and the accuracy of the employed functional can be assessed by increasing the size of the one-particle basis set until it is practically complete and then compare the obtained excitation energies with available experimental data. Such a benchmarking procedure for linear silanes showed that the optical gaps calculated at the CC2 level agree well with experimental values for Si_2H_6 and Si_3H_8 measured in the gas phase, whereas for SiH_4 the discrepancy between calculated and measured values is as much as 0.7 eV.[47] An improved treatment of electron correlation effects was obtained using coupled-cluster calculations considering triple excitations (CC3).[46] The CC3 calculations yielded excitation energies in close agreement with the CC2 values suggesting that the reported experimental excitation threshold of 8.8 eV for SiH_4 does not correspond to a vertical excitation.[48] The best calculated value for the vertical excitation is 9.45 eV.[47] The close agreement between excitation energies obtained at the CC2 and CC3 levels shows that the CC2 model works rather well also for SiH_4. For Si_4H_{10}, Si_5H_{12}, and neo-Si_5H_{12}, the optical gaps calculated at the CC2 level are on the other hand 0.46 eV, 0.37 eV and 0.34 eV larger than the excitation threshold measured in the liquid phase.[49] The major part of the difference between measured and CC2 excitation energies is most likely due to solvent effects which usually redshift the excitation energies. The excitation energies calculated at the CC2 level for linear silanes are compared to available experimental data in Figure 1. The CC2 calculations on the silanes also showed that there is a linear relation between the length of the molecule and the oscillator strength of the lowest excitation; the longer the silane chain the stronger is the absorption.

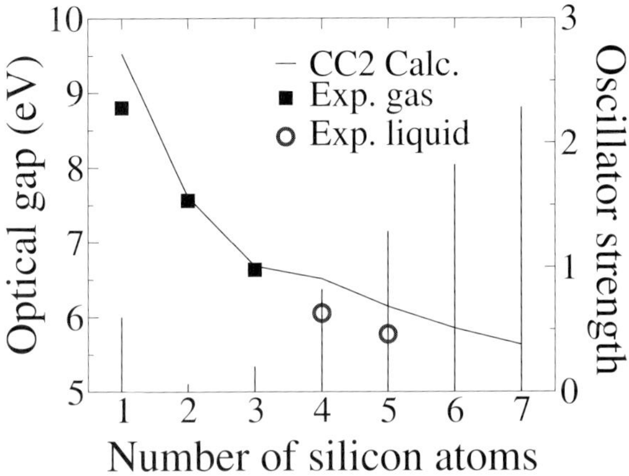

Fig. 1. The optical gap (in eV) of linear silanes calculated at the CC2 level with Dunning's quadruple ζ basis set augmented with diffuse and polarization functions (aug-cc-pV(Q+d)Z)[50] are compared to experimental values. The calculated oscillator strengths are given as histograms.

Coupled-cluster calculations on silicon nanoclusters at the CC2 level are computationally expensive, especially when large basis sets are employed. A reasonable alternative is to use DFT methods in studies on large silicon nanoclusters. The accuracy of DFT calculations depends on the size of the one-particle basis set and the functional employed. The optical gap usually decreases with increasing size of the basis set. The gap tends to be larger with hybrid functionals which account partly for the Hartree-Fock exchange contributions than with local density approximation (LDA) and generalized gradient approximation (GGA) functionals. Benchmark studies on linear silanes showed that the optical gaps calculated at the LDA and GGA levels are systematically 0.7–0.8 eV smaller than obtained in the CC2 calculations.[47] The only exception is SiH_4 which apparently is not very well described at the DFT level. Hybrid functionals such as the Perdew-Burke-Ernzerhof hybrid functional (PBE0) perform better.[51] At the PBE0 level, the optical gaps of the linear silanes are about 0.3 eV smaller than the corresponding CC2 values. As CC2 calculations slightly overestimate the optical gaps, the PBE0 functional and Becke's three parameter functional with the Lee-Yang-Parr correlation functional (B3LYP) seem to be a good compromise as far accuracy and computational costs are concerned.[52,53] The optical gaps are compared in Figure 2. They were obtained using quadruple ζ quality basis sets augmented with polarization functions.[47] The optical gap of the linear silanes decreases with increasing chain length. The excitation thresholds calculated at the CC2 level agree well with available experimental values and they seem to be an upper bound for the optical gap. DFT calculations employing hybrid functionals yield slightly to small excitation thresholds. The optical gaps calculated at the DFT level using GGA and LDA type functionals are almost identical but they are systematically too small as compared to experiment. Good estimates for the optical gaps can therefore be deduced from the GGA or LDA energies by adding a constant shift.

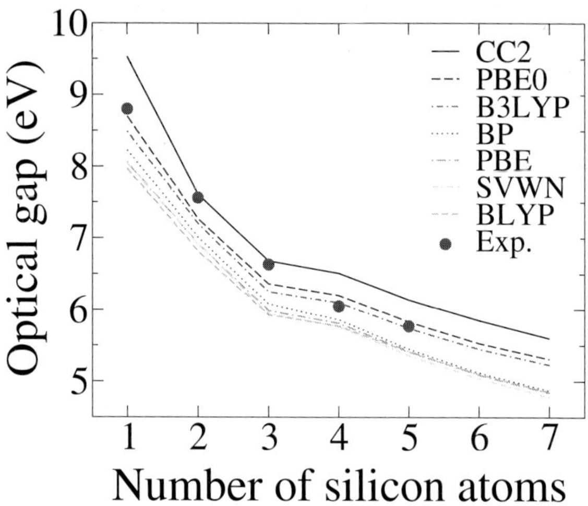

Fig. 2. The optical gap (in eV) of linear silanes calculated at the CC2 and DFT levels using the aug-cc-pV(Q+d)Z basis sets.

In the silicon nanocluster studies, smaller basis sets have to be employed because otherwise the computational costs become huge. The use of smaller basis sets usually blueshifts the lowest electronic transitions. PBE0 calculations with triple ζ quality basis sets augmented with polarization functions (TZVP) were found to yield optical gaps for the silanes in close agreement with experiment due to cancellation of errors.[22,47] The basis-set convergence of the optical gap of $Si_{10}H_{16}$ is shown in Figure 3. The optical spectrum for the recently synthesized sila-adamantane cluster capped with methyl and trimethylsilyl groups ($Si_{14}C_{24}H_{72}$) has a strong absorption maximum at 5.59 eV.[54] DFT calculations at the PBE0/TZVP level yielded an optical gap of 5.06 eV. As the first few transitions are rather weak, the observed transition probably corresponds to the two strong bands at 5.54 eV and 5.60 eV with oscillator strengths of 0.42 and 0.69, respectively.[22]

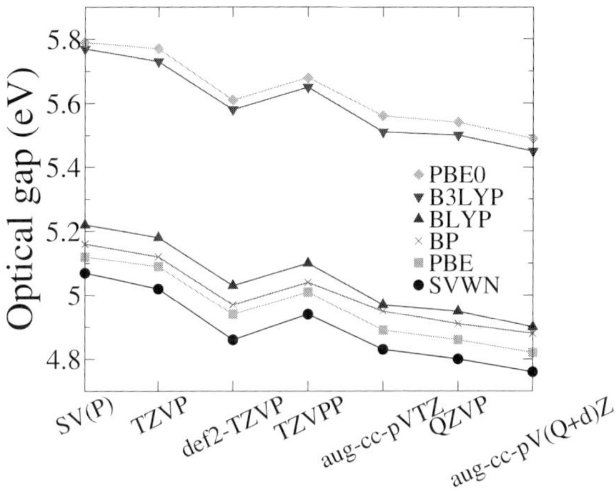

Fig. 3. The optical gap (in eV) of sila-adamantane calculated at DFT levels using different basis sets. SV(P) = split valence basis set with one set of polarization functions on Si.[55] TZVP = triple ζ valence basis set with one set of polarization functions.[56] def2-TZVP = the newly optimized TZVP.[57] TZVPP = triple ζ valence basis set with two sets of polarization functions.[56] aug-cc-pVTZ = correlation consistent basis set of valence triple ζ quality augmented with diffuse functions.[58] QZVP = quadruple ζ valence with one set of polarization functions.[59] aug-cc-pV(Q+d)Z = correlation consistent basis set of valence quadruple ζ quality augmented with diffuse functions and with an extra *d* shell on Si.[50]

The molecular structure of the silicon nanocluster also affects the calculated optical gap. The first excitation energy of SiH_4 calculated using the experimental structure is 0.1 eV larger than obtained with the structure optimized using the Becke-Perdew (BP) functional and TZVP basis sets.[60–62] This implies that a better precision than 0.1 eV cannot be expected for calculated excitation energies.

Many other computational methods have been used to predict the optical gap of silicon nanoclusters. These studies comprise highly accurate calculations at the many-body GW-Bethe-Salpeter-equation (GW-BSE)[63–65] and quantum Monte Carlo (QMC) levels.[64,66–68] The QMC is in principle a very accurate many-body approach that can be applied to silicon nanoclusters consisting of hundreds of atoms.[69,70] However, the QMC approach is not quite unproblematic for excited state studies because the construction of the trial function is not trivial and the obtained results might depend on that choice.[68] For SiH_4, the best optical gap obtained with the QMC is 9.47(2) eV[66] which is in excellent agreement with the best *ab initio* value calculated at the CC3 level.[47] For SiH_4, a smaller optical gap of 9.1(1) eV has also been obtained with the QMC.[64] For larger clusters, the choice of the trial function is unfortunately not easier than for a small molecule. As a consequence, optical gaps calculated with the QMC for silanes are not always mutually concordant.[65–68] However, the QMC optical gaps for the neo-Si_5H_{12} molecule possessing T_d symmetry and for the sila-adamantane ($Si_{10}H_{16}$) cluster calculated by Benedict *et al.*[65] agree well with the best *ab initio* excitation energies and with estimated experimental gas-phase values.[22,47] The assignment of the electronic excitation spectra using QMC calculations is difficult because the oscillator strengths are not readily obtained with the QMC method.

The optical gap calculated at the TDDFT level using the BP GGA functional becomes equal to the HOMO-LUMO splitting in the limit of infinitely large clusters. For clusters larger than 0.8 nm the difference between the excitation threshold and the HOMO-LUMO gap is less than 0.1 eV and for silicon nanoclusters larger than 1.7 nm in diameter, the difference is less than 0.01 eV.

Figure 4 shows a combination of optical gaps calculated at the BP TDDFT and effective mass approximation (EMA) levels. The EMA data are taken from Ref. 71. The gaps calculated at the BP TDDFT level using TZVP basis sets are shifted by 0.8 eV. This shift is chosen such that the first excitation energy for $Si_{10}H_{16}$ obtained at the BP TDDFT level agrees with the accurate optical gap of 5.95 eV estimated using CC2 and QMC calculations. The dashed line in Figure 4 represents the optical gaps calculated at the QMC level using the diffusion Monte-Carlo (DMC) approach.[67]

The optical gaps calculated at the BP TDDFT level using def2-TZVP basis sets for hydrogen-capped silicon nanoclusters are summarized in Figure 5. In the DFT and TDDFT calculations, the TURBOMOLE split valence basis set augmented with polarization functions were employed.[57] The analytical expression for the size dependence of the optical gap of silicon nanoclusters was obtained from the BP TDDFT optical gaps (E_g) of hydrogen-capped nanoclusters larger than 1 nm by using a power-law fit.

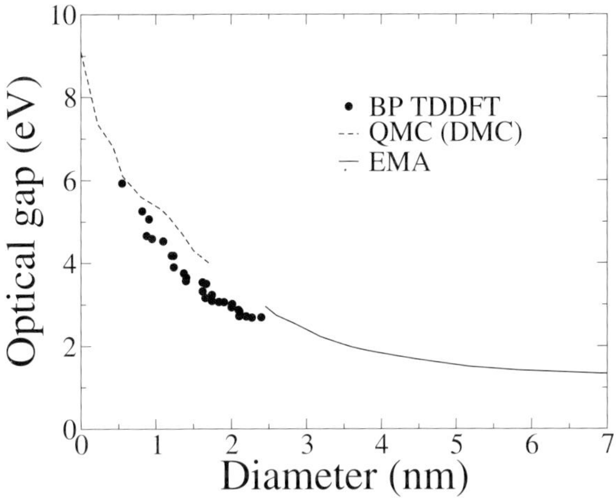

Fig. 4. The optical gap (in eV) of silicon nanoclusters as obtained at the BP TDDFT and EMA levels. The BP TDDFT optical gaps are shifted by 0.8 eV to yield the correct optical gap for $Si_{10}H_{16}$. The QMC and EMA optical gaps are taken from Refs. 67 and 71.

The fitted expression for the size dependence of the optical gap is

$$E_g[\text{eV}] = 1.12 + 3.00/d[\text{nm}]^{1.575}, \tag{1}$$

where d is the cluster diameter in nanometers and 1.12 eV is the assumed band gap for bulk silicon. Belomoin *et al.*[6] reported a size dependence of $E_g[\text{eV}]=3.44/d[\text{nm}]^{0.5}$ which was obtained by fitting to experimental PL energies of four nanoclusters. Delerue *et al.* proposed a size-dependence of $E_g[\text{eV}]=1.12+3.73/d[\text{nm}]^{1.39}$ based on PL results and calculations.[72] Meier *et al.*[73] determined the size-dependence of the oscillator strength by measuring the PL decay rates of silicon nanoclusters of different size. The cluster sizes were estimated from the size-dependence of the optical gap using the fit of Delerue *et al.*

The oscillator strengths calculated at the BP TDDFT level for the first electronic transition are shown in Figure 5. The oscillator strengths decrease less systematically with increasing cluster diameter than the excitation energy. However, a clear trend is seen in the graph. The curve describing the size dependence of the oscillator strengths was obtained by first determining the energy dependence of the oscillator strengths (f_{osc}) using a similar analytical expression as used by Meier *et al.*[73] The obtained expression is

$$f_{osc} = 0.1351 \cdot 10^{-6} \exp\left(E_g[\text{eV}]/0.3036\right). \tag{2}$$

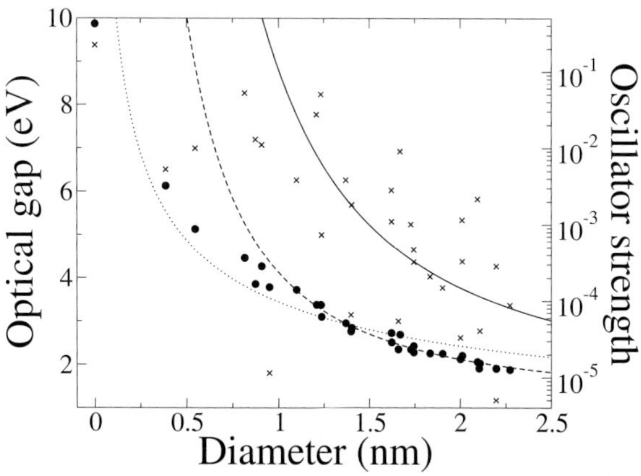

Fig. 5. Optical gaps (•, in eV) and oscillator strengths (×) for silicon clusters of different size as a function of the cluster diameter calculated at the BP TDDFT level using the def2-SV(P) basis set. The size dependence of the optical gap fitted using the BP TDDFT data is shown as a dashed line. The experimental fit[6] is represented with the dotted line. The solid line shows size dependence of the calculated oscillator strength obtained using the fitting procedure described in the text.

Combining Eq. (2) with the fitted expression in Eq. (1) yields the following expression for the size dependence of the oscillator strength

$$f_{osc} = 0.5406 \cdot 10^{-5} \exp\left(9.881/d[\mathrm{nm}]^{1.575}\right). \tag{3}$$

The curve representing Eq. (3) is drawn as a solid line in Figure 5. A similar expression with the coefficients of $1.4 \cdot 10^{-6}$, 11.24, and 1.39 was obtained by Meier *et al.*[73] using experimental PL life times in combination with the size dependence of the absorption energy calculated by Delerue *et al.*[72] The analytical expressions obtained using the two fitting procedures agree well. The oscillator strengths obtained for the larger hydrogen-capped silicon nanoclusters are indeed very small but they agree qualitatively with the experimental data of Meier *et al.*[73] The size-dependence function for the oscillator strength indicates that the large oscillator strengths of about unity obtained for the very bright silicon nanoclusters are most likely not merely due to confinement effects.

A surface dimer model suggested by Allan *et al.* has often been used to explain the luminescence of silicon nanoclusters.[16] According to this model, the silicon dimers on the cluster surface have a double-well potential in the excited state. The observed blue emission would take place from the outer well which is the global minimum on the excited state potential energy surface (PES). To assess the validity of this model, the PESs of the ground state and the first few excited states of one single surface dimer on the surface of a $Si_{29}H_{34}$ (C_{2v}) cluster was studied at the BP TDDFT level using the def2-TZVPP basis sets. No double-well potential was obtained for any of the excited states considered. The study indicates that the surface dimer model[16] used for explaining the bright luminescence of hydrogen-capped silicon nanoclusters can be questioned.

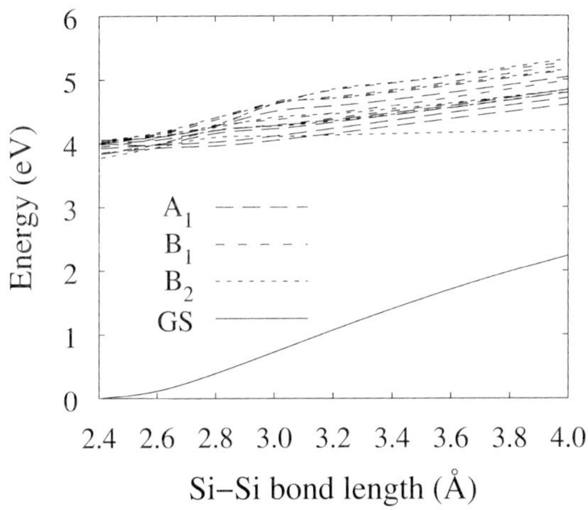

Fig. 6. The ground state (GS) and the excited state potential energy surfaces for a surface dimer in $Si_{29}H_{34}$ as a function of dimer elongation. A_1, B_1, and B_2 denote the irreducible representation of the excited states.

4. Absorption and Luminescence Spectra

A deeper understanding of light-absorption and light-emission processes requires knowledge about the potential-energy surfaces of both the ground and the involved excited states. The photon excites the cluster in the ground state equilibrium geometry to an excited state and during the excitation process the molecular structure is more or less unchanged; the excitation is usually thought to be vertical. However, the ground-state structure is not necessarily optimal for the excited state resulting in a relaxation of the molecular structure of the excited state. The de-excitation giving rise to the luminescence can again be considered to occur vertically from the relaxed excited state structure. The optimized molecular structure of the excited state is not completely optimal for the ground state.

The ground-state energy for the excited state structure is therefore somewhat higher than for the ground-state geometry resulting in a reduction in the emission energy; the strain of the ground-state structure leads to an additional contribution to the redshift of the de-excitation energy.

Calculation of luminescence properties is much more involved than computing the absorption spectrum. In simulations of the luminescence, a variety of relaxation processes have to be considered because the observed light emission is the outcome of several competing and coupled processes. The emission intensity and the energy of the emitted light depend on the de-excitation route and populations of the levels. Both radiative and non-radiative processes might play a role. The luminescence can be simulated by solving the rate equations for the involved relaxation pathways. However, Stokes shifts and some conception of the luminescence intensities can be obtained by comparing excitation energies and oscillator strengths calculated for the molecular structures of the ground and first excited states, respectively. The difference between absorption and emission spectra is then assumed to be merely a consequence of the changes in the molecular structure.

Stimulated emission and optical gain necessary for laser action can be observed when population inversions between the ground and a radiating excited state occurs. Optical gain was recorded for the first time in silicon nanoclusters by Pavesi *et al.*[13] They proposed a three-level model to explain it. They suggested that an inversion of the population of a radiative state at the silicon-nanocluster silicon-dioxide interface lying energetically between the ground state and the absorbing excited states is responsible for the optical gain. The prerequisites for a silicon nanocluster based laser is strong absorption of light, a fast relaxation of the excited state, and a recombination with a smaller transition probability from the relaxed excited state than for the absorption process.[4,12,13]

5. Hydrogen-Capped Silicon Nanoclusters

Silicon nanoclusters covered by hydrogens can be synthesized through electrochemical etching with hydrofluoric acid.[6,9,74–77] By using this technique, Nayfeh *et al.* dispersed bulk Si into nanoparticles of about 1 nm in diameter. The obtained silicon nanoclusters were found to be strongly luminescent exceeding the luminescence activity of fluorescein.[75,78] The nanoparticles were also found to possess directed light beam emission[77] and second harmonic generation.[79] Luminescent silicon nanoclusters can also be synthesized from NaSi.[80,81] By letting NaSi react with NH_4Br, Zhang *et al.* manufactured slightly larger hydrogen-capped silicon nanoclusters with an average diameter of 3.9 ± 1.3 nm exhibiting strong PL at 438 nm.[81] An alternative route to synthesize hydrogen-capped silicon nanoclusters is by reducing Si^{4+} ions in micelles by $LiAlH_4$.[82] The growth of the clusters is restricted by the micelle size. The smallest clusters are 1.8 nm in diameter and consist of ~150 Si atoms covered by hydrogens.

Even though the hydrogen-capped silicon nanoclusters have been studied computationally at many levels of theory,[6,19–22,65–68,77–79,83–100] the reason for their strong PL emission is still unresolved. A consensus concerning the molecular structure of the manufactured 1 nm nanoclusters has not been reached either.

The discrepancies between the optical gaps of silanes and sila-adamantane calculated at the *ab initio*, DFT, and QMC levels are more or less understood. By combining the values obtained at these levels one can estimate rather accurately the optical gaps for the 1 nm and larger nanoclusters. The absorption strengths, which are readily obtained at the *ab initio* and DFT levels, provide additional information about the excited states. The optimization of the cluster structure for the excited state yields luminescence energies and a conception of the PL strength.

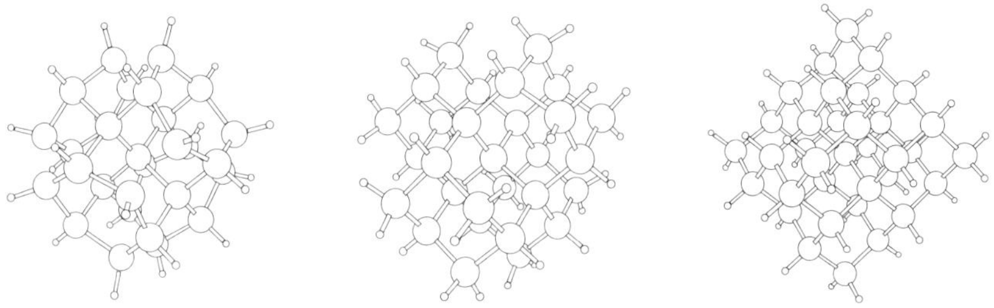

(a) (b) (c)

Fig. 7. The ground state structures of the hydrogen-capped silicon nanoclusters (a) $Si_{29}H_{24}$, (b) $Si_{29}H_{36}$ and (c) $Si_{35}H_{36}$.

The most popular candidates for the structure of the 1 nm silicon nanocluster are $Si_{29}H_{24}$, $Si_{29}H_{36}$, and $Si_{35}H_{36}$ shown in Figure 7. However, their absorption and emission energies as well as the oscillator strengths do not agree well with experimental observations. The absorption thresholds for $Si_{29}H_{24}$, $Si_{29}H_{36}$, and $Si_{35}H_{36}$ calculated at the DFT level using the BP functional are 2.70 eV, 3.72 eV, and 3.60 eV, respectively.[21] The B3LYP

DFT, BP TDDFT and CC2 calculations yield an estimated optical gap of about 4.5 eV for $Si_{29}H_{36}$ which is significantly larger than the experimental excitation threshold of 3.7 eV.[84] The corresponding QMC value is 5.3 eV.[67] A similar extrapolation procedure for $Si_{29}H_{24}$ and $Si_{35}H_{36}$ as done for $Si_{10}H_{16}$ in Section 3 yields excitation thresholds of 3.6 eV and 4.3 eV as compared to the QMC optical gaps of 3.6 eV and 5.1 eV, respectively.[65,67]

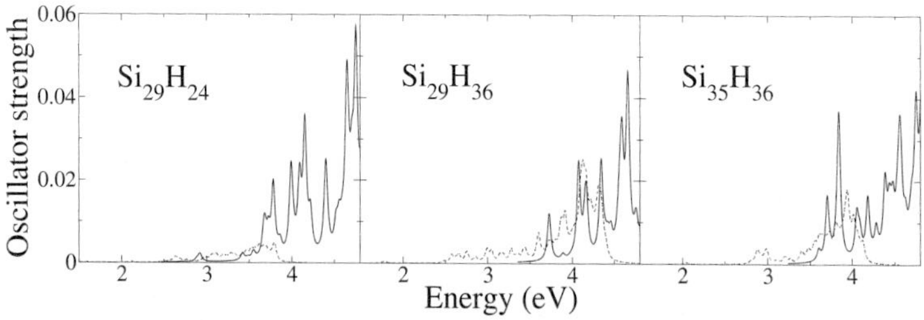

Fig. 8. The electronic excitation spectra of $Si_{29}H_{24}$, $Si_{29}H_{36}$ and $Si_{35}H_{36}$ calculated at the BP TDDFT level using the def2-TZVP basis set. The solid line shows the absorption spectra obtained using the optimized ground-state structure and the dashed line is obtained using the optimized geometry of the first excited state. The spectra using the excited state structure were not calculated for the whole energy range shown.

The electronic excitation spectra for $Si_{29}H_{24}$, $Si_{29}H_{36}$ and $Si_{35}H_{36}$ in Figure 8 were obtained by plotting each excitation energy using a Lorentzian line shape function. The calculated oscillator strengths determine the height and a constant full width at half maximum (FWHM) of 50 meV was assumed. In the absorption spectra one sees that the first transitions are very weak and that at higher energies states with stronger absorption intensity appear. However, no bright states were obtained in the calculations. The oscillator strengths of 10^{-3}–10^{-2} for these transitions are two to three orders of magnitude smaller than those observed experimentally.[23]

The PL energy and Stokes shifts can be obtained in TDDFT calculations using the cluster structure of the excited state. The corresponding oscillator strengths provide an estimate for the PL strength. For $Si_{29}H_{36}$, the optimization of the excited state structure in D_{2d} symmetry yielded a Stokes shift of 0.7 eV,[20,21] whereas a more recent calculation without any symmetry constraints resulted in a totally unsymmetric structure (C_1) and a large Stokes shift of 2 eV. For the two other clusters, the Stokes shifts are 1.1 eV and 1.6 eV, respectively. The vertical excitation and de-excitation energies for $Si_{29}H_{24}$, $Si_{29}H_{36}$ and $Si_{35}H_{36}$ are summarized in Table 1. The

Stokes shifts and the two relaxation contributions to are also given. The oscillators strengths obtained for the excited state structures are significantly smaller than obtained for the ground-state structures, especially for the two larger clusters.

Table 1. The excitation energies (in eV) for the Si nanoclusters calculated at the BP/def2-TZVP level using the ground-state (E_{gs}) and the first excited state (E_{es}) structures. The Stokes shift (ΔE_{Stokes}) and its contributions from the strain of the ground-state (ΔE_{gs}) and the relaxation of the excited state (ΔE_{es}) are also given. f_{gs} and f_{es} denote the oscillator strengths of the corresponding transitions.

	$Si_{29}H_{24}$	$Si_{29}H_{36}$	$Si_{35}H_{36}$
E_{gs}	2.70	3.72	3.60
E_{es}	1.61	1.79	2.04
ΔE_{Stokes}	1.10	1.94	1.56
ΔE_{gs}	0.70	1.33	1.20
ΔE_{es}	0.40	0.61	0.36
f_{gs}	0.000089	0.011	0.0025
f_{es}	0.00015	0.00052	0.00045

Computational studies on sila-adamantane based silicon nanoclusters showed that the oscillator strengths are practically independent of the interior structure of the cluster.[22] Clusters with fused sila-adamantane cages as core have very weak transitions[22] even though sila-adamantane was found to have strong absorption at 5.6 eV.[22,54] Reboredo *et al.*[97] also reported that the surface chemistry of silicon nanoclusters is independent of the core structure. Thus, most likely none of these nanoclusters is responsible for the observed strong blue PL corresponding to an oscillator strength of 0.92.[23]

Optimizations of the cluster structures for the first excited state show that many local minima with completely different Stokes shifts can occur on the PES of the excited states. The large experimentally observed Stokes shift has been explained by localized surface states in the silicon nanocluster.[16] Allan *et al.* proposed that such states exist under the form of self-trapped excitons at the Si–Si dimers on the nanocluster surface. However, in the optimizations of the cluster structures for the excited state, we did not obtain any significantly elongated Si–Si bonds. Thus, our silicon nanocluster studies do not support the notion that surface dimers are the light emitters of hydrogen-capped silicon clusters.

Figure 9 illustrates changes in the Si–Si bond lengths of the $Si_{29}H_{36}$ structure upon excitation. The largest structural changes (0.050–0.095 Å)

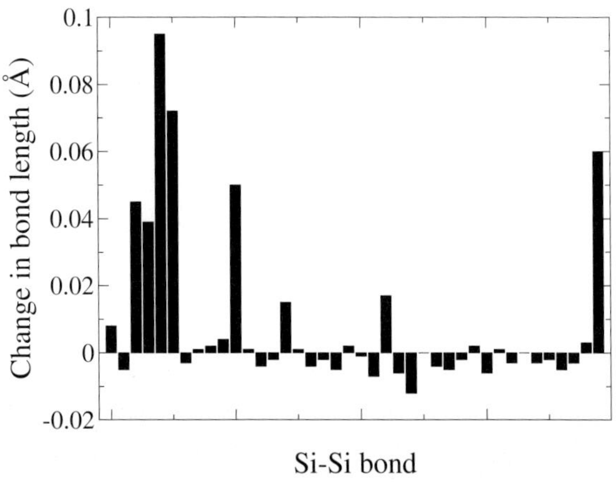

Fig. 9. Changes in Si–Si bond lengths due to structural relaxations of the ground-state structure upon excitation to the lowest excited state of the $Si_{29}H_{36}$ cluster calculated at BP TDDFT and BP TDDFT levels using the def2-TZVP basis set.

occur at the cluster surface while changes inside the cluster core are slightly smaller (0.039–0.045 Å). The largest decrease in the bond lengths of -0.012 Å occurs also at the cluster surface.

6. Oxidized Silicon Nanoclusters

Silicon nanoclusters prepared through electrochemical dispersion are mainly capped by hydrogens after the etching.[9,76,101] They also contain a small amount of impurity atoms[9,78] and the surface is partly oxidized.[76] The synthesized silicon nanoclusters react in the presence of molecular oxygen increasing the silicon oxide coverage of the cluster surface.[12] The oxidation leads to a significant redshift of the PL.[102,103] The quantum yield is also much lower for the oxidized clusters.[12] Measurements of the PL of oxidized silicon nanoclusters display emission contributions that can be assigned to Si-O-Si and Si=O groups.[101,103,104] The spontaneous oxidation also forms -OH groups on the surface.[12]

The oxidized silicon nanoclusters studied at the BP TDDFT level are shown in Figures 10 and 11. The silicon nanocluster in Figure 10a is obtained by replacing the twelve $=SiH_2$ groups of $Si_{29}H_{36}$ with Si=O groups. The calculated electronic absorption spectrum in Figure 12 shows that the optical gap for the oxidized cluster is redshifted by about 1.5 eV as compared to the hydrogen-capped one. The first band of the calculated absorption spectrum consists of several transitions and it has an oscillator strength maximum of 0.06. The oscillator strength for the oxidized cluster with

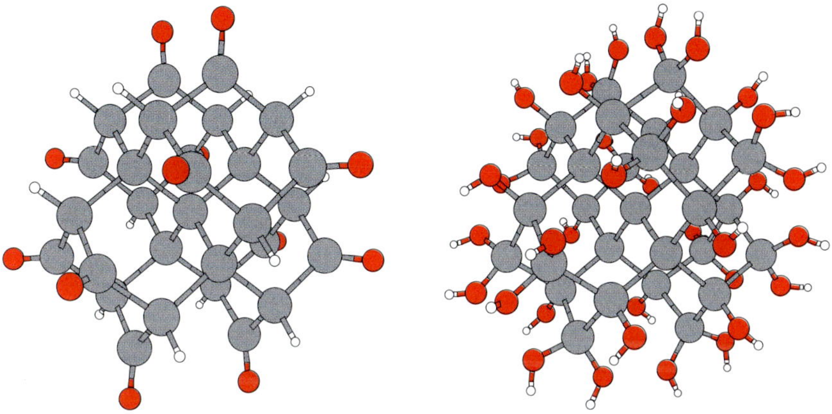

(a) (b)

Fig. 10. (a) $Si_{29}O_{12}H_{12}$ is a silicon nanocluster with twelve Si=O groups. (b) $Si_{29}(OH)_{36}$ is a silicon nanocluster covered with hydroxyl groups

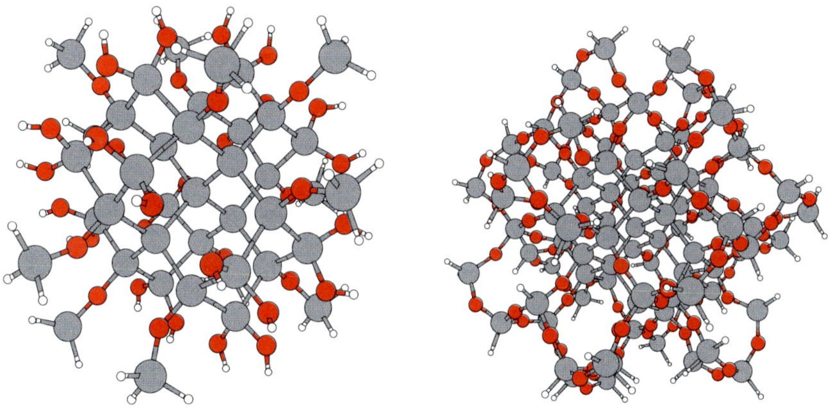

(a) (b)

Fig. 11. (a) $Si_{29}(OH)_{24}(OSiH_3)_{12}$ with twelve Si-O-Si bonds at the surface. (b) $Si_{29}@Si_{58}O_{96}H_{76}$ is a silicon nanocluster embedded in silicon dioxide.

Si=O groups on the surface is thus larger than obtained for the hydrogen-capped ones. The oscillator strength is still about 15 times smaller than obtained experimentally for the hydrogen-capped silicon nanocluster samples.[23] The optical gap calculated for $Si_{29}(OH)_{36}$, which is a silicon nanocluster capped by 36 hydroxyl (-OH) groups, is about 1 eV smaller than for the corresponding hydrogen-capped one.

The oscillator strengths for the hydroxyl-capped silicon nanocluster are of the same size as obtained for the hydrogen-capped ones. Vasiliev *et al.*[102] obtained similar results in their computational study. Figure 11a shows a silicon nanocluster terminated with 24 -OH and 12 -Si-O-SiH_3 groups. The BP TDDFT calculation shows that the optical gap is shifted by about 1 eV as compared to the corresponding hydrogen-capped one. The oscillator strengths for the lowest transitions of the absorption spectrum of $Si_{29}(OH)_{24}(OSiH_3)_{12}$ shown in Figure 12 are also of the same size as obtained for the hydroxyl terminated cluster.

The silicon nanocluster shown in Figure 11b consists of a Si_{29} core embedded in silicon dioxide. The silica is modelled by a $Si_{58}O_{96}H_{76}$ shell surrounding the Si_{29} core. Its electronic absorption spectrum shown in Figure 12 is redshifted by about 1 eV as compared to hydrogen-terminated Si_{29}. The obtained oscillator strengths for the lowest transitions of $Si_{29}@Si_{58}O_{96}H_{76}$ are as weak as obtained for the oxidized silicon nanoclusters.

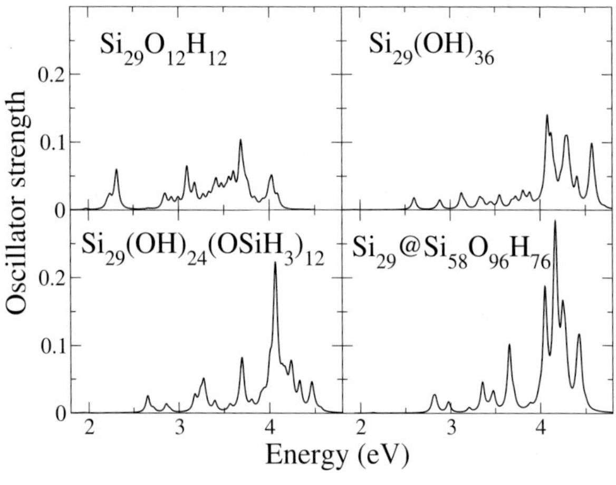

Fig. 12. The electronic excitation spectra of the oxidized silicon nanoclusters calculated at BP TDDFT level using the TZVP basis set.

7. Silane-Capped Silicon Nanoclusters

The computational studies on hydrogen-capped silicon nanoclusters show that the optical gaps for them are somewhat larger than obtained in the experiment. The calculated Stokes shifts are also larger than the experimental values. The calculations also show that strong ground-state transitions are lacking among the low-lying excited states. The benchmark studies on the linear silanes[47] as well as the calculated and measured absorption spectra of silyl and methyl capped sila-adamantane[22,54] suggest that strongly emitting silicon nanoclusters can be designed by attaching molecular functional groups to the cluster surface where the substituents function as light emitters. Based on semiempirical calculations, Takeda *et al.* suggested that oligosilane bridged silicon nanoclusters might be the origin of the PL of porous silicon.[105] Luterova *et. al.*,[106,107] and Schauer *et al.*[15] studied experimentally the PL of porous silicon samples and found that -$(SiH_2)_n$-units might form efficient luminescence centers in these materials. Mu *et. al.*[108] also reported strong blue PL from large silica nanospheres covered by chlorotrimethylsilane.

Recent DFT calculations on silicon nanoclusters covered by linear silane groups support this conception.[99] The silane-capped $Si_{29}H_{35}(SiH_2)_{n-1}SiH_3$ with n=10 is shown in Figure 13a. The calculations showed that these structures have significantly larger oscillator strengths for the lowest electronic transitions than obtained for the hydrogen-capped ones. The obtained optical gaps are rather independent of the length of the silane chain, whereas the oscillator strength increases linearly with its length. Figure 14 shows the excitation energies of the two lowest excited states and the corresponding oscillator strengths for $Si_{29}H_{36}$ substituted with one silane.

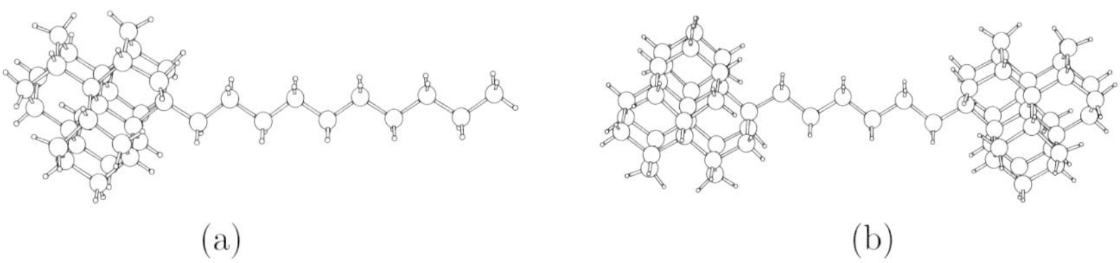

Fig. 13. (a) $Si_{29}H_{35}(SiH_2)_9SiH_3$ is an example of a silane substituted silicon nanocluster. (b) $Si_{29}H_{35}(SiH_2)_6Si_{29}H_{35}$ is an example of a bridged silicon nanocluster dimer.

The BP TDDFT calculations on the dimers consisting of two $Si_{29}H_{36}$ clusters bridged by a silane chain indicate that such structures might be responsible for the strong light emission observed for nano-sized silicon

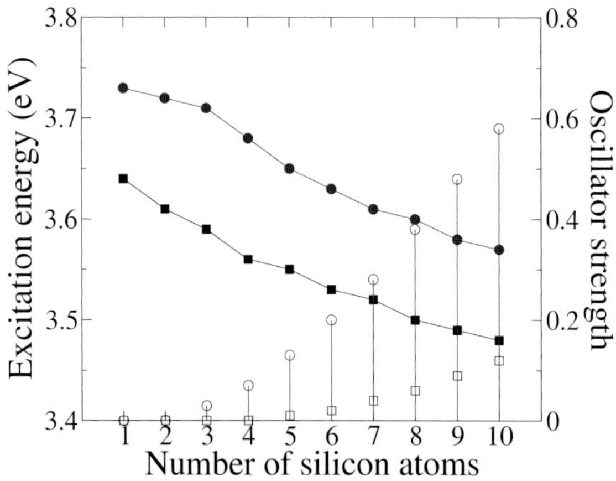

Fig. 14. The excitation energies (in eV) of the two lowest transitions and the corresponding oscillators strengths for the single silane-capped silicon nanoclusters (C_s) obtained at the BP TDDFT level shown as a function of the number of Si atoms of the substituent. The filled symbols show the excitation energies and the histograms the oscillator strengths.

structures.[99] The dimers have low-lying strong transitions, the energy of which is almost independent of the length of the silane bridge. On the other hand, the oscillator strength is linearly proportional to the number of Si atoms in the chain. The size of the cluster tunes the absorption wavelength, whereas the silane bridge functions as an 'antenna'. The excitation energies and the corresponding oscillator strengths for $Si_{29}H_{35}(SiH_2)_nSi_{29}H_{35}$ are shown in Figure 15. The dimer with n=6 is depicted in Figure 13b.

The calculations on the silane-capped silicon nanoclusters show that variations in the cluster structure do not significantly affect the wavelength of the PL, whereas the emission strengths are strongly dependent on the presence and character of the substituents. This could explain why only a few percent of the nanoclusters are found to be luminescent in experimental studies.[101,109,110] The PL properties of a silicon nanocluster could even be tailored by attaching appropriate substituents to the cluster surface.[8,101] Light emitting substituents could also explain the dissimilar results obtained for the relation between the size of the silicon nanocluster and the observed luminescence energies.[82,90]

Hydrogen-capped silicon nanoclusters oxidize easily in the presence of molecular oxygen.[8,12,103,111] However, they can be stabilized against oxidation by terminating them with alkyl groups.[8,12,111,112] The alkyl ter-

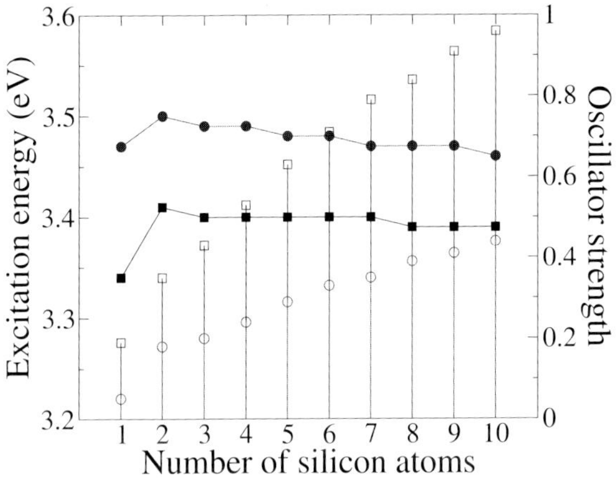

Fig. 15. The excitation energies (in eV) of the two lowest transitions and the corresponding oscillators strengths for the silane-bridged silicon-nanocluster dimers obtained at the BP TDDFT level shown as a function of the number of -SiH_2- units in the bridge. The filled symbols show the excitation energies and histograms the oscillator strengths.

mination does not have a large effect on the size of the optical gap, because the alkyl groups function as capping groups and do not contribute significantly to the electron delocalization path. For example, for the methyl and silyl capped sila-adamantane, the methyl termination decreased the optical gap by 0.17 eV. At the BP TDDFT level using the def2-TZVP basis set, the oscillator strength for $Si_{29}H_{24}[(CH_2)_2CH_3]_{12}$ is 10^4 times smaller than for the corresponding silane-capped silicon nanocluster, i.e. $Si_{29}H_{24}[(SiH_2)_2SiH_3]_{12}$. The optical gap calculated for the alkyl-capped silicon nanocluster is 3.57 eV as compared to the 3.20 eV obtained for the silane-capped one. The optical gap for the corresponding hydrogen-capped silicon nanocluster calculated at the same level is 3.72 eV.

8. Conclusions

The current status of *ab initio* and density functional theory methods for studies of optical properties of silicon nanoclusters has been reviewed. The calculated excitation energies are qualitatively in good agreement with experimental data. However, identification of cluster structures based on comparisons of calculated and measured optical gaps is still difficult. The computational studies indicate that the strong PL observed for some silicon nanoclusters is not merely due to quantum confinement effects. The size dependences of the oscillator strength deduced from PL lifetimes and obtained in calculations agree qualitatively, but cannot explain why some silicon nanoclusters have oscillator strengths close to unity. Most likely, the very bright silicon nanoclusters have functional groups at the cluster surface emitting the light. According to the calculations, silane-capped silicon nanoclusters have bright low-lying transitions and they might be a good starting point for development of material for optical devices. The size of the silicon nanocluster determines the size of the optical gap and the emission properties of the silane substituent adjust the oscillator strengths.

Acknowledgments

This research has been supported by the Academy of Finland through its Centers of Excellence Programme 2006–2011 and the OPNA research project (118195).

References

1. P. Y. Yu and M. Cardona, *Fundamentals of Semiconductors: Physics and Materials Properties* (Springer, 2005), 3rd edition.
2. C. Pickering, M. I. J. Beale, D. J. Robbins, P. J. Pearson and R. Greef, *J. Phys. C* **17**, 6535–6552 (1984).
3. L. T. Canham, *Appl. Phys. Letters* **57**, 1046–1048 (1990).
4. H. Wong, V. Filip, C. K. Wong and P. S. Chung, *Microelectron. Reliab.* **47**, 1–10 (2007).
5. L. T. Canham. Visible photoluminescence from porous silicon. In ed. L. T. Canham, *Properties of Porous Silicon*, pp. 249–255. Institution of Engineering and Technology (1997).
6. G. Belomoin, J. Terrien, A. Smith, S. Rao, R. Twesten, S. Chaieb, L. Wagner, L. Mitas and M. H. Nayfeh, *Appl. Phys. Letters* **80**, 841–843 (2002).
7. Y. Kanemitsu, *Phys. Rep.* **263**, 1–91 (1995).
8. J. G. C. Veinot, *Chem. Commun.*, pp. 4160–4168 (2006).
9. M. H. Nayfeh and L. Mitas. Silicon Nanoparticles: New Photonic and Electronic Material at the Transition Between Solid and Molecule. In ed. V. Kumar, *Nanosilicon*, pp. 3–78. Elsevier, Amsterdam (2007).
10. L. T. Canham. Visible photoluminescence from porous silicon. In ed. L. T. Canham, *Properties of Porous Silicon*, pp. 247–255. INSPEC, The Institution of Electrical Engineers, London, UK (1997).
11. C. Marschner, J. Baumgartner and A. Wallner, *Dalton Trans.*, pp. 5667–5674 (2006).
12. D. Jurbergs, E. Rogojina, L. Mangolini and U. Kortshagen, *Appl. Phys. Letters* **88**, 233116 (2006).

13. L. Pavesi, L. Dal Negro, C. Mazzoleni, G. Franzò and A. Priolo, *Nature* **408**, 440–444 (2000).
14. H. Suzuki, S. Hoshino, K. Furukawa, K. Ebata, C. H. Yuan and I. Bleyl, *Polym. Adv. Technol.* **11**, 460–467 (2000).
15. F. Schauer, I. Kuritka, N. Dokoupil and P. Horváth, *Physica E* **14**, 272–276 (2002).
16. G. Allan, C. Delerue and M. Lannoo, *Phys. Rev. Letters* **76**, 2961–2964 (1996).
17. P. Roura and J. Costa, *Eur. J. Phys.* **23**, 191–203 (2002).
18. D. Kovalev, H. Heckler, G. Polisski and F. Koch, *Phys. Stat. Sol. (b)* **215**, 871–932 (1999).
19. D. Sundholm, *Nano Letters* **3**, 847–849 (2003).
20. D. Sundholm, *Phys. Chem. Chem. Phys.* **6**, 2044–2047 (2004).
21. O. Lehtonen and D. Sundholm, *Phys. Rev. B* **72**, 085424 (2005).
22. O. Lehtonen and D. Sundholm, *Phys. Rev. B* **74**, 045433 (2006).
23. A. Smith, Z. H. Yamani, N. Roberts, J. Turner, S. R. Habbal, S. Granick and M. H. Nayfeh, *Phys. Rev. B* **72**, 205307 (2005).
24. R. Ahlrichs, M. Bär, M. Häser, H. Horn and C. Kölmel, *Chem. Phys. Lett.* **162**, 165–169 (1989). current version: see http://www.turbomole.com.
25. P. Hohenberg and W. Kohn, *Phys. Rev.* **136**, B864–B871 (1964).
26. W. Kohn and L. J. Sham, *Phys. Rev.* **140**, A1133–A1138 (1965).
27. E. Runge and E. K. U. Gross, *Phys. Rev. Letters* **52**, 997–1000 (1984).
28. W. Koch and M. C. Holthausen, *A Chemist's Guide to Density Functional Theory* (Wiley-VCH, Weinheim, 2001).
29. J. Olsen and P. Jørgensen, *J. Chem. Phys.* **82**, 3235–3264 (1985).
30. M. E. Casida. Time-dependent density-functional response theory for molecules. In ed. D. P. Chong, *Recent Advances in Density Functional Methods, Part I*, p. 155. World Scientific, Singapore (1995).
31. R. Bauernschmitt and R. Ahlrichs, *Chem. Phys. Lett.* **256**, 454–464 (1996).
32. R. van Leeuwen, *Int. J. Mod. Phys. B* **15**, 1969–2023 (2001).
33. F. Furche and R. Ahlrichs, *J. Chem. Phys.* **117**, 7433–7447 (2002).
34. F. Furche and R. Ahlrichs, *J. Chem. Phys.* **121**, 12772–12773 (2004).
35. F. Furche and D. Rappoport. Density functional methods for excited states: equilibrium structure and electronic spectra. In ed. M. Olivucci, *Ch. III of Computational Photochemistry, Vol. 16 of Computational and Theoretical Chemistry.* Elsevier, Amsterdam (2005).
36. D. Rappoport and F. Furche. Excited states and photochemistry. In eds. M. A. L. Marques, C. A. Ullrich, F. Nogueira, A. Rubio, K. Burke and E. K. Gross, *Time-Dependent Density Functional Theory*, number 706 in Lecture Notes in Physics, pp. 337–354. Springer-Verlag (2006).
37. K. Burke, J. Werschnick and E. K. U. Gross, *J. Chem. Phys.* **123**, 062206 (2005).
38. F. Furche and K. Burke. Time-dependent density functional theory in quantum chemistry. In ed. D. Spellmeyer, *Annual Reports in Computational Chemistry, 1*, pp. 19–30. Elsevier, Amsterdam (2005).
39. D. Rappoport and F. Furche, *J. Chem. Phys.* **122**, 064105 (2005).

40. T. Helgaker, P. Jørgensen and J. Olsen, *Molecular Electronic-Structure Theory* (Wiley, Chichester, 2000).
41. C. Hättig. Beyond Hartree-Fock: MP2 and Coupled-Cluster Methods for Large Systems. In eds. J. Grotendorst, S. Blugel and D. Marx, *Computational Nanoscience: Do It Yourself!*, vol. 31, pp. 245–278. NIC series, Jülich (2006). ISBN 3-00-017350-1.
42. O. Christiansen, H. Koch and P. Jørgensen, *Chem. Phys. Lett.* **243**, 409–418 (1995).
43. C. Hättig and F. Weigend, *J. Chem. Phys.* **113**, 5154–5161 (2000).
44. A. Köhn and C. Hättig, *J. Chem. Phys.* **119**, 5021–5036 (2003).
45. C. Hättig, *Adv. Quantum Chem.* **50**, 37–60 (2005).
46. O. Christiansen, H. Koch and P. Jørgensen, *J. Chem. Phys.* **103**, 7429–7441 (1995).
47. O. Lehtonen and D. Sundholm, *J. Chem. Phys.* **125**, 144314 (2006).
48. U. Itoh, Y. Yasutake, H. Onuki, N. Washida and T. Ibuki, *J. Chem. Phys.* **85**, 4867–4872 (1986).
49. F. Fehér, *Molekülspektroskopische Untersuchungen auf dem Gebiet der Silane und der Heterocyclischen Sulfane* (Forschungsbericht des Landes Nordrhein-Westfalen (Westdeutscher Verlag), Köln, 1977).
50. T. H. Dunning Jr, K. A. Peterson and A. K. Wilson, *J. Chem. Phys.* **114**, 9244–9253 (2001).
51. J. P. Perdew, K. Burke and M. Ernzerhof, *J. Chem. Phys.* **105**, 9982–9985 (1996).
52. A. D. Becke, *J. Chem. Phys.* **98**, 5648–5652 (1993).
53. C. Lee, W. Yang and R. G. Parr, *Phys. Rev. B* **37**, 785–789 (1988).
54. J. Fischer, J. Baumgartner and C. Marschner, *Science* **310**, 825–825 (2005).
55. A. Schäfer, H. Horn and R. Ahlrichs, *J. Chem. Phys.* **97**, 2571–2577 (1992).
56. A. Schäfer, C. Huber and R. Ahlrichs, *J. Chem. Phys.* **100**, 5829–5835 (1994).
57. F. Weigend and R. Ahlrichs, *Phys. Chem. Chem. Phys.* **7**, 3297–3305 (2005).
58. D. E. Woon and T. H. Dunning Jr, *J. Chem. Phys.* **98**, 1358–1371 (1993).
59. F. Weigend, F. Furche and R. Ahlrichs, *J. Chem. Phys.* **119**, 12753–12762 (2003).
60. S. H. Vosko, L. Wilk and M. Nusair, *Can. J. Phys.* **58**, 1200–1211 (1980).
61. J. P. Perdew, *Phys. Rev. B* **33**, 8822–8824 (1986).
62. A. D. Becke, *Phys. Rev. A* **38**, 3098–3100 (1988).
63. M. Rohlfing and S. G. Louie, *Phys. Rev. Letters* **80**, 3320–3323 (1998).
64. J. C. Grossman, M. Rohlfing, L. Mitas, S. G. Louie and M. L. Cohen, *Phys. Rev. Letters* **86**, 472–475 (2001).
65. L. X. Benedict, A. Puzder, A. J. Williamson, J. C. Grossman, G. Galli, J. E. Klepeis, J. Y. Raty and O. Pankratov, *Phys. Rev. B* **68**, 085310 (2003).
66. A. R. Porter, M. D. Towler and R. J. Needs, *Phys. Rev. B* **64**, 035320 (2001).
67. A. Williamson, J. C. Grossman, R. Q. Hood, A. Puzder and G. Galli, *Phys. Rev. Letters* **89**, 196803 (2002).
68. J. E. Vincent, J. Kim and R. M. Martin, *Phys. Rev. B* **75**, 045302 (2007).

69. J. C. Grossman, *J. Chem. Phys.* **117**, 1434–1440 (2002).
70. A. J. Williamson, R. Q. Hood and J. C. Grossman, *Phys. Rev. Letters* **87**, 246406 (2001).
71. G. Nesher, L. Kronik and J. R. Chelikowsky, *Phys. Rev. B* **71**, 035344 (2005).
72. C. Delerue, G. Allan and M. Lannoo, *Phys. Rev. B* **48**, 11024–11036 (1993).
73. C. Meier, A. Gondorf, S. Lüttjohann and A. Lorke, *J. Appl. Phys.* **101**, 103112 (2007).
74. M. H. Nayfeh, O. Akcakir, J. Terrien, Z. Yamani, N. Barry, W. Yu and E. Gratton, *Appl. Phys. Letters* **75**, 4112–4114 (1999).
75. O. Akcakir, J. Terrien, G. Belomoin, N. Barry, J. D. Muller, E. Gratton and M. H. Nayfeh, *Appl. Phys. Letters* **76**, 1857–1859 (2000).
76. G. Belomoin, J. Terrien and M. H. Nayfeh, *Appl. Phys. Letters* **77**, 779–781 (2000).
77. M. H. Nayfeh, N. Barry, J. Terrien, O. Akcakir, E. Gratton and G. Belomoin, *Appl. Phys. Letters* **78**, 1131–1133 (2001).
78. M. H. Nayfeh. Stimulated Blue Emission from Ultrasmall Si Nano Particles (A New Phase of Silicon). In eds. N. Bloembergen, N. Rahman and A. Rizzo, *Conference Proceedings Vol. 71: Atoms, Molecules and Quantum Dots in Laser Fields: Fundamental Processes*, pp. 83–96. Societa' Italiana di Fisica, Bologna (2001).
79. M. H. Nayfeh, O. Akcakir, G. Belomoin, N. Barry, J. Therrien and E. Gratton, *Appl. Phys. Letters* **77**, 4086–4088 (2000).
80. S. Lee, W. J. Cho, Y. D. Kim, E. K. Kim and J. G. Park, *Jap. J. Appl. Phys.* **44**, 5843–5846 (2005).
81. X. Zhang, D. Neiner, A. Wang, A. Y. Louie and S. M. Kauzlarich, *Nanotechnology* **18**, 095601 (2007).
82. J. P. Wilcoxon, G. A. Samara and P. N. Provencio, *Phys. Rev. B* **60**, 2704–2714 (1999).
83. C. S. Garoufalis, A. D. Zdetsis and S. Grimme, *Phys. Rev. Letters* **87**, 276402 (2001).
84. L. Mitas, J. Terrien, R. Twesten, G. Belomoin and M. H. Nayfeh, *Appl. Phys. Letters* **78**, 1918–1920 (2001).
85. G. Belomoin, E. Rogozhina, J. Terrien, P. V. Braun, L. Abuhassan, M. H. Nayfeh, L. Wagner and L. Mitas, *Phys. Rev. B* **65**, 193406 (2002).
86. A. Puzder, A. J. Williamson, J. C. Grossman and G. Galli, *J. Chem. Phys.* **117**, 6721–6729 (2002).
87. I. Vasiliev, S. Ogut and J. R. Chelikowsky, *Phys. Rev. B* **65**, 115416 (2002).
88. E. W. Draeger, J. C. Grossman, A. J. Williamson and G. Galli, *Phys. Rev. Letters* **90**, 167402 (2003).
89. A. Franceschetti and S. T. Pantelides, *Phys. Rev. B* **68**, 033313 (2003).
90. A. Puzder, A. J. Williamson, J. C. Grossman and G. Galli, *J. Am. Chem. Soc.* **125**, 2786–2791 (2003).
91. Z. Zhou, L. Brus and R. A. Friesner, *J. Am. Chem. Soc.* **125**, 15599–15607 (2003).
92. Z. Zhou, R. A. Friesner and L. Brus, *Nano Letters* **3**, 163–167 (2003).

93. E. Degoli, G. Cantele, E. Luppi, R. Magri, D. Ninno, O. Bisi and S. Ossicini, *Phys. Rev. B* **69**, 155411 (2004).
94. E. Luppi, E. Degoli, G. Cantele, S. Ossicini, R. Magri, D. Ninno, O. Bisi, O. Pulci, G. Onida, M. Gatti, A. Incze and R. Del Sole, *Opt. Mater.* **27**, 1008–1013 (2004).
95. S. Rao, J. Sutin, R. Clegg, E. Gratton, M. H. Nayfeh, S. Habbal, A. Tsolakidis and R. M. Martin, *Phys. Rev. B* **69**, 205319 (2004).
96. Y. B. Ge and J. D. Head, *Mol. Phys.* **103**, 1035–1045 (2005).
97. F. A. Reboredo and G. Galli, *J. Phys. Chem. B* **109**, 1072–1078 (2005).
98. C. S. Garoufalis and A. D. Zdetsis, *Phys. Chem. Chem. Phys.* **8**, 808–813 (2006).
99. O. Lehtonen and D. Sundholm, *Phys. Chem. Chem. Phys.* **8**, 4228–4232 (2006).
100. F. Pichierri, *Chem. Phys. Letters* **421**, 319–323 (2006).
101. D. A. Eckhoff, J. D. B. Sutin, R. M. Clegg, E. Gratton, E. V. Rogozhina and P. V. Braun, *J. Phys. Chem. B* **109**, 19786–19797 (2005).
102. I. Vasiliev, J. R. Chelikowsky and R. M. Martin, *Phys. Rev. B* **65**, 121302 (2002).
103. M. V. Wolkin, J. Jorne, P. M. Fauchet, G. Allan and C. Delerue, *Phys. Rev. Letters* **82**, 197–200 (1999).
104. X. D. Pi, L. Mangolini, S. A. Campbell and U. Kortshagen, *Phys. Rev. B* **75**, 085423 (2007).
105. Y. Takeda, S. Hyodo, N. Suzuki, T. Motohiro, T. Hioki and S. Noda, *J. Appl. Phys.* **73**, 1924–1928 (1993).
106. K. Luterová, I. Pelant, P. Fojtík, M. Nikl, I. Gregora, J. Kočka, J. Dian, J. Valenta, P. Malý, J. Kudrna, J. Štepánek, A. Poruba and P. Horváth, *Philos. Mag. B* **80**, 1811–1832 (2000).
107. K. Luterová, A. Poruba, J. Dian, O. Salyk, P. Horváth, P. Knápek, J. Valenta, J. Kočka and I. Pelant, *J. Porous Mater.* **7**, 135–138 (2000).
108. J. Mu, L. Xu, X. Gao and Y. Zhang, *J. Disp. Sci. Tech.* **26**, 483–486 (2005).
109. G. M. Credo, M. D. Mason and S. K. Burattoa, *Appl. Phys. Letters* **74**, 1978–1980 (1999).
110. M. D. Mason, D. J. Sirbuly, P. J. Carson and S. K. Burattoa, *J. Chem. Phys.* **114**, 8119–8123 (2001).
111. Y. Chao, A. Houlton, B. R. Horrocks, M. R. C. Hunt, N. R. J. Poolton, J. Yang and L. Siller, *Appl. Phys. Letters* **88**, 263119 (2006).
112. K. A. Pettigrew, Q. Liu, P. P. Power and S. M. Kauzlarich, *Chem. Mater.* **15**, 4005–4011 (2003).

CHAPTER 4

OPTICAL GAIN IN SILICON NANOCRYSTAL WAVEGUIDES MEASURED BY THE VARIABLE STRIPE LENGTH TECHNIQUE

Hui Chen, Jung H. Shin*, Philippe M. Fauchet

Department of Physics and Astronomy and Department of Electrical and Computer Engineering, University of Rochester, Rochester, NY 14627, USA

**Permanent address: Department of Physics, Korea Advanced Institute of Science and Technology, 373-1 Guseong-dong, Yuseong-Gu, Daejeon 305-701, Republic of Korea*

In this chapter, time-resolved photoluminescence and optical gain in nanocrystalline silicon prepared by ion implantation, magnetron sputtering and PECVD, with both oxide and nitride passivations, are investigated using the Variable Stripe Length technique. The results indicate that oxide passivation plays an important role in providing optical gain from silicon nanocrystals.

1. Introduction

1.1. *Silicon photonics: optical interconnects*

Silicon photonics is the "technology of making optical devices using silicon and standard CMOS manufacturing techniques".[1] Although silicon is the most widely used semiconductor for building integrated circuits, including microprocessors, it is not common to make silicon photonic devices, such as light sources and modulators. Because silicon is an indirect bandgap semiconductor, light emission in bulk silicon is not an efficient process due to competing non-radiative processes such as free carrier absorption and Auger recombination. This strongly limits silicon's role as a light source. Unstrained pure crystalline silicon has a zero second-order optical susceptibility due to its symmetric crystal structure, and therefore exhibits no linear electro-optic (EO) effect. For this reason a silicon based modulator needs to rely on other effects such as thermo-optical or free-carrier plasma effects, which makes it less efficient compared to traditional EO modulators using lithium niobate (LiNbO3) or polymers.[2,3] These intrinsic characteristics have made silicon a less interesting material for photonic applications.

However the situation is changing rapidly. A multi-billion dollar industry, silicon microelectronics has been a great success for many years. However, its growth has started to slow down due to the aggressive scaling down of the CMOS transistors. Scaling is a key factor for the increase in transistor speed, but it is also responsible for the significant performance degradation of the interconnect system. Conventional electrical interconnects are a network of metal wires that distributes power, clock and signal among different functional blocks on a chip. As the size of the metal wires continues to scale down, their resistance and capacitance increase significantly. Eventually the RC delay and power dissipation associated with the interconnects will increase to a point where they no longer satisfy the system performance requirements. At the same time, the denser packing of the wires also leads to degradation of the signal integrity due to severe electromagnetic interference and electromigration effects. These factors act together causing the electrical interconnect bottleneck to emerge. Novel interconnect technologies have been proposed to tackle this problem. Optical interconnects are among one of them.

The concept of optical interconnects was first proposed by Goodman in 1984.[4] He described the possibility of using optics to replace the electrical interconnects for backplane-to-backplane and chip-to-chip communications in VLSI systems. The motivation for using optics to solve the electrical interconnect bottleneck comes from the pervasive success of optics in the telecommunication industry. The idea is to combine the unique advantages of optics for communication with the advanced CMOS technology to drive the silicon microelectronic industry forward. Compared to traditional electrical interconnects, optical interconnects have many advantages, such as large bandwidth, low crosstalk, no distance dependent loss and essentially no heat dissipation in the interconnect itself. It has been predicted that within 5 to 10 years, optics will appear in board-to-board and chip-to-chip level interconnects.[5] However, for optical interconnects to migrate to the chip level, several advances need to take place. Recent system-level analyses of on-chip optical interconnects have identified the performance gains and limitations associated with global optical interconnects.[6,7,8] A study comparing high speed on-chip electrical and optical interconnects has shown that under certain assumptions an average performance speedup of 55% can be achieved using on-chip optical interconnects.[9] Although these results are very promising, the real implementation of chip-level all silicon based optical interconnects has turned out to be a grand challenge in part due to the lack of an electrically injected silicon based laser.

Like any other optical communication system, an on-chip optical interconnect system requires light sources, modulators, waveguides and photo-detectors. The main difference is that all the components have to be monolithically integrated on a silicon chip using CMOS compatible technology. This requirement has ruled out most of the III-V semiconductors due to difficulties in integrating them with silicon. Therefore developing silicon based optical components becomes extremely important. While significant progress has been achieved with many building blocks including silicon-based waveguides, modulators and detectors over the past few years,[10,11,12] an electrically injected silicon laser is still lacking. The development of a silicon laser poses an enormous challenge to silicon photonics researchers because it is both an engineering problem and a fundamental physics issue. Is it feasible to achieve any form of lasing in silicon? Until recently the answer would have been "No". However, this is no longer true since the first ever demonstration of a silicon laser in 2004,[13] using Stimulated Raman Scattering (SRS). Although this achievement itself has marked a major milestone in the development of silicon photonics, its role in on-chip optical interconnects, the main motivation for silicon photonics, is doubtful. Raman scattering relies on the inelastic scattering of an incident photon to produce a Stokes (or anti-Stokes) photon. Therefore the silicon Raman laser needs to be optically pumped by another laser. This makes it impractical in on-chip optical interconnects. The ideal solution would be to have a silicon based laser that can be electrically pumped. This requires other means to extract light from silicon. One of the solutions is to use low dimensional silicon.

1.2. *Physics of silicon nanocrystal light emission*

The fast development of nanoscience and nanotechnology has created a new family of materials: low dimensional materials such as quantum wells, quantum wires and quantum dots. In these materials the charge carriers lack one or more degrees of freedom, i.e., they are spatially confined along certain directions. Therefore the optical and electrical properties of these structures are very different from those of their bulk counterpart.

In 1990 bright photoluminescence in the visible range was observed at room temperature in porous silicon, a form of low dimensional silicon, under UV excitation.[14] This discovery has opened up a new route to extract light from silicon. Since then intensive research efforts have been devoted both experimentally and theoretically to understand the origin of this light emission from porous silicon.[15,16] Porous silicon is a spongy network formed by void spaces and silicon clusters of various shapes and sizes, from the nanometer to the micron range depending on the electrochemical etching condition.[34] It has been well accepted that the existence of nanometer-sized silicon crystallites in porous silicon is responsible for the observed visible luminescence that is well above bulk silicon bandgap energy. These nanometer-sized silicon crystallites are called silicon nanocrystals or quantum dots since the carriers are spatially confined in all three dimensions. The fact that the emission energy is strongly size dependent provides direct evidence of the involvement of the quantum confinement effect.

When the diameter of the nanocrystals decreases below 10 nm, the spatial extent of the electron wave function becomes comparable to the size of the dot. The confined electrons feel the nanocrystal boundaries and their electronic states are subject to quantum confinement. The uncertainty principle teaches us that when carriers are strongly confined spatially, they experience uncertainty in momentum. The spread of electron and hole wave functions in momentum space should make it possible for the electrons and holes to recombine directly and emit a photon without the assistance of phonons.[35] This causes a significant increase in the radiative recombination efficiency in silicon nanocrystals, whereas in bulk silicon, the radiative recombination of an electron and a hole is always an indirect process which requires phonon participation and exhibits very low efficiency.

These discoveries have been quite intriguing and promising since they open a new route towards efficient light emission in silicon. Advances in material science have provided us with many different fabrication techniques for forming silicon nanocrystals, including ion-implantation, reactive ion deposition, plasma enhanced chemical vapor deposition, magnetron sputtering and evaporation etc.[17-23] These methods provide silicon nanocrystals with narrower size distributions and better controlled shapes and morphologies.

1.3. *Review of optical gain in silicon nanocrystals*

In 2000, Pavesi et al. reported stimulated emission and optical gain in silicon nanocrystals using the Variable Stripe Length (VSL) method.[17] The VSL method was first proposed by Shaklee et al. in 1971 as an experimental technique for the direct determination of optical gain in semiconductor materials.[24] In the VSL method, the intensity of the edge photoluminescence emission is correlated with the length of an optically excited area in a single-pass geometry. This method has been used to establish stimulated emission and quantify optical gain in many direct bandgap semiconductors. The strength of this technique is that it is based on single pass amplification to establish gain; therefore no fabrication of a resonant optical cavity is required. This makes the VSL technique very suitable for screening novel materials. However, this technique also has its weakness: several artifacts may complicate the interpretation of the results, if precautions are not taken.

Table 1. Partial list of published results on optical gain observed in silicon nanocrystals measured by VSL technique.[17-23]

Sample	Pumping condition	Gain number reported	References
Ion implantation	Pulsed	up to 100 cm^{-1}	*Nature* **408**, 440, 2000
Reactive ion deposition	Pulsed	6 cm^{-1}	*APL* **79**, 1249, 2001
PECVD	Pulsed	10 cm^{-1}	*APL* **82**, 4636, 2003
Magnetron sputtering	Pulsed	up to 50 cm^{-1}	*APL* **83**, 5479, 2003
Evaporation	Pulsed	26 cm^{-1}	*JAP* **96**, 3164, 2004
Sol gel PSi	Pulsed	25 cm^{-1}	*APL* **84**, 3280, 2004
Oxidized PSi	Pulsed	up to 74 cm^{-1}	*PRL* **93**, 207402, 2004

Since 2000, the VSL method has become the major technique to study optical gain in silicon nanocrystals and has yielded many promising results. Table 1 summarizes the current status of optical gain studies on silicon nanocrystals.

Although these results are very promising, it is important to note the following concerns. Some samples experienced light amplification and some did not. There were discrepancies in the reported gain coefficients, from as high as 100 cm^{-1} (Ref. 17) to less than 10 cm^{-1}.[19] Meanwhile, the exact mechanism and physical origin of the optical gain in silicon nanocrystals remain unclear. In this chapter, we will provide some insight into these issues and gain a better understanding of the conditions leading to optical gain by investigating a wide range of nanocrystal films.

2. Sample Preparation

There are many ways to fabricate nanocrystalline silicon, including ion implantation, reactive ion deposition, evaporation, plasma enhanced chemical vapor deposition (PECVD), magnetron sputtering, etc.[18-21,25] In this work we mainly focus on nanocrystalline silicon prepared by ion implantation, magnetron sputtering and PECVD. Both silicon oxide and

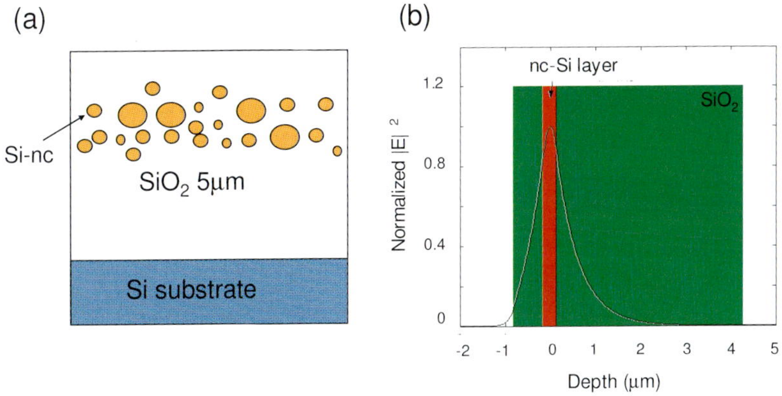

Fig. 1. (a) Cross-sectional view of the ion-implanted nc-Si sample. (b) Simulation results of the normalized electric field distribution (obtained from Dr. N. Smith at ANU).

silicon nitride are used to passivate the nanocrystal surfaces so the role of surface passivations on optical gain can be studied.

Ion implanted nanocrystal samples were produced at the Australian National University (ANU), Canberra, Australia. Si ions of 600 keV energy were implanted into 5 μm thick SiO_2 layers grown on (100) Si wafers by high temperature, high pressure oxidation. After annealing at 1100°C for one hour in flowing N_2 ambient nanocrystals were formed with a size distribution between 3 and 5nm and a peak nanocrystal density of around 5×10^{18} cm^{-3}.[26] Figure 1a shows a sketch of the cross section of the film, and Figure 1b a calculation (obtained from ANU) of the normalized electric field distribution inside this waveguide structure.

Nc-Si superlattices were fabricated by radio frequency magnetron sputtering at the University of Rochester. A 5 μm thick SiO_2 layer was thermally grown on silicon wafers to form a bottom cladding layer. Alternating nanometer-thick layers of amorphous silicon (a-Si) and SiO_2 were then deposited in the same chamber to form a superlattice (SL) structure. The a-Si layers were then transformed into dense arrays of nanocrystals by a two-step annealing procedure: a rapid thermal annealing at 950°C for 30 seconds followed by a high temperature furnace annealing at 1100°C for 1 hour, both in N_2 ambient. Under these

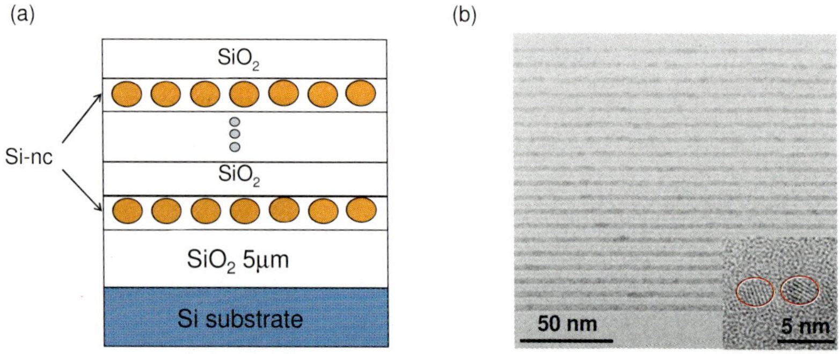

Fig. 2. (a) Drawing of the cross section of the nc-Si superlattices. (b) Cross-sectional scanning electron microscope (SEM) image. Inset: a high resolution transmission electron microscope (TEM) image.

conditions, the entire a-Si layer is transformed into densely-packed nanocrystals with a very tight size distribution.[27,28] The diameter of the nanocrystals is given by the thickness of the initial a-Si layer. The sample studied here had 80 periods of nc-Si and SiO_2 layers, the thicknesses of which were 2 nm and 5 nm, respectively. Figure 2a is a sketch of the sample cross section and Figure 2b shows the cross-sectional scanning electron microscope (SEM) image, the inset of which is a high resolution transmission electron microscope (TEM) image showing the nanocrystals.

Nanocrystalline silicon films fabricated by PECVD were produced at the Korea Advanced Institute of Science and Technology (KAIST). A 200 nm thick amorphous silicon-rich oxide film was deposited on a quartz wafer using plasma enhanced chemical vapor deposition. After deposition, annealing for 1 hr at 1100°C in flowing Ar led to nanocrystal growth. Figure 3 shows the schematic view of the cross section of the PECVD nc-Si film.

The above samples were all passivated with oxide. Silicon nanocrystals with nitride passivation were also grown. The samples were produced at the Electronics and Telecommunications Research Institute in Korea. Nitride-passivated nc-Si thin films were grown *in-situ* on a Si wafer covered with a 15 μm thick thermal oxide layer using plasma enhanced chemical vapor deposition with SiH_4 and N_2 as sources gases. The film thickness was about 150 nm, and the flow rates of SiH_4 and N_2 were modulated to control the size of the silicon nanocrystals.[29] Two samples, with average nanocrystal sizes of 4.6 and 3.1 nm, showing red (800 nm) and green (550 nm) luminescence, respectively, were studied.

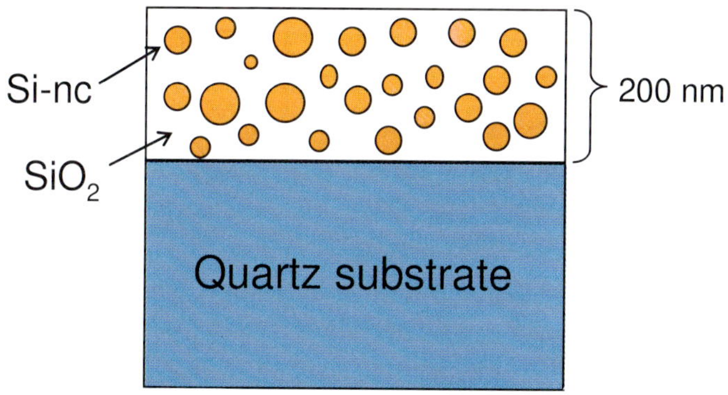

Fig. 3. Schematic view of the cross section of the PECVD nc-Si samples obtained from KAIST.

3. The VSL Method

Figure 4 illustrates the VSL method concept. A sketch of the setup is shown in graph (a). The sample is optically excited using a thin stripe-shape laser beam. An adjustable blade is mounted in front of the sample surface to vary the length of the excited area. The photoluminescence emitted along the length of the stripe is collected from the sample edge. Figure 4b plots the edge emission intensity I versus the excitation length l. If the emission intensity increases superlinearly with the excitation length, optical gain is present. If the emission increases sublinearly with the length, optical loss dominates. If the relationship is

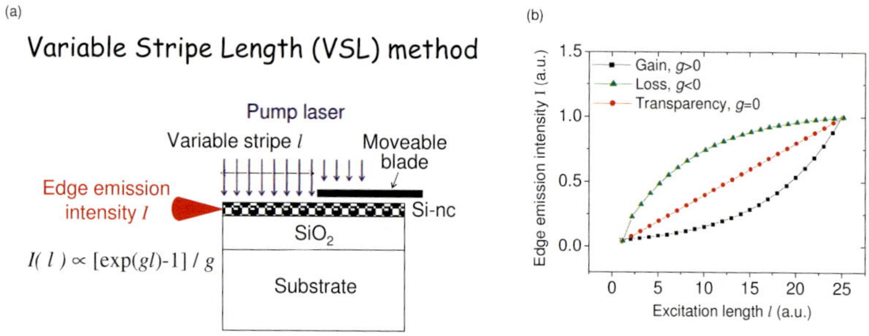

Fig. 4. (a) Variable stripe length method. (b) Simulated VSL curves: Edge emission intensity versus excitation length.

linear, light is neither absorbed nor amplified, indicating optical transparency.

The strength of the VSL technique is that it is based on single-pass amplification to establish gain; therefore no complicated fabrication of a resonant cavity is required. This makes the VSL technique very suitable for screening gain in novel materials. However, this technique also has its weakness: as artifacts could distort the results, if precautions are not taken.

There have been several reports focusing on the variety of artifacts that may occur during the measurement and their effects on VSL results.[30,31] The most commonly recognized ones are caused by the un-optimized distance between the colleting microscope objective and the sample edge, and by the non-uniformity of the pump beam. The focal spot of the microscope objective used to collect the edge emission should be set exactly at the sample edge where the waveguided light exits the film. But in reality this may not be the case due to misalignment. If this happens, then as the blade moves away from the sample edge, the excited spot could move into or out of focus with respect to the objective, depending on whether the focal spot of the objective is inside or outside the sample edge. This will cause an artificial increase or decrease of the VSL signal. Detailed discussions on this effect and possible solutions were given by Valenta *et al.*[30] and Dal Negro *et al.*[31] They suggested using the Shifted Excitation Spot (SES) method to correct for this effect. Instead of using a stripe-shaped beam to excite the sample, the pump beam is focused into a small circular spot and gradually shifted from the sample edge while the edge emission signal is collected. If the objective is focused exactly on the sample edge, the edge emission versus excited spot position should be an exponential decay curve, due to the increased signal absorption when the excitation spot is moved away from edge. SES measurements were performed before VSL measurements. The second well documented artifact is caused by the non-uniformity of the pump beam, which could result from the Gaussian profile of the pump beam itself, and also the diffraction introduced by the edge of the moving blade. The former one can be avoided by using only the center portion of the beam where the intensity is nearly constant, whereas the latter one requires more careful treatment. Diffraction occurs when the pump beam hits the edge of the blade. This generates an oscillating pump intensity pattern on the exposed sample surface. According Fresnel diffraction law, the closer the blade is to the sample surface, the shorter the length scale is for the diffraction pattern. Therefore, in all VSL measurements reported later, the separation between the sample surface and the blade was kept around 500 μm. This resulted in an artificial exponential increase of the signal only in the first 10 μm or less.

One more artifact is usually not addressed carefully. It regards the correct detection of the VSL signal by the microscope objective. VSL theory is based on the one dimensional amplifier model, which assumes that the detected signal is waveguided in the active medium for a distance *l*. This causes light to be attenuated or amplified by an exponential factor. However, the light collected by the objective may not be the desired guided signal coming from the nanocrystal waveguide core, but from other pathways, such as the spontaneous PL scattered off the top surface or the bottom nc-Si/SiO_2 interface. To form a waveguide, a thick layer of SiO_2 was used as the bottom cladding in all the samples described later. The thickness of the nanocrystal core (~300 nm) is small compared to that of the SiO_2 layer (several microns). Therefore it is very difficult to ensure that the signal is indeed the waveguided signal. The light that is not guided gets reflected or scattered by the sample surface or the bottom SiO_2/Si interface into all directions. For specific angles, it may be collected by the objective and mixed into the VSL signal. In addition, there is little attenuation of visible light by the thick SiO_2 layer

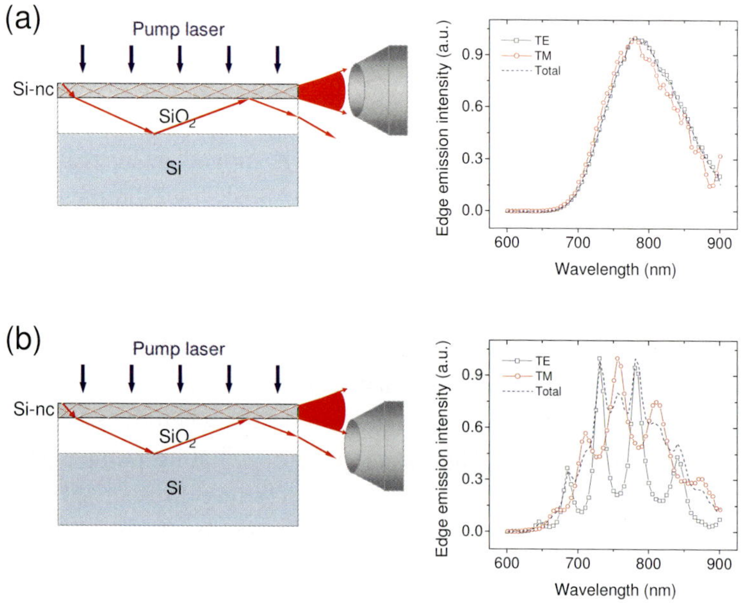

Fig. 5. (a) Edge emission PL spectra when the guided signal is detected by the objective. (b) Edge emission spectra when the stray PL signal is detected by the objective.

in contrast to what may take place in the nc-Si core. Therefore the contributions from the stray PL can be relatively large compared to the VSL signal. If the detection optics is aligned by maximizing the signal intensity, as is often done, artifacts are very likely to occur.

In Figure 5 we give an example. By slightly shifting the objective away from the plane of the nanocrystal core towards the bottom SiO_2 layer, strikingly different PL spectral shapes are observed. In Figure 5a when the objective is aligned with the waveguide core, a smooth Gaussian curve is detected, which is characteristic of nanocrystal emission and very little difference is noticed between the normalized TE and TM spectra. However, if the objective is slightly off alignment, as shown in graph (b) (displacement exaggerated for demonstration), the spectrum is drastically altered. If we use a polarizer to separate the TE and TM polarized light, regularly distributed peaks are observed in the spectra with the peaks in one polarization corresponding to the valleys in the other. This phenomenon was observed before and was attributed to modes of the waveguiding layer.[30] However, we believe this is simply caused by light interference due to the layered structure. These fringes are equally spaced in energy and fitting shows their spacing corresponds well with the film thickness. This strongly suggests that the signal detected in this case is not from the waveguiding layer, but dominated by stray PL.

In order to make sure the true VSL signal is collected, preceding the VSL measurements the detection optics was first aligned using a reference laser beam. A prism coupler was introduced to the VSL setup for alignment purpose only. A schematic view of the setup is shown in Figure 6. The sample was mounted on the prism coupler stage. An 810 nm laser beam was coupled into the nanocrystal film through the prism. We used it as a reference to align our detection optics to the guided signal. Once the alignment was completed, this reference beam was removed and the pump beam was introduced for VSL measurements without further modifying the alignment. By doing so, we confirmed the existence of guided modes in the Si-nc layer as well as the detection of the mode by the objective.

Time-resolved PL and VSL measurements were performed using a frequency doubled amplified Ti-sapphire system generating ultra-short laser pulses (~200 fs, 1 kHz, 405 nm) focused into a 50 μm wide stripe and incident normally upon the film surface. The energy per pulse and pump fluence were up to 20 μJ and 40 mJ/cm^2, respectively. The guided signal collected from the sample edge using a microscope objective was analyzed by a scanning monochromator equipped with a PMT and a gated photon counter. In all measurements the photon counter was gated separately from 0 to 10 nsec and from 50 nsec to 5 μsec to record the fast and slow PL and VSL signals.

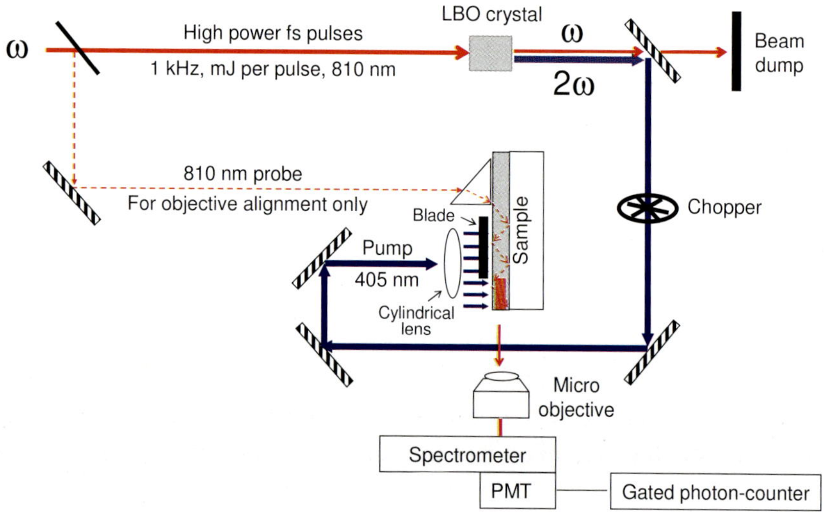

Fig. 6. Schematic view of the experimental setup for VSL measurements.

4. Results and Discussion

4.1. *Oxide passivated silicon nanocrystals*

Oxide passivated nc-Si films prepared by ion-implantation, magnetron sputtering and plasma enhanced chemical vapor deposition were studied.

4.1.1. *Ion implanted nanocrystal system*

Figure 7a shows the PL spectra of this sample. The photon counter was gated separately from 0 to 10 nsec and from 50 nsec to 5 μsec to record the fast (square) and slow (circle) PL signals. The PL signal collected within the first 10 nsec is several orders of magnitude smaller than that collected with the 5 μsec gate. The normalized PL spectra are shown in Figure 7b. After normalization, the two spectra are found to be identical, indicating a common origin.

As mentioned in the previous section, an SES measurement was performed preceding the VSL measurements to eliminate artifacts.

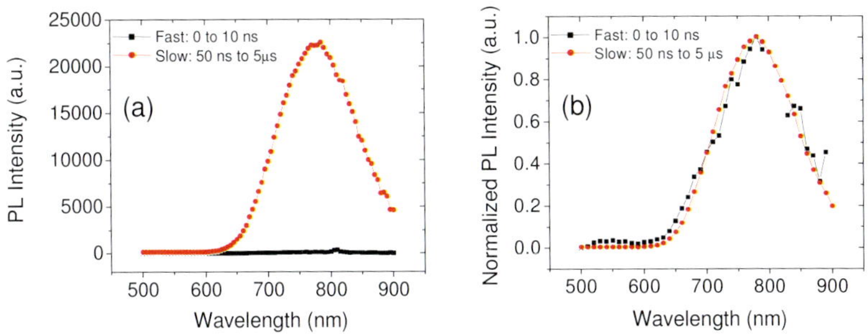

Fig. 7. (a) Fast (square) and slow (circle) PL spectra. (b) Normalized PL spectra: Fast (square) and slow (circle).

Figure 8 shows the SES curve taken at 700 nm. An exponential decay function fits the data points and a loss coefficient of 32 cm^{-1} can be extracted.

Figures 9a and 9b show the fast (first 10 ns after excitation) and slow (50 ns to 5 μs) VSL measurements at different wavelengths. The solid lines are fitting curves using the one dimensional amplifier model, $I(l) \propto [\exp(gl)-1] / g$. The pump fluence is 40 mJ/cm^2. Both fast and slow VSL signals show a sublinear increase with increasing pumping length, indicating optical loss, throughout the whole spectral region (600 to 780 nm) that was investigated. Table 2 lists the loss coefficient for the fast and slow components obtained at different wavelengths from curve fitting. There is no clear difference between the loss numbers of the fast

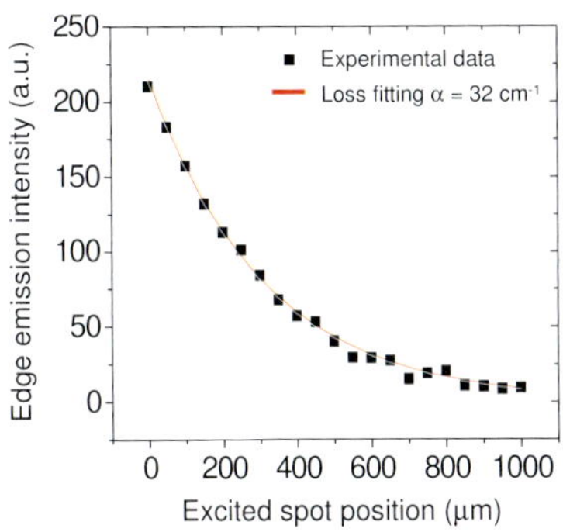

Fig. 8. Shifted Excited Spot (SES) data taken at 700 nm. Solid line: exponential decay fitting.

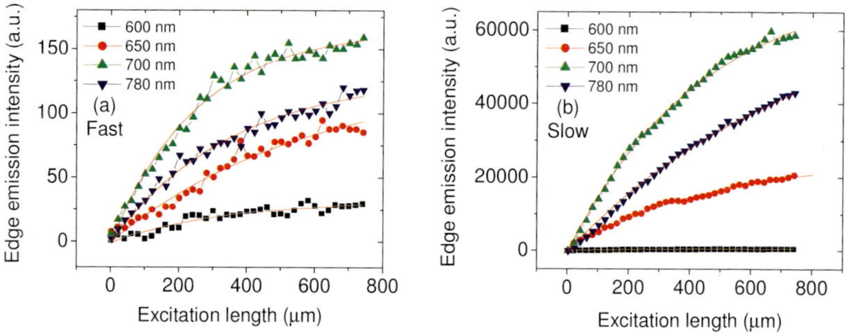

Fig. 9. (a) Fast component (from 0 to 10 ns) VSL data taken at 600 nm (square), 650 nm (circle), 700 nm (up triangle), and 780 nm (down triangle). (b) Slow component (from 50 ns to 5 μs) VSL data taken at 600 nm (square), 650 nm (circle), 700 nm (up triangle), and 780 nm (down triangle). Solid lines are fitting curves with the one dimensional amplifier model.

Table 2. Loss coefficient of the fast and slow components at different wavelengths.

Wavelength (nm)	Energy (eV)	Loss coefficient (cm^{-1}) Fast component	Loss coefficient (cm^{-1}) Slow component
600	2.1	40	34
650	1.9	39	31
700	1.8	37	24
780	1.6	29	14

and slow components. They vary in the range from 10 to 40 cm^{-1} depending on the wavelength. The shorter the wavelength the higher the loss tends to be.

To summarize, time-resolved photoluminescence measurements revealed almost identical PL spectra for the fast and slow emission components after normalization. VSL measurements performed on both fast and slow components show a sublinear increase of the signal with increasing pumping length, indicating optical loss.

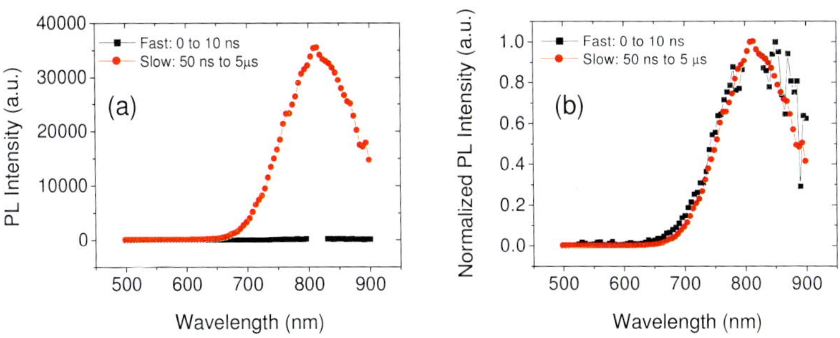

Fig. 10. (a) Fast (square) and slow (circle) PL spectra. (b) Normalized PL spectra: Fast (square) and slow (circle).

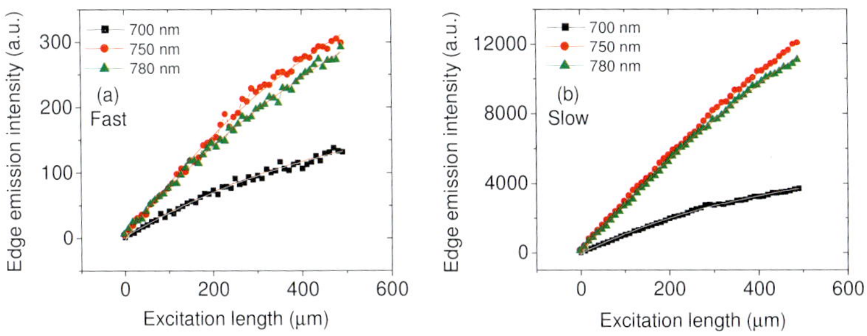

Fig. 11. (a) Fast component (from 0 to 10 ns) VSL data taken at 700 nm (square), 750 nm (circle), and 780 nm (up triangle). (b) Slow component (from 50 ns to 5 μs) VSL data taken at 700 nm (square), 750 nm (circle), and 780 nm (up triangle). Solid lines are fitting curves with the one dimensional amplifier model.

4.1.2. *Magnetron sputtered Si/SiO$_2$ superlattices*

Nanocrystalline silicon / silicon oxide superlattices were prepared by radio frequency magnetron sputtering. A total of 80 periods of nc-Si and SiO_2 layers were deposited, the thicknesses of which were 2 and 5 nm, respectively.

Figure 10a shows the fast and slow PL spectra of this sample. Again, due to the significant difference in the signal intensity, the signal counts in the fast PL spectra are almost negligible compared to that of the slow PL spectrum. The normalized PL spectra shown in Figure 10b almost completely overlap each other.

Table 3. Loss coefficient of the fast and slow components at different wavelengths.

Wavelength (nm)	Energy (eV)	Loss coefficient (cm^{-1}) Fast component	Loss coefficient (cm^{-1}) Slow component
700	1.77	19	22
750	1.65	16	17
780	1.59	12	11

Figures 11a and 11b show the fast (first 10 ns after excitation) and slow (50 ns to 5 μs) VSL measurements at different wavelengths. The solid lines represent the fitting curves. The pump fluence was 40 mJ/cm^2. In the spectral region studied (700 to 780 nm), both fast and slow VSL signals show a sublinear increase with increasing pumping length, indicating optical loss. Due to the extremely low signal level below 700 nm, no measurement was performed for wavelengths shorter than 700 nm. Table 3 shows the loss coefficients vs. photon energy for fast and slow components. The loss coefficients vary from 11 to 22 cm^{-1} and the numbers obtained for the fast and slow components are almost the same.

To conclude, time-resolved photoluminescence and VSL measurements were performed on the nc-Si/SiO_2 superlattices. The results were very similar to that observed from the ion-implanted nanocrystal film. We found that the spectral shapes of the fast and slow emission components are almost identical. VSL measurements performed on both fast and slow components showed a sublinear increase of the signal with pumping length, indicating optical loss.

This result is in contrast to what has been reported previously.[20] In those experiments, gain was observed in some samples immediately after optical pumping. There were some differences in the waveguide geometry used in these experiments, but they seem unlikely to be the cause for the discrepancy. A more likely cause is that the PL collected within 10 nsec after pumping is not blue shifted in the present study, whereas a strong blue shift was observed in those experiments where gain was achieved.[20] The origin of this blue PL with nsec or shorter time

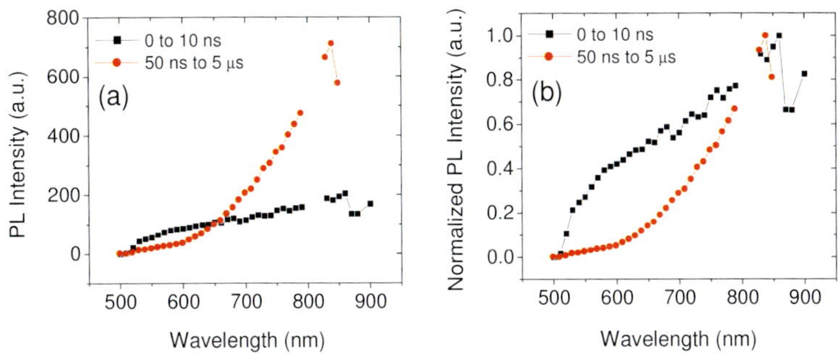

Fig. 12. (a) Fast (square) and slow (circle) PL spectra. (b) Normalized PL spectra: Fast (square) and slow (circle).

decay is not established and may be related to the presence or absence of Si=O at the interface of the quantum dots,[15] as will be mentioned in the next section.

4.1.3. *PECVD nanocrystal system*

An oxide-passivated nc-Si thin film was also fabricated by depositing an amorphous silicon-rich oxide film on a quartz wafer using plasma-enhanced chemical vapor deposition. After deposition, annealing for 1 hr at 1100°C in flowing Ar produced nanocrystals.

Figure 12a shows the fast and slow PL spectra of this sample. The solid squares and circles represent the fast and slow signals, respectively. The slow component has much higher signal counts than that of the fast, but the shapes of the two spectra are different. Once we normalize the spectra, this difference becomes much clearer, as shown in Figure 12b. The fast PL spectrum is strongly blue shifted and also much broader compared to the slow PL spectrum, which peaks in the near IR region. This is very different from what was observed in the previous two samples, where the fast and slow PL spectra almost completely overlap each other. The difference in the spectral shape indicates that the fast and the slow emission components may have different physical origins. It is possible that the fast emission is due to recombination between quantized states, whereas the slow one involves surface states, such as the Si=O

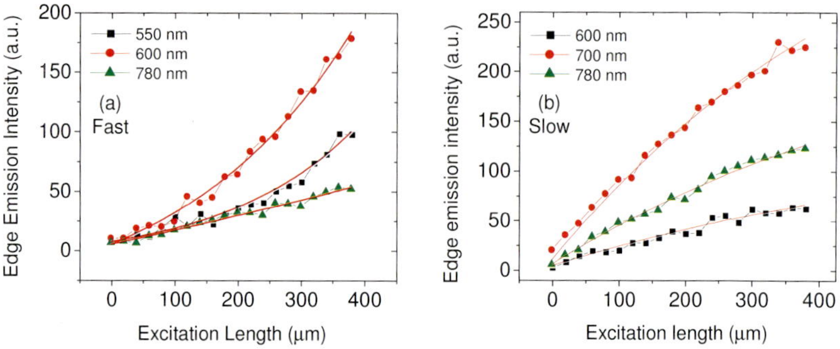

Fig. 13. (a) Fast component (from 0 to 10 ns) VSL data taken at 550 nm (square), 600 nm (circle), and 780 nm (up triangle). (b) Slow component (from 50 ns to 5 μs) VSL data taken at 600 nm (square), 700 nm (circle), and 780 nm (up triangle). Solid lines are fitting curves with the one dimensional amplifier model.

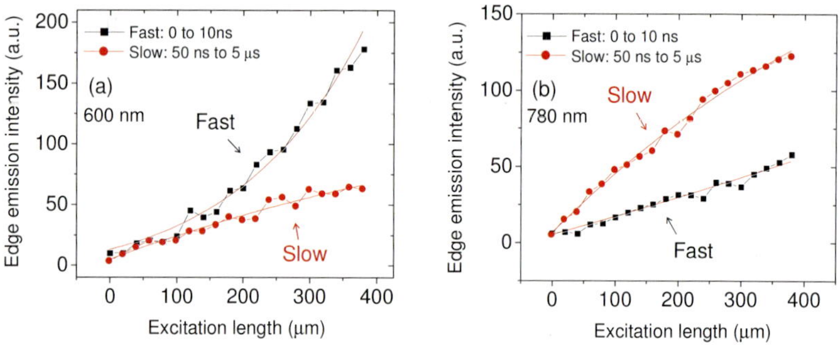

Fig. 14. (a) VSL data taken at 600 nm: Fast component (square), slow component (circle). (b) VSL data taken at 780 nm: Fast component (square), slow component (circle).

bonds identified by Wolkin *et al.*[15] This would explain the higher photon energy of the fast emission component.

Figures 13a and 13b show the fast (first 10 ns after excitation) and slow (50 ns to 5 μs) VSL measurements at different wavelengths. The pump fluence was 40 mJ/cm^2. In the spectral region from 550 to 780 nm, the fast VSL signal exhibits a strong superlinear increase with increasing pumping length, indicating the presence of optical gain. Although the trend is similar at all wavelengths studied, the exponential increase is stronger for shorter wavelengths, indicating that the gain is higher in the

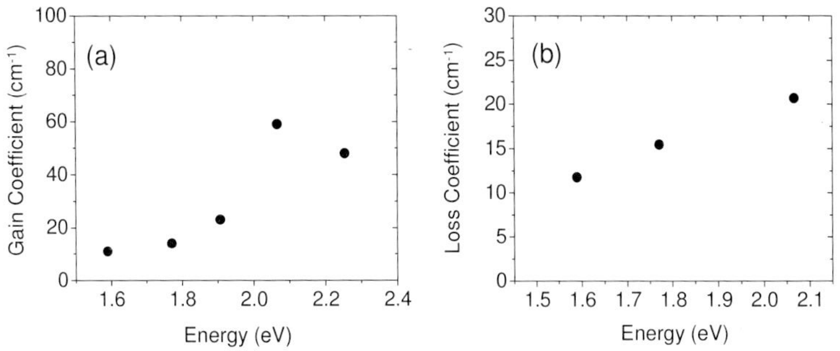

Fig. 15. (a) Gain coefficient vs. photon energy for fast component (from 0 to 10 ns); (b) Loss coefficient vs. photon energy for slow component (from 50 ns to 5 μs).

bluer region of the spectrum. Figures 13b shows the VSL measurements of the slow component. The VSL signal increases sublinearly with the pumping length throughout the spectral region from 600 to 780 nm, indicating optical loss. For comparison, the fast and slow VSL data are plotted on the same graph in Figure 14. At 600 nm, the shape of VSL signal changes from sublinear (slow component) to superlinear (fast component). A similar trend is found at 780 nm too, shown in Figure 14b; however the exponential increase of the fast component is not as strong as that observed at 600 nm. These results, including the blue shift of the fast PL component, are similar to that reported by Ruan *et al.*[20]

Figure 15a plots the gain coefficient vs. photon energy for the fast component, and Figure 15b plots the loss coefficient vs. photon energy for the slow component. The gain peaks around 600 nm when the gain coefficient is 59 cm^{-1}. The gain value gradually decreases at longer wavelengths and it is only about 11 cm^{-1} at 780 nm. The loss coefficients vary from 21 cm^{-1} at 600 nm to 12 cm^{-1} at 780 nm.

To conclude this section, a blue shifted fast emission component was observed in the PECVD nc-Si sample. Optical gain was measured for the fast component in the spectral region from 500 to 780 nm. The peak of the gain occurs near 600 nm with a gain coefficient of 59 cm^{-1}. Optical loss was measured in the slow emission component throughout the spectral region from 600 to 780 nm.

4.2. *Nitride passivated silicon nanocrystals*

All the research on optical gain of silicon nanocrystals relied on nanocrystals with oxide passivation. However, as the optical properties of the silicon nanocrystals are very sensitive to surface passivation, studying nanocrystals passivated with a different material can help us gain a better understanding of the origin of gain, whether it is intrinsic to nanocrystals or associated with specific defects related to the oxide passivation.

In oxide passivated nc-Si, when optical gain is observed, a set of similar properties are nearly always observed, independent of the preparation methods. First, there exists a fast, nsec-range component that is blue-shifted compared to the omnipresent "slow", μsec-range nc-Si luminescence peak in the near infra-red region. Second, optical gain is observed only within the fast component, while the slow component always shows optical loss. The above measurements on oxide passivated nanocrystals support this argument: the presence of a blue shifted fast emission component is a necessary condition leading to optical gain.

In this context, nitride passivated nc-Si may be a very good candidate for optical gain for several reasons. First, silicon nitride is also a widely used dielectric in the CMOS industry. Second, since silicon nitride has a smaller bandgap than that of silicon oxide, carrier injection should be easier, producing more efficient electroluminescence. More importantly, there have been recent reports that the luminescence lifetime in nitride-passivated nc-Si is in the nsec-range,[32,33] comparable to the fast-component in the oxide-passivated nc-Si luminescence that appears to be responsible for optical gain.

Nitride passivated silicon nanocrystal films were grown *in-situ* on a Si wafer covered with a 15 μm thick thermal oxide layer using plasma-enhanced chemical vapor deposition with SiH_4 and N_2 as the sources gases. The film thickness was about 150 nm, and the flow rates of SiH_4 and N_2 were modulated to control the size of the silicon nanocrystals. Two samples average nanocrystal sizes of 4.6 and 3.1 nm, showing red (800 nm) and green (550 nm) luminescence, respectively, were studied. Henceforth, they will be referred to as sample R (red) and G (green), respectively.

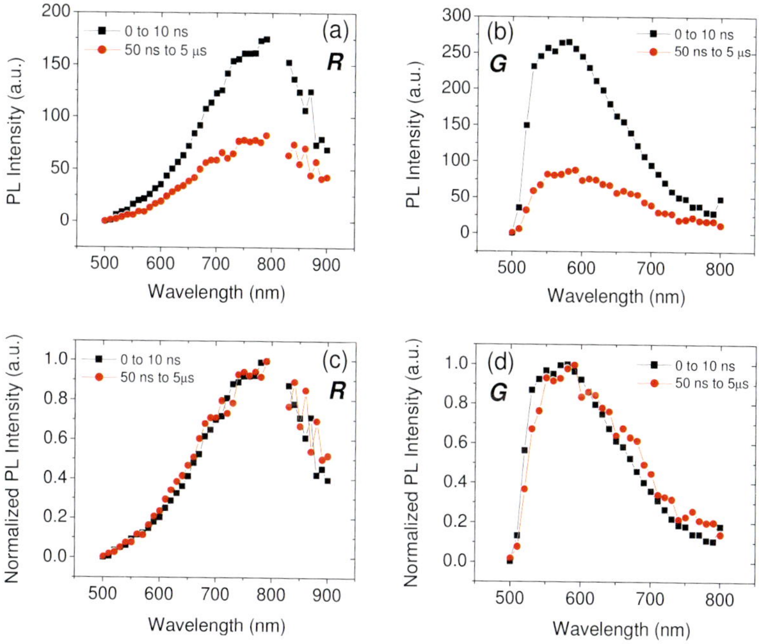

Fig. 16. (a) Fast (square) and slow (circle) PL spectra of sample R. (b) Fast (square) and slow (circle) PL spectra of sample G. (c) Normalized fast (square) and slow (circle) PL spectra of sample R. (d) Normalized fast (square) and slow (circle) PL spectra of sample G.

Figures 16a and 16b show the PL spectra of sample R and G, respectively. The photon counter was gated separately from 0 to 10 nsec and from 50 nsec to 5 μsec to record the fast (square) and slow (circle) PL signals. The normalized PL spectra are shown in Figures 16c and 16d.

The PL spectral shape of the nitride passivated nanocrystals closely resembles those of the oxide passivated ones. However, unlike in the oxide passivated nanocrystals where the slow PL intensity is much higher than the fast PL intensity, in the case of samples R and G, the intensity of fast PL is more than twice that of slow PL. Furthermore, after normalization the two spectra are identical. This is very different from what has been observed in the samples that show optical gain. In those samples a clear difference exists between the fast and the slow PL

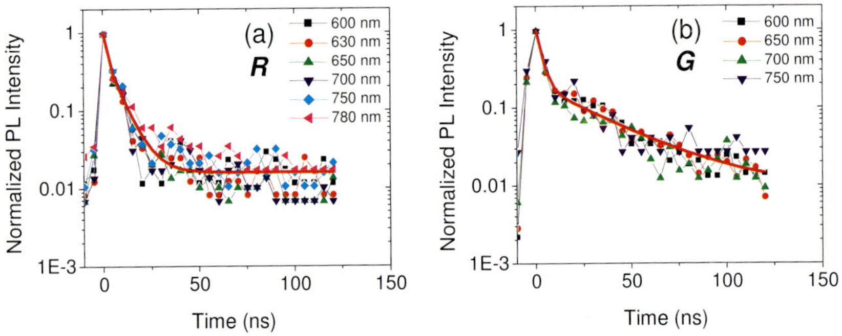

Fig. 17. (a) Normalized PL decay traces at different wavelengths for sample R. (b) Normalized PL decay traces at different wavelengths for sample G. Solid lines are double exponential fitting curves.

spectra – a broad blue-shifted PL peak that dominates the fast PL spectrum, whereas a PL peak in the near IR that dominates the slow PL spectrum.

Figures 17a and 17b show the normalized PL decay curves of sample R and G at various emission wavelengths. These normalized traces almost overlap each other, indicating a uniform decay throughout the spectral region from 600 nm to 780 nm. In sample R a faster decay with a time constant $\tau_1 = 0.5$ nsec and a slower decay with constant $\tau_2 = 7$ ns were obtained. In sample G, the faster decay constant $\tau_1 = 2$ ns and the slow decay constant $\tau_2 = 33$ ns.

It should be pointed out here that the signal intensities of the R and G samples are comparable to that of the blue shifted fast emission component from the gain sample. This suggests that the fast emission from R and G samples is very likely due to a faster radiative recombination mechanism in nitride-passivated nc-Si, and not due to shortening of the conventional, slow-PL component of nc-Si caused by defect-related non-radiative decay. This is in good agreement with recent theoretical investigations that have suggested that nitride-passivated nc-Si can have radiative lifetimes in the nsec range.[32,33]

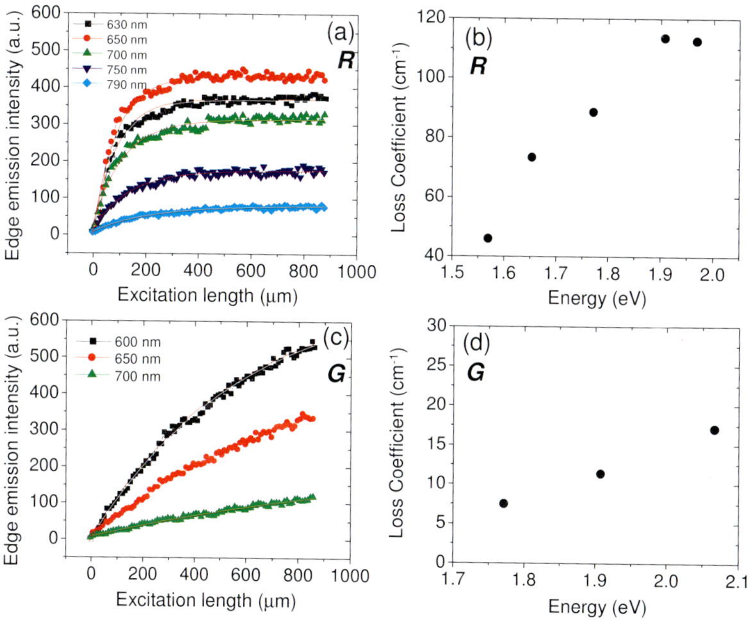

Fig. 18. (a) VSL data for sample R taken at 630 nm (square), 650 nm (circle), 700 nm (up triangle), 750 nm (down triangle) and 790 nm (diamond). The solid lines are the fitting curves with the one dimensional amplifier model. (b) Optical loss coefficients (obtained from fitting) versus photon energy. (c) VSL data for sample G taken at 600 nm (square), 650 nm (circle), 700 nm (up triangle). (d) Optical loss coefficients versus photon energy. The pump fluence is 40 mJ/cm^2. Time gate was set to detect only the fast component (from 0 to 10 ns).

It has been established that optical gain in oxide-passivated nc-Si is associated with a fast, blue-shifted component of the luminescence. Since in nitride-passivated nc-Si all the luminescence is fast, this material should be ideal for obtaining strong optical gain. However, this is not the case. Figures 18a and 18c show the VSL measurements of the fast PL component at different wavelengths for samples R and G, respectively. Figures 18b and 18d plot the loss coefficients obtained from curve fitting versus photon energy. The signal with nitride-passivated films shows a strong sub-linear increase with increasing pumping length, indicating optical loss, throughout the entire spectral region (600 to 800 nm).

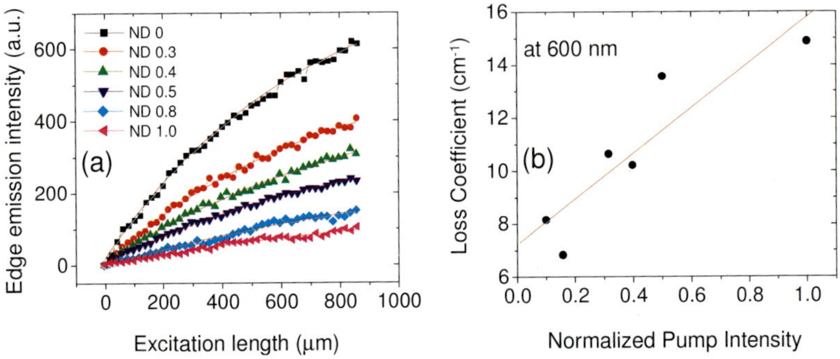

Fig. 19. (a) Sample G: VSL data taken at 600 nm with different pump fluences, 40 mJ/cm^2 (square), 20 mJ/cm^2 (circle), 16 mJ/cm^2 (up triangle), 13 mJ/cm^2 (down triangle), 6 mJ/cm^2 (diamond) and 4 mJ/cm^2 (left triangle). (b) Optical loss coefficient versus normalized pump intensity.

Fig. 19 shows the dependence of optical loss with increasing pump fluence in sample G. A set of neutral density filters was used to attenuate the pump intensity. ND 0 represents the unattenuated beam, which is 40 mJ/cm^2. ND 0.3, ND 0.4, ND 0.5, ND 0.8 and ND 1.0 filters attenuated the pump fluence to 50%, 40%, 32%, 16% and 10% of its original value. Graph (a) shows the VSL measurements at 600 nm taken under different pump fluences. Graph (b) plot the loss coefficient obtained from curve fitting versus normalized pump intensity. The optical loss at 600 nm increases with increasing pump fluence, indicating that a substantial part of the optical loss was due to photo-injected carrier absorption. At very low pump fluence, the optical loss approaches 7 cm^{-1} in this sample.

Thus, nitride passivation of nc-Si does not lead to optical gain despite the fast luminescence decay in the nsec range and the possibility of tuning the peak luminescence in the blue. In fact, the overall luminescence dynamics, including the stronger optical loss for the R sample and the increasing optical loss at higher energies, is consistent with what had been expected of an indirect semiconductor, and argues against the possibility of obtaining optical gain from nitride-passivated nc-Si.[36]

5. Conclusions

To summarize, time-resolved photoluminescence and VSL measurements have been performed on nc-Si samples prepared by ion-implantation, magnetron sputtering and PECVD, with both oxide and nitride passivations. In the oxide passivated nanocrystal samples, optical gain is observed only in samples that exhibit a clear blue shifted fast emission component, independent of the details of the sample preparation. Gain is only observed immediately after carrier injection, which suggests it might be impossible to achieve CW lasing. In samples that have identical fast and slow PL spectral shapes, optical loss is observed in both fast and slow components.

In nitride passivated nc-Si samples, a fast luminescence decay is observed, but no gain is achieved even if the PL spectrum peaks at short wavelengths. Thus, taken together, the data suggest that optical gain from nc-Si is not likely to be intrinsic to nc-Si alone, but instead requires the additional presence of oxide passivation. The presence of a fast blue shifted emission component seems necessary for achieving optical gain. The role oxide passivation plays in providing this blue-shifted, fast luminescence component associated with optical gain is not established. Although it is possible that this fast emission is due to specific recombination from quantized states, it could also be caused by a surface defect, analogous to the Si=O bond that leads to red-shift of nc-Si PL peak.

Acknowledgements

Special thanks go to Dr. Jinhao Ruan who performed the first optical gain measurements at the University of Rochester. The authors thank Dr. Robert G. Elliman, Dr. Jung H. Shin's research group, and Dr. Jae-Heon Shin and Dr. Gun Yong Sung who provided different nc-Si samples. This work was supported by grants from the US National Science Foundation, the Semiconductor Research Corporation and the Air Force Office of Scientific Research through a MURI grant (G. Pomrenke).

References

1. http://www.intel.com/technology/silicon/sp/glossary.htm.
2. M. Minakata, *Proceedings of SPIE* **4532**, 16 (2001).
3. Y. Shi, W. Lin, D. J. Olson, J. H. Bechtel, H. Zhang, W. H. Steier, C. Zhang and L. R. Dalton, *Appl. Phys. Lett.* **77**, 1 (2000).
4. J. W. Goodman, *Proceedings of the IEEE* **72**, 850 (1984).
5. N. Savage, "Linking with light," *IEEE Spectrum* **39**, 32 (2002).
6. M. J. Kobrinsky, B. A. Block, J. Zheng, B. C. Barnett, E. Mohammed, M. Reshotko, F. Robertson, S. List, I. Young and K. Cadien, *Intel Technology Journal* **8**, 129 (2004).
7. M. Haurylau, G. Chen, H. Chen, J. Zhang, N. A. Nelson, D. H. Albonesi, E. G. Friedman and P. M. Fauchet, *IEEE Journal of Selected Topics in Quantum Electronics* **12**, 1699 (2006).
8. G. Chen, H. Chen, M. Haurylau, N. A. Nelson, D. H. Albonesi, P. M. Fauchet and E. G. Friedman, *Integration-the VLSI journal* **40**, 434 (2007).
9. N. Nelson, G. Briggs, M. Haurylau, G. Chen, H. Chen, D. H. Albonesi, E. G. Friedman and P. M. Fauchet, *Proceedings of the Workshop on Unique Chips and Systems* **45** (2005).
10. H. Yamada, T. Chu, S. Ishida and Y. Arakawa, *IEEE Photon. Tech. Lett.* **17**, 585 (2005).
11. A. Liu, R. Jones, L. Liao, D. Samara-Rubio, D. Rubin, O. Cohen, R. Nicolaescu and M. Paniccia, *Nature* **427**, 615 (2004).
12. B. Yang, J. D. Schaub, S. M. Csutak, D. L. Rogers and J. C. Campbell, *IEEE Photon. Tech. Lett.* **15**, 745 (2003).
13. O. Boyraz and B. Jalali, *Optics Express* **12**, 5269 (2004).
14. L. T. Canham, *Appl. Phys. Lett.* **57**, 1046 (1990).
15. M. V. Wolkin, J. Jorne, P. M. Fauchet, G. Allan and C. Delerue, *Phys. Rev. Lett.* **82**, 197 (1999).
16. C. Delerue, G. Allen and M. Lannoo, *Phys. Rev. B* **48**, 11024 (1993).
17. L. Pavesi, L. Dal Negro, C. Mazzoleni, G. Franzò and F. Priolo, *Nature* **408**, 440 (2000).
18. L. Khriachtchev, M. Räsänen, S. Novikov and J. Sinkkonen, *Appl. Phys. Lett.* **79**, 1249 (2001).
19. L. Dal Negro, M. Cazzanelli, L. Pavesi, S. Ossicini, D. Pacifici, G. Franzò, F. Priolo and F. Iacona, *Appl. Phys. Lett.* **82**, 4636 (2003).
20. J. Ruan, P. M. Fauchet, L. Dal Negro, M. Cazzanelli and L. Pavesi, *Appl. Phys. Lett.* **83**, 5479 (2003).
21. M. Cazzanelli, D. Navarro-Urriós, F. Riboli, N. Daldosso, L. Pavesi, J. Heitmann, L. X. Yi, R. Scholz, M. Zacharias and U. Gösele, *J. Appl. Phys.* **96**, 3164 (2004).
22. K. Luterová, K. Dohnalová, V. Svrcek, I. Pelant, J.-P. Likforman, O. Crégut, P. Gilliot and B. Hönerlage, *Appl. Phys. Lett.* **84**, 3280 (2004).

23. M. Cazzanelli, D. Kovalev, L. Dal Negro, Z. Gaburro and L. Pavesi, *Phys. Rev. Lett.* **93**, 207402 (2004).
24. K. L. Shaklee and R. F. Leheny, *Appl. Phys. Lett.* **18**, 475 (1971).
25. R. G. Elliman, M. J. Lederer, N. Smith and B. Luther-Davies, *Nucl. Instr. and Meth. B* **206**, 427 (2003).
26. P. M. Fauchet, J. Ruan, H. Chen, L. Pavesi, L. Dal Negro, M. Cazzaneli, R. G. Elliman, N. Smith, M. Samoc and B. Luther-Davies, *Optical Materials* **27**, 745 (2005).
27. L. Tsybeskov, K. D. Hirschman, S. P. Duttagupta, M. Zacharias, P. M. Fauchet, J. P. McCaffrey and D. J. Lockwood, *Appl. Phys. Lett.* **72**, 43 (1998).
28. G. F. Grom, D. J. Lockwood, J. P. McCaffrey, H. J. Labbé, P. M. Fauchet, B. White, J. Diener, D. Kovalev, F. Koch and L. Tsybeskov, *Nature* **407**, 358 (2000).
29. T.-Y. Kim, N.-M. Park, K.-H. Kim, G. Y. Sung, Y.-W. Ok, T.-Y. Seong and C.-J. Choi, *Appl. Phys. Lett.* **85**, 5355 (2004).
30. J. Valenta, I. Pelant and J. Linnros, *Appl. Phys. Lett.* **81**, 1396 (2002).
31. L. Dal Negro, P. Bettotti, M. Cazzanelli, D. Pacifici and L. Pavesi, *Optics Communications* **229**, 337 (2004).
32. L. Dal Negro, J. H. Yi, L. C. Kimerling, S. Hamel, A. Williamson and G. Galli, *Appl. Phys. Lett.* **88**, 183103 (2006).
33. L. Dal Negro, J. H. Yi, J. Michel and L. C. Kimerling, T.-W. F. Chang, V. Sukhovatkin and E. H. Sargent, *Appl. Phys. Lett.* **88**, 233109 (2006).
34. P. M. Fauchet, "Silicon: Porous" in *Encyclopedia of Applied Physics, Update 2*, Wiley-VCH Verlag, 249 (1999).
35. D. Kovalev, H. Heckler, M. Ben-Chorin, G. Polisski, M. Schwartzkopff and F. Koch, *Phys. Rev. Lett.* **81**, 2803 (1998).
36. H. Chen, J. H. Shin, P. M. Fauchet, J. Y. Sung, J. H. Shin and G. Y. Sung, *Appl. Phys. Lett.* **91**, 173121 (2007).

CHAPTER 5

SI-NC BASED LIGHT EMITTERS AND ER DOPING FOR GAIN MATERIALS

Olivier Jambois, Se-Young Seo, Paolo Pellegrino and Blas Garrido

EME, Departament d'Electrònica, IN2UB, Universitat de Barcelona
Barcelona, 08028, Spain
E-mail: blas@el.ub.es

We present an overview of light emitting devices based on Si-nc, and eventually their doping to shift the emission to the UV or near infrared. Several materials and two basic device structures will be presented to test the emission properties. For visible emission the focus will be on the electroluminescence (photoluminescence is widely tackled in other chapters of this book), while for materials doped with Er attention will be paid to material properties and simple waveguide structures, that should prove their suitability as optical gain media.

1. Introduction

A wealth of combination of Si-based materials and fabrication methods have been proposed and demonstrated so far for electroluminescence (EL). The basic material is a high quality, non-defective silicon-rich oxide or nitride, or a combination of them (SiO_x, SiN_x, SiO_xN_y).

Depending on the the emission pursued, doping with luminescent centers can be performed. Of particular interest is the doping with Er, as it allows an emission at the strategic 1.5 μm telecom window. Efficient excitation of Er ions through Si-nc can provide population inversion and thus optical amplification. An optical modeling of the system is required, including Si-nc excitation, energy transfer from Si-nc to Er, density of Er ions coupled to Si-nc and control of detrimental effects.

Consequently, this chapter is organized into two main sections. Section I deals with Si-nc based light emitting devices (LEDs). It shows devices with Si-nc emitting in the red, while by C doping they shift the emission to the green. Both DC and AC excitation are considered and eventually the realization of a tunable source is shown. Section II deals with Er doping that places the emission in the near infrared. Firstly, Er excitation dynamics and luminescence properties are considered. Afterwards the main limiting factors for the Er luminescence are analysed. Finally, a glimpse is given on the realization of waveguide amplifiers.

2. Si Nanocluster Based Light Emitters

2.1. *Brief review and perspective*

To overcome the inability of bulk Si to efficiently emit light because of its indirect gap leading to dominant non-radiative recombination, several Si-based materials for light emission and amplification have been proposed with different strategies. For example, by using high purity bulk Si, non-radiative recombination paths are strongly reduced, allowing to obtain LEDs with 1% power efficiency.[1] Another approach to obtain Si band-edge EL is by introducing dislocation loops, in order to spatially confine the carriers and reach a power efficiency of 10^{-2}%.[2] Intentional introduction of defects in SiO_2 by ion implantation has allowed to attain 0.5% power efficiency.[3] On the other hand, the high turn-on voltage required for the polarization of these LEDs does not make these materials suitable for long time use.

Several Si nanoparticles-based materials have been proposed as an alternative for EL emission. Their emission is mostly based on: i) the creation of an exciton inside a nanoparticle (mostly of Si, Si-nc), with a significant increase of its radiative recombination due to quantum confinement; or ii) the localization of the injected carriers at luminescent centers such as dopants elements (C, rare earths such as Er^{3+}, Tm^{3+} or Gd^{3+}), that allows to shift the emission up to the UV or down to the near infrared.

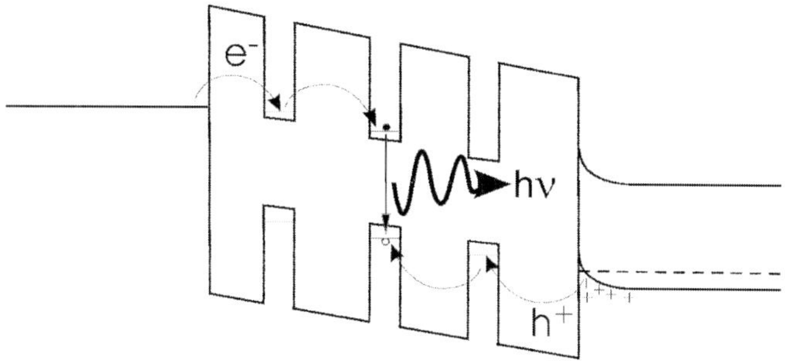

Fig. 1. Forward bias injection of carriers and light emission in bipolar devices.

Most of the LEDs proposed so far have an architecture that resembles that of a MIS (Metal-Insulator-Semiconductor) capacitor, where luminescent centers are located inside the dielectric. The electrical excitation is provided by carrier injection -either unipolar of bipolar- from the top electrode and the substrate. Electrodes can be a ultrathin semi-transparent metallic layer, such as Al (Ref. 4) or Au,[5] Indium Tin Oxide (ITO),[3] or n-type polycrystalline silicon.[6] A 200 nm thick doped polysilicon layer is the usual choice, as it is a standard contact in microelectronics and provides large light extraction in the near infrared and most of the visible spectral range.

Devices based on bipolar injection rely on the recombination of injected electron-hole pairs in the luminescent centers, and/or transfer of the excitation to dopants (see Figure 1). Under forward bias, excitons are created in the luminescent centers by injection of electrons from the electrode and of holes from the p-type substrate. Different kinds of injection mechanisms such as thermoionization effect assisted by the electric field or direct tunnelling injection may occur. Conduction mechanisms can be due to tunnel effects or thermoionic conduction enhanced by the electric field (Poole Frenkel law).[7] Devices working under DC bias and with limited EL efficiency (< 0.01%) and reliability (few hours) have been reported.[4]

Recently demonstrated is the sequential injection, under AC polarization, of electrons and holes from the inverted substrate into the Si-nc close to the interface (see Figure 2). Once the gate gets positively biased, an inversion layer appears in the p-type Si substrate and electrons

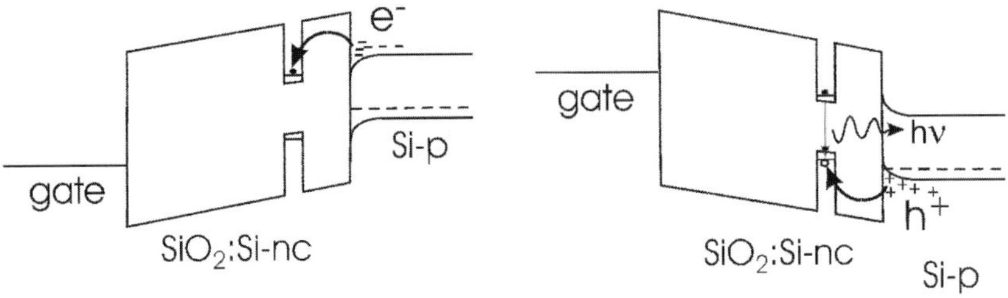

Fig. 2. Mechanism of sequential injection of carriers from the p-type Si substrate into the nanocrystals embedded at the close interface of the dielectric.

are injected in the Si-nc. As the voltage gate gets negative, an accumulation region of holes appears in the substrate. Holes are then injected inside the Si-nc, already negatively charged, and electron-hole pairs can be created by Coulomb attraction.

So far most of the reported devices are based on unipolar transport of electrons injected under F-N mechanism. The application of strong electric fields allows injection of hot electrons (Fowler-Nordheim tunnel injection, F-N) that, accelerated by the field, result into impact excitation of luminescent centers and/or field ionization, as can be seen in Figure 3. In devices with only Si-nc, power efficiency of 0.01 - 0.03% has been reported.[8] On the other hand, for devices doped with rare earths higher values, of about 0.3% have been estimated.[9] While a bipolar or unipolar injection in DC mode provides a conduction current flow across the dielectric, most of it is actually leakage current that decreases both

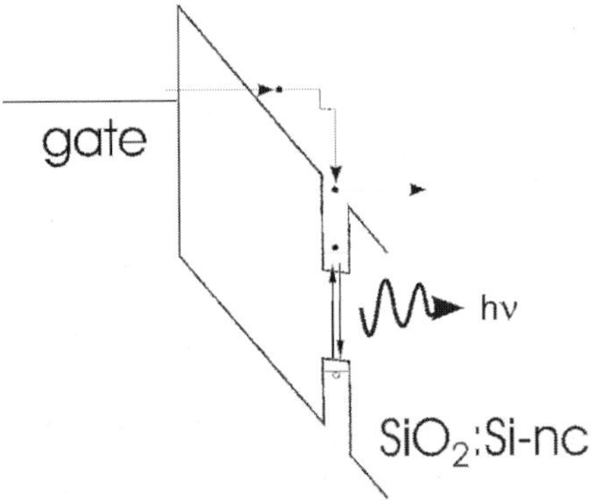

Fig. 3. Fowler Nordheim injection of carriers and excitation of luminescent centers by impact ionization.

Table 1. Power efficiency of the electroluminescent devices based on silicon technology.

Material	**Power efficiency**
Pure bulk Si	1% (ref. 1)
Dislocations in bulk Si	~ 10^{-2}% (ref. 2)
Defects in SiO_2	0.5% (ref. 3)
Porous Si	0.37% (ref. 11)
Si-nc:SiO_2	~ 10^{-2}% (ref. 8), 10% (theoretical value, ref.12)
Si-nc:SiN_y	10^{-3}% (ref. 13)
RE:SiO_2 (RE = Tb, Yb)	0.2% (ref. 14), 0.3% (ref. 9)
Er:SiO_2	0.2% (ref. 14)
Er:SRO	0.02% (ref. 14)
C:Si-nc:SiO_2	> 10^{-3}% (ref. 15)

efficiency and reliability. On the contrary, alternate injection deals with a capacitive-like kind of devices with the absence of any leakage current or wear-out. Power efficiencies of 10% have been predicted for this kind of devices.[12]

Er doping is of particular interest for telecoms applications, as its emission at 1.55 μm corresponds to the minimum of absorption in silica. LED's of SiO_2 doped with Er have been demonstrated to show 0.2% power efficiency. Si nanoparticles are able to transfer its excitonic energy to Er ions, and LEDs using this principle have been demonstrated.[14]

The different Si-based materials proposed in the literature and their power efficiency are summarized in Table 1. Pure bulk Si gives the highest power efficiency, followed by porous Si. But pure bulk Si is expensive, and porous Si needs wet processing, not compatible with CMOS technology. Only few studies report values of power efficiency of Si-nc fabricated by dry processing (PECVD, implantation, etc.). Nevertheless, excitation by sequential injection of electron and holes brings a new interest in the study of Si-nc based materials.

2.2. *Si-nc embedded in SiO_2 for red emitters*

LEDs based on Si-nc embedded in SiO_2 present broad luminescence in the red-infrared region. The fabrication of Si-nc / SiO_2 composites requires the creation of a silicon-rich silicon oxide (SRSO) that, after annealing at high temperature, will become the final active layer. Several approaches have been tested to fabricate these SiO_x layers. We will focus only on the two most popular techniques, as to say, plasma-enhanced chemical-vapour-deposition (PECVD) and ion implantation.

In order to choose the most suitable parameters for device fabrication, different sets of test samples with different Si amounts have been fabricated. PECVD layers with a thickness ranging from 40 nm to 100 nm were deposited by reaction of SiH_4 and NO_2 precursors, decomposed by the action of a RF (radio frequency) field. This mechanism accelerates the layer growth but at the cost of a high amount of N and H impurities incorporated inside the matrix. The ultimate Si-nc growth is obtained by a thermal treatment at about 1250ºC. During this process all H atoms are deasorpted reducing the porosity of the layers. N atoms, instead, get strongly bound to the matrix.

The implanted layers have been fabricated by double implantation of Si atoms into 40 nm thick thermally grown oxide. In order to obtain a constant Si excess in the matrix, a 30-nm thick nitride top-mask had been previously deposited, enabling to place the implantation peak into the oxide and the tail in the nitride. The post-implantation annealing temperature commonly employed is around 1100ºC. Energy deposition and strong ionization during implantation greatly enhance Si diffusion in the oxide enabling Si-nc precipitation at a lower temperature than in PECVD layers.

In the deposited layers strong N incorporation can be detected, comparable to the Si excess concentration, while N in-diffusion is apparent in the implanted oxides. In the implanted layers the Si-nc distribution is uniform all through the oxide, while in the deposited layers two regions depleted of Si-nc are observable: one at the outer surface, probably due to ambient oxidation, the other one at the interface with the substrate. This latter is formed by Si diffusion to the substrate during annealing, attracted by its chemical potential, or by instabilities during the first stage of the deposition.

A comparison of the PL spectra from the layers reveals that independently of the fabrication method, the emission is optimized for a Si excess of about 17%. Beyond that composition the loss in quantum confinement is noticeable and, due to the large density and size of the nanoparticles, their mutual interaction is also favoured. In order to study the carrier injection, layers of about 40-50 nm have been elaborated with this Si content.

2.3. *Electroluminescence mechanisms*

The majority of the authors reported electroluminescence under DC (direct current) polarization. In this case, when the emitting layer is quite thick the carrier injection occurs through Fowler-Nordheim (F-N) tunneling under relatively high voltages. In this regime, the charge transport in the oxide takes place via hot-electrons and the electron-hole pairs are generated by impact ionization of the Si-nc. Hot electron transport leads to a fast degradation of the oxide matrix and a low device endurance. On the contrary, when the SRSO oxide is very thin, impact ionization is suppressed while creation of excitons is favoured by simultaneous injection of electrons and holes from the opposite contacts. The drawback of these devices lies in the very low efficiency, due to a strong dissipation of electrical power.

The AC injection is relatively new and has been reported the first time by Walters *et al.*[6] This mechanism consists in the sequential injection of electron and holes into Si-nc from the substrate under an alternate bias. The switching of the polarization signal has to be very sharp in order to avoid the discharge of the previously charged nanocrystals. For this the most suitable polarization is a square signal, which can be either bipolar or unipolar.[9] The advantage of this mechanism is the reduction of the current flow across the structure, that highly improves the reliability. In our PECVD samples the onset of the emission lies between 15 and 18 V and takes place under Fowler-Nordheim injection, as revealed by current-voltage characteristics.

The principles of this mechanism can be visualized in Figure 2. At the onset of the negative pulse an accumulation of holes takes place at the interface substrate / SiO_x. Some holes get to the closest nanocrystals and remain stored for a lapse that depends on the density of available leakage paths. When the voltage switches back to zero (i.e. slightly above the flat-band voltage) a depletion of holes is generated at the interface and electrons are injected, attracted by the electric field from the positive charged nanocrystals. A transient emission is then observed by recombination of the created excitons, until the total exhaustion occurs of electron-hole pairs. The measured decay-time is about 10 μs, in good agreement with the typical values reported for this kind of structure.

If the applied signal is bipolar some additional considerations should be made. In this case negative charge is also accumulated during the positive cycle of the signal and a second transient is observed when the signal switches to negative voltages. The emission observed is stronger than that observed under unipolar pulses as electron injection is enhanced by the applied field. For p-type substrates the second transient is weaker due to the faster migration of the stored electrons and the higher barrier for holes injection.

In a similar manner the layer has been polarized by a unipolar negative signal. The threshold voltage ranges then from 3 V to 5 V. This operation regime not only generates pulsed emission but also a continuous component. The DC emission, not observed in PECVD devices, takes place while the negative voltage is applied and gives evidence of a net current flow across the structure.

For all the implanted layers the decay time is few ms, a value remarkably larger than the one observed in deposited devices. This could be due to the high amount of charge present in the whole oxide when the voltage drops back to zero. A part of the accumulated charge migrates to the contacts in a process that is longer than the AC recombination. The application of bipolar pulses slightly increases the AC emission and, in the same way than in PECVD devices, the second transient appears.

In the deposited samples the depletion of nanocrystals close to the substrate hinders the continuous current flow, so that only the transient emission is allowed, with efficiency up to ~0.3%. In implanted samples most of the large current flow, instead, is lost by direct conduction mechanisms between the electrodes, and the efficiency does not exceeed $10^{-4}\%$. These results highlight the important role played by the oxide sublayer observed in PECVD devices and the suitability of the alternate injection. Nevertheless the improvement in emission efficiency is reached at the cost of rising the threshold voltage up to ~10 V. Eventually the best solution resides in a compromise between efficiency and threshold voltage. For this matter the thickness of the oxide sublayer depleted of Si-nc appears as a critical parameter.

2.4. *C-rich nanoparticles for white emitters*

Although emission from Si-nc in SiO_2 is confined to the red part of the visible range,[16] few studies have recently reported broad photoluminescence in the whole visible domain, by doping Si-rich SiO_2 layers with carbon.[17,18] This material has shown intense, white to the eye PL coming from the convolution of the red band from Si-nc, and a blue-green band from C-rich nanoparticles, with a composition close to graphitic C and SiC.

This intense white luminescence could be useful for the realization of white LEDs. Moreover, by adding on top of the LED a filter, such as a Bragg mirror, it would allow to filter the emitted light to a single wavelength. The large spectral emission could also be used for bio- or chemical- sensors. Moreover, it has been demonstrated that PL from C-rich nanoparticles show a very fast decay time, typically between 100 ps and few ns.[19] This very fast decay would allow to fabricate very fast modulators in the visible range. This material appears thus very interesting from the point of view of its luminescence properties. By tailoring the Si and C distribution across the dielectric, it is possible to selectively excite one or both families of nanoparticles.[7,15]

Sequential low energy implantation of Si^+ and C^+ was performed into a 40 nm thick SiO_2 layer grown by dry oxidation at 1000°C on a p-type <100> Si wafer. Sample 1 has been designed to have a single C peak in the middle of the film and a uniform profile of Si obtained by means of a four step implantation. This favours injection and conduction of charges in the layer, in order to excite the luminescent centers in DC excitation conditions. C ions have been introduced in the middle of the layer to

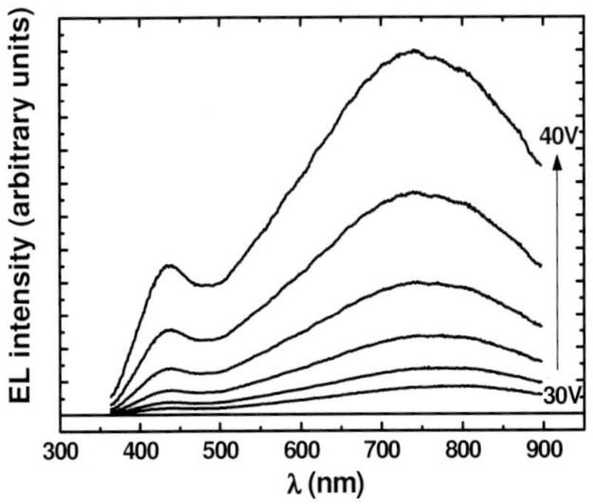

Fig. 4. EL spectra of sample 1 in DC excitation for an increasing applied voltage.

prevent their out-diffusion at high temperature.[18] The second sample, labelled sample 2, has been designed to have C and Si peak distribution closer to the substrate, in order to favour the sequential injection of electrons and hole from the substrate to the luminescent nanoparticles. The thin oxide buffer depleted of Si-ncs at the interface with the gate contact prevents any leakage current.

Figure 4 shows the EL spectra of sample 1 under DC bias for different applied voltages. A broad white EL can be observed. The electrode is a 100 nm Si polycrystalline layer, which shows a significant reduction of the optical transmission in the blue spectral region.

The origin of EL can be determined by looking at the I-V characteristics plotted in Figure 5 in a Fowler-Nordheim representation. A linear trend is observed above 25 V, suggesting that above this voltage electrons are injected in the conduction band of SiO_2. On the right scale, the corresponding EL integrated intensity has been included, also showing a linear trend. Moreover, the onset for EL corresponds to the onset of Fowler-Nordheim regime. These evidences strongly suggest that hot electrons coming by Fowler-Nordheim injection give raise to radiative impact ionization of the luminescent centers.

By calculating the optical power and the electrical power at 40 V, a power efficiency of 10^{-4}% has been estimated. This result is comparable to the highest power efficiency reported so far,[20] with the added value that the emission here spans the whole visible range. An increase in

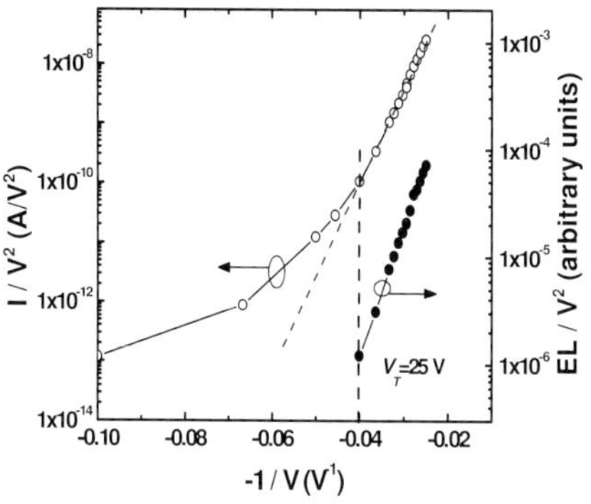

Fig. 5. I-V and EL-V characteristics of sample 1 in Fowler-Nordheim representation.

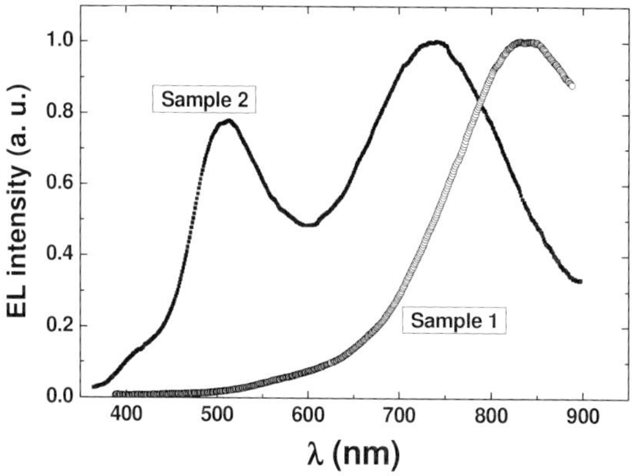

Fig. 6. EL spectra of sample 1 and sample 2 in AC mode at 40 Hz with ±20 V (sample 1), and ±14 V amplitude (sample 2).

efficiency is possible in AC polarization, preventing any leakage current and suppressing the dissipative losses.

By exciting sample 1 with a square pulse of ±20 V amplitude, it is apparent that the white emission is lost (see Figure 6), and that only a band at 800 nm is present. In fact, only the central region of the oxide layer contains C-related nanoparticles, sandwiched between two outer regions containing Si-nc. This implies that under pulsed excitation the C-rich nanoparticles are too far to be excited and only Si-nc close to the

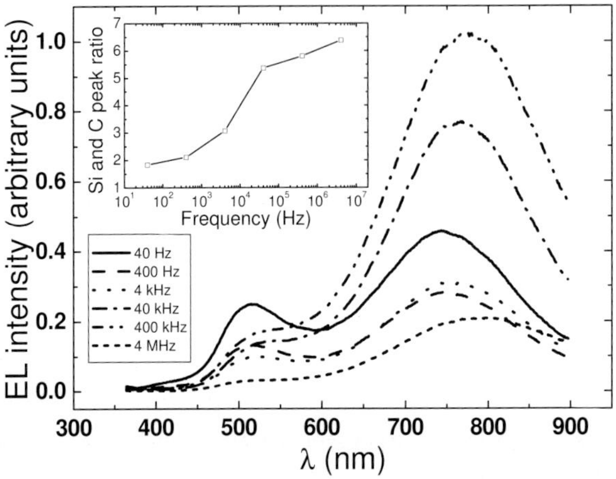

Fig. 7. Evolution of EL spectra with frequency of the square pulse for an applied voltage of ±20 V. Inset shows how the ratio of red to blue-green peak increases with the applied frequency.

substrate are excited leading to a red EL. In sample 2 instead, C-rich nanoparticles are located closer to the substrate, so that the white EL is recovered (see Figure 6). This strongly suggests that under AC excitation a mechanism of sequential injection of electrons and holes occurs into the luminescent centers located very close to the substrate. A lower limit for power efficiency of 10^{-3}% has been estimated, implying an improvement in efficiency by the use of field-effect EL.

By increasing the applied voltage for a constant frequency of 40 Hz, no change of the shape is observed, as to say, both red and blue-green peak increase in intensity by almost the same amount. However, by changing the frequency of the pulse, while keeping the same voltage, i.e. ±20 V, the relative contribution from each peak is changed, as can be seen in Figure 7. In particular, the contribution of the red band is favoured at higher frequency. This results into a tuneable EL, from white at low frequency to a red-dominant one as the frequency is increased as shown in the inset of Figure 7, where the ratio between Si and C peaks is shown. This observation opens interesting perspectives on the colour tuneability of field effect EL devices.

3. Er Doping for Gain Materials with Si Nanoclusters

For the operation in the *C*–band of optical telecommunication at 1.54 μm, Si based light sources and/or amplifiers have been regarded as one of most crucial topics in Si nanophotonics. For optical gain materials based on silicon, Er-doping has been thus the natural choice, tracing the previous technical success of Er-doping into silica fiber. The initial studies had focused on the Er doping of bulk silicon.[21] Even though successful Er^{3+} luminescence was widely reported, some drawbacks of this approach were found, as to say a strong thermal quenching and nearly no Er^{3+} luminescence at room-temperature, together with very short Er^{3+} luminescence lifetimes.[22] This weak efficiency is linked to the presence of strong de-excitation processes of Er^{3+} ions, which has darken their usefulness as gain materials. Therefore the interest turned to alternatives to bulk Si, still compatible with the mainstream Si technology.

In 1997, it was discovered that the Er^{3+} emission is strongly enhanced by using crystalline Si-nc as sensitizers in silica when co-doped with Er.[23] Er-doped Si-nc, independently of its aggregation phase, have been widely regarded as one of the most promising Si based gain materials. The advantages of Si-nc sensitizing include efficient room-temperature Er^{3+} luminescence, due to the lack of de-excitation processes,[24] and larger optical activation of Er^{3+} ions thanks to oxygen-rich environment, compared to bulk Si. An additional noticeable one is Er^{3+} luminescence lifetime sometimes as long as ~ 10 msec, comparable to that in silica.

In addition, Er-doped Si-nc are even more advantageous than Er-doped silica. For visible pump wavelength, Er^{3+} effective excitation cross-section is more than four orders of magnitude higher than an optical absorption cross-section of Er^{3+} in silica. The broad absorption band of the Si clusters, used in this case to indirectly excite Er, enable the use of cheap flash-lamps or light-emitting-diodes (LEDs) for side-pumping, efficiently replacing the expensive pumping lasers which have been used in EDFA. Furthermore, the higher refractive index produced by the Si-nc can be used to provide the index contrast necessary for wave-guiding. These properties offer the possibility to fabricate a compact, Si-compatible, integrated and cost-effective erbium-doped

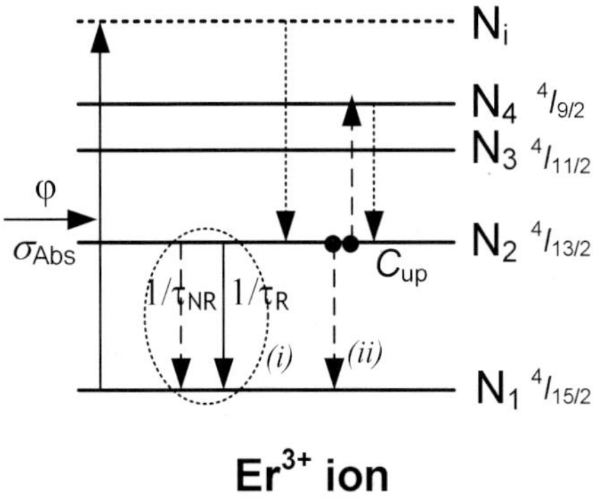

Fig. 8. Schematic descriptions 4f levels of Er^{3+} ion, and the excitations and de-excitations by direct absorption.

waveguide amplifier (EDWA), where the active layer is composed by silica co-doped with Si-nc and Er.

H. Han *et al.*, firstly reported on the signal-enhancement of 4 dB / cm and attainable gains of up to 7 dB/cm from EDWA adopted Si-nc sensitizing and a vertical pump laser 476 nm (non-resonant to Er^{3+} 4*f* levels).[25] The pump flux was as low as 4×10^{18} cm^{-2} / s, already viable with commercial LEDs. The same group has recently claimed the signal enhancement by top-pumping with an array of GaN LEDs.[25]

Despite of such promising results, just scarce results about the signal enhancement followed by other groups,[26] and little progress on the device development using Er doped Si-nc has been made so far. On the contrary, nearly none or even negative SE were reported by some groups. The reason they pointed out includes strong carrier absorption,[27] no-increase of emission cross-section,[28] and the low fraction Er^{3+} ions coupled to Si-nc.[29]

3.1. *Resonant excitation by direct absorption*

Figure 8 shows the schematic description of Er^{3+} ion level and resonant excitation process of Er^{3+} ions which are common in most insulators. Only specific photons, whose energy matches the corresponding energy difference between 4*f* levels of Er^{3+} ions, can be resonantly absorbed.

These excited Er^{3+} ions mostly non-radiatively relax to $^4I_{13/2}$ level, and then can radiatively decay, with a decay time τ_R, emitting photons at ~1.54 μm, or non-radiatively decay with a decay time τ_{NR}, due to quenching centers. Such combined decay (dotted circle marked (*i*)) is called an intrinsic decay of Er^{3+} ions. Compared to the relaxation rate of Er^{3+} from $^4I_{13/2}$ level to the ground state ($^4I_{15/2}$), usually as long as 10^5 s^{-1}, Er^{3+} ions in other higher excited states are expected to be quickly relaxed to the first excited state; so that Er^{3+} ions are assumed to be either on the first excited state or on the second excited one.

In addition, one of the most frequently reported de-excitation processes is the co-operative up-conversion (CUC). It involves two Er ions, both in the first excited level, and results in the promotion of one ion to the third excited level ($^4I_{9/2}$) and the de-excitation of the other to the ground state (arrow marked (*ii*)). Subsequently, the excited ion decays rapidly to the second level and then non-radiatively to the first excited level through the emission of phonons in a lapse of few μs, or radiatively to the fundamental state with a low branching ratio (weak luminescence from the second excited level can be detected at 980 nm).

The rate equation of Er^{3+} ions in the first excited state is

$$\frac{dN_2}{dt} = \sigma_{abs}\,\varphi\,N_1 - \frac{N_2}{\tau_2} - C_{up}N_2^2 \qquad (1)$$

where, N_1, N_2, and C_{up} is the Er^{3+} population in the ground state, in the first excited state, and the CUC coefficient, respectively.

In the low pump flux regime most Er^{3+} ions are on the ground state and related de-excitation processes of excited Er^{3+} are small. Then excited Er^{3+} ions are expected to monotonically decay with the experimentally measured PL lifetime τ_{PL} that includes as a second order term the CUC (we used τ_{PL} in order to distinguish it from the intrinsic decay τ_2). In steady state conditions N_2 is

$$N_2 = \frac{N_{Er}\varphi}{\varphi + 1/\sigma_{abs}\tau_{PL}}, \qquad (2)$$

where $N_{Er} = (N_1 + N_2)$ is the population of optically active Er^{3+} ions.

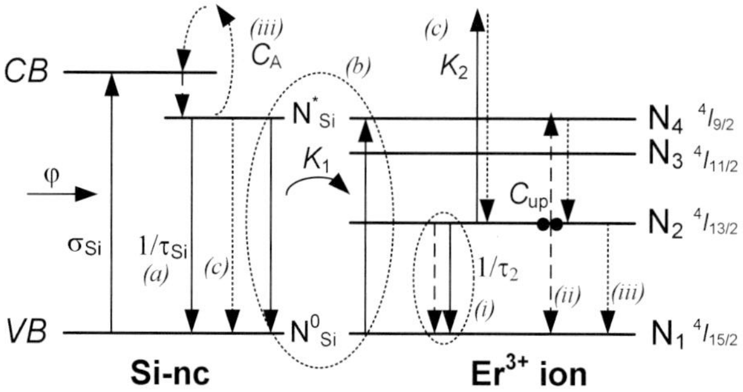

Fig. 9. Schematic descriptions for the interaction model between Si-nc and Er^{3+} ion, and the excitations and de-excitations of them.

3.2. *The interaction between silicon nanoclusters and erbium ions*

In the case of Er^{3+} excitation via Si-nc the excitation/de-excitation of Si-nc and the interaction between Si-nc and Er^{3+} ions should be taken into account, in addition to Er^{3+} excitation.[29] Figure 9 schematically illustrates the interaction between Si-nc and Er^{3+} ions, and indicates the paths for excitations and de-excitations. As soon as a Si-nc, with its absorption cross-section σ_{Si}, absorbs incident pump photons with a flux φ, carriers are generated, and eventually trapped at interfacial states.

The energy scheme of Si-nc can be regarded as a quasi two-level system, and the exciton generation rate can be defined as $\sigma_{Si}\varphi$. When Si-nc are coupled to Er^{3+} ions, the recombination energy of excitons can be transferred to the Er^{3+} ions in the ground state promoting them to higher excited states (the arrow with (*b*)). An additional mechanism that depletes the Si-nc is the excited-state-absorption (ESA), where an excited Er is pumped to a higher excited level by the energy transfer from the recombination of an excited Si-nc (the arrow with *(c)* in Figure 9).

The rate equation for the Si-nc subsystem in the coupled system is:

$$\frac{dN^*_{Si}}{dt} = \sigma_{Si}\varphi\, N^0_{Si} - \frac{N^*_{Si}}{\tau_{Si}} - K_1 N_1 N^*_{Si} - K_2 N_2 N^*_{Si}, \tag{3}$$

where N^*_{Si} is the density of excited Si-nc, N^0_{Si} the Si-nc density in the fundamental state, $N_{Si} = N^*_{Si} + N^0_{Si}$ is the total number of Si-nc; K_1 is the coupling constant between Si-nc and Er in the ground state, and K_2 is the coefficient for ESA.

The N_1 ions in the ground state are excited by energy-transfer from excitons. Together with the intrinsic decay and CUC introduced in the previous section, the de-excitation of N_2 by Auger excitation of excitons (AEE) comes to play a role. This de-excitation involves N_2 and N^*_{Si} similarly to ESA, but with the difference that the energy of N_2 is transferred to N^*_{Si} promoting them to a higher energy level (arrow marked (*iii*)). Similarly to CUC effect, ESA fills other higher excited states of Er^{3+} ions. (i.e., $\rightarrow$ $^2H_{11/2}$). Additionally, it is still unclear to which state Er will be excited through the energy transfer from Si-nc, considering the controversy about the quasi-resonant excitation of Er^{3+}.[31]

Despite a lower accuracy than the more detailed modeling that includes higher energy levels, the quasi two-level system of Er^{3+} ions as introduced for resonant excitation model will be again used, due to comparable transition rates of Er^{3+} ions in Si-nc system.

As a result, the rate equation for Er^{3+} ions in the first excited state in the coupled system becomes:

$$\frac{dN_2}{dt} = K_1 N^*_{Si} N_1 - \frac{N_2}{\tau_2} - C_{up} N_2^2 - C_A N^*_{Si} N_2 , \qquad (4)$$

where C_A is the AEE coefficient, N_0 the number of Er ions coupled to Si-nc ($N_0 = N_1 + N_2$). It should be noted that N_0 is used instead of N_{Er}, since a clear evidence of having all N_{Er} coupled to Si-nc is lacking, at this stage. And this will be discussed more in details afterwards.

Moreover, by comparing the dynamics of excited Si-nc and excited Er^{3+} ions, the overall process of photo-generation of excitons in Si-nc plus energy transfer to Er^{3+} ions is much faster than the excitation and de-excitation of Er^{3+} ions. When the pump flux is raised, Si-nc will reach very quickly the steady state equilibrium. Thus, it is more convenient to handle the dynamics of Er^{3+} by introducing their own effective excitation cross-section σ_{eff} instead of dealing with Si-nc and Er^{3+}, individually. This assumption allows writing:

$$\frac{dN_2}{dt} = \sigma_{\mathrm{eff}}\varphi N_1 - \frac{N_2}{\tau_2} - C_{\mathrm{up}}N_2^2 - C_A N_{\mathrm{Si}}^* N_2, \tag{5}$$

where

$$\sigma_{\mathrm{eff}} \approx \sigma_{\mathrm{Si}} \frac{K_1 N_{\mathrm{Si}}}{\sigma_{\mathrm{Si}}\varphi + \frac{1}{\tau_{\mathrm{Si}}} + K_1 N_1 + K_2 N_2}. \tag{6}$$

The last three terms in Eq. (3) can be rewritten as

$$\cdots - N_2\left(\frac{1}{\tau_2} + C_{\mathrm{up}}\, N_2 + C_A\, N_{\mathrm{Si}}^*\right). \tag{7}$$

and this reflects different de-excitation processes for an excited Er^{3+} ion: the intrinsic decay, CUC, and AEE, with de-excitation rate of $1/\tau_2$, $C_{up} N_2$ and $C_A N_{Si}^*$, respectively.

By analyzing the low pump flux regime, it is apparent that most Er^{3+} ions are on the ground state so that de-excitation processes of excited Er^{3+} are small. Then excited Er^{3+} ions are expected to monotonically decay with the experimentally measured PL lifetime τ_{PL} that includes as second order terms the CUC and AEE (i.e., $1/\tau_2 + C_{up}N_2 + C_A N_{Si}^* \rightarrow 1/\tau_{PL}$). In such a case, Eq. (3) is simplified to:

$$\frac{dN_2}{dt} = \sigma_{\mathrm{eff}}\, \varphi N_1 - \frac{N_2}{\tau_{\mathrm{PL}}}. \tag{8}$$

In steady state conditions, the population of excited Er^{3+} in the low pump regime is finally given by

$$N_2 = \frac{N_0 \varphi}{\varphi + 1/\sigma_{\mathrm{eff}}\tau_{\mathrm{PL}}}, \tag{9}$$

that is the typical sigmoidal curve that defines the population of a simplified two levels system. Note that even at high pump flux σ_{eff} has the physical meaning of a cross-section for Er excitation and τ_{PL} of Er de-excitation time, although they will be flux-dependent when other de-excitation processes start to affect N_2.

3.3. *Limiting factors for Er luminescence*

The precise comparison to a well-characterized reference sample (Er-doped silica glass) allows to estimate the absolute density of excited Er^{3+} ions from the experimental measurements of Er^{3+} luminescence intensity.[32,33] The detailed calibration procedure is well described in Ref. 33.

For Er-doped Si-nc films with a Si excess of 7 and 12 at.% the ratio between N_2 and N_{Er} has been estimated for a wide range of pumping powers.[32,34] Surprisingly the ratio hardly exceeds 3% for a flux of almost 10^{21} cm^{-2} sec^{-1}. This low ratio of the luminescent Er^{3+} to N_{Er} clearly indicates that a strong limiting phenomenon is taking place in Si-nc sensitizing system. By looking at Eq. (9) where N_2 is calculated, the effective excitation cross-section σ_{eff}, the deexcitation rate of Er^{3+} ions $1/\tau_{PL}$, and the number N_0 of Er^{3+} ions coupled to Si-nc are three major parameters which can play a role in limiting N_2.

3.4. *The effective excitation cross-section*

Equation (6) tells that σ_{eff} is a function of τ_{Si}, σ_{Si}, K_1, K_2, N_{Si}, as well as φ. It is clear that a large effective cross-section is expected for a long intrinsic Si-nc lifetime as described in Eq. (6). If the exciton can stay in the upper state as long as possible, more chances will be for the Si-nc to transfer its energy to Er^{3+} ions.

C. Garcia *et al.* reported on Si-nc size dependence of the optical absorption cross-section.[35] It has been found that σ_{Si} is an increasing function of the energy of the excitation light in the whole visible - UV range. Even though the increase in pump photon energy is less than 13% (from 2.4 eV for 514 nm to 2.7 eV for 457 nm), the maximum value of σ_{Si} increases more than twice, from 4×10^{-17} cm^2 to ~8×10^{-17} cm^2. This result is explained by an exponential-like increase of the distribution of exciton states from the bottom of Si-nc conduction band. . As Si-nc size increases above 2.5 nm (and the electronic bandgap of Si-nc decreases), it is expected that the density of states at a given energy increases, while at the same time the oscillator strength drops. The net balance of both contributions is that a maximum in σ_{eff} appears for Si-nc with a size of ~3 nm.

The value of the coupling constant between Si-nc and Er^{3+} ions was found to be 3×10^{-15} cm^3/s by Pacifici *et al.*[30] Nonetheless, the quasi resonant energy transfer behavior from the band around 1.6 eV of Si-nc to the $^4I_{9/2}$ 4*f* state of Er^{3+} ion has to be discussed.[31] In such process, Si-nc would excite Er^{3+} ions on the first excited state up to the $^2H_{11/2}$ 4f level. Some recent reports have emphasized the importance of such ESA of excitation energy of Si-nc.[36] A difference of more than four orders of magnitude in the value of K_2 has been reported.[30,31] While a K_2 of less than 3×10^{-19} cm^3/s has been reported, the highest value has been $K_2 = 1\times10^{-15}$ cm^3/s, similar to $K_1 = 3\times10^{-15}$ cm^3/s, implying that the ESA rate can be as fast as the energy transfer from Si-nc to Er^{3+} ions.

N_{Si} is just proportional to the effective cross-section, implying that an increase of N_{Si} is the most straightforward way to achieve large σ_{eff}. In fact, even though ~ 3 nm seems to be most desirable size for σ_{Si}, similar or even larger energy transfer efficiency has also been reported for smaller but denser Si-nc doped with Er ions.[30]

Apart from N_{Si}, σ_{eff} is mainly determined by the competition between the absorption of pump photon, the intrinsic decay rate, the energy transfer rate to Er, and ESA process, as it appears in the denominator in Eq. (6). A value lower than 10^4 sec^{-1} for σ_{Si} φ is expected as far as φ does not exceed 10^{20} cm^{-2} sec^{-1}, while the intrinsic decay rate seems not to exceed 10^5 / sec. Energy transfer rate was found to be ~10^5 sec^{-1}. This comparison implies that ESA starts to become significant once the excited population population N_2 exceeds 10^{20} cm^{-3}, a not realistic value in most cases. Hence, the impact of ESA in the effective excitation cross-section can be considered as negligible.

σ_{eff} is a function of φ indicating that the excitation efficiency will be affected by large values of φ. In fact, at low pump fluxes σ_{eff} is assumed to be almost constant as $\sigma_{eff} \approx \sigma_{Si}K_1N_{Si}/(1/\tau_{Si} + K_1N_1)$. On the other hand, as φ increases, the exact correlation between σ_{eff} and φ appears unclear, since both N_1 and N_2 are the values at steady state and thus are also φ-dependent, while the first two terms in the denominator of Eq. (4) are straightforward.

Using experimental values of N_0 and τ_{PL}, the effective excitation cross-section can be de-convoluted together with the help of Eq. (9). To

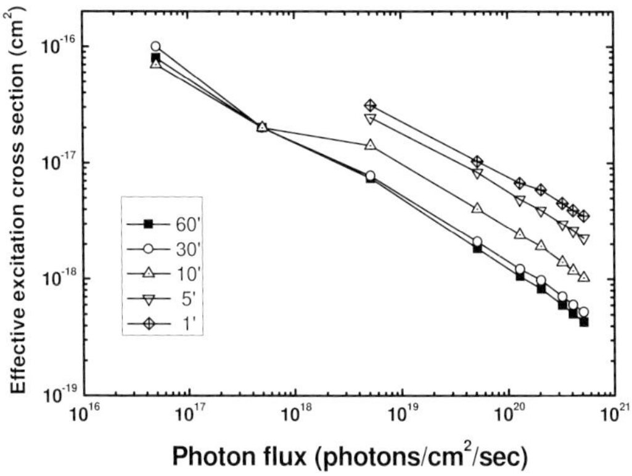

Fig. 10. Effective excitation cross-section of Er^{3+} ions versus pump photon flux for Er-doped Si-nc films.

evaluate N_0 one needs some care, as it will be explained later on, and more in details in Ref. 31.

Garrido *et al.* reported the values of $\sigma_{\rm eff}$ for the Er-doped Si-nc film with silicon excess of 7 at.%.[32] The effective excitation cross-section gets $\sigma_{\rm eff} = 7.34\times10^{-18}$ cm^2 for the longest annealed (60 min.) film, pumped at 476 nm with $\varphi = 5\times10^{18}$ cm^{-2}s^{-1}.[32] This value is more than 10 times smaller than the expected excitation cross-section of Si-nc. In general, $\sigma_{\rm eff}$ increases from 3.1×10^{-17} cm^2 as the annealing time increases from 1 to 60 min.

Another and simpler method to evaluate the value of $\sigma_{\rm eff}$ is deduced from Eq. (8). From time-resolved luminescence traces, where the pump flux is turned on and off, one can evaluate the rise rate R (φ) and the decay rate of PL $W_{\rm PL}(\varphi)$ ($= 1/\tau_{\rm PL}(\varphi)$), that are related by:

$$R(\varphi) - W_{\rm PL}(\varphi) = \sigma_{\rm eff}\ \varphi. \quad (10)$$

Figure 10 shows how $\sigma_{\rm eff}$ depends on φ for films treated at 900°C for different times. The results, as expected from the data above, show that $\sigma_{\rm eff}$ is similar for all the samples investigated, but it roughly decreases one order of magnitude per decade of increasing flux. Thus, the Si-nc to Er energy transfer quenches with increasing flux.

3.5. *De-excitation processes*

Similarly to the intrinsic Si-nc lifetime, a long Er^{3+} intrinsic lifetime is also favorable in order to achieve higher Er^{3+} luminescence. Therefore a rather similar strategy for achieving a longer lifetime is applicable similarly as for the Si-nc intrinsic lifetime. As previously described, the intrinsic lifetime consists of radiative and non-radiative lifetimes ($1/\tau = 1/\tau_R + 1/\tau_{NR}$).

PL lifetime at 1.54 μm has been recorded for all the Er-doped Si-nc layers annealed at 900°C.[32] The used pump flux of 10^{18} $cm^{-2}s^{-1}$ is relatively low, so that decay-processes such as CUC and EEA are negligible. PL lifetime can be regarded then as the intrinsic lifetime of Er^{3+} ions, and it increases from 0.5 to 4.5 msec as the annealing time rises from 1 to 60 min. At the same time, the refractive index of the layer is also found to increase with the annealing time.[32] For this reason the increase in lifetime cannot be assigned to the change in radiative lifetime, which is expected to decrease with the annealing time, but more reasonably to the suppression of non-radiative processes.

In steady state conditions, under a constant illumination, excitons always exist and an Auger effect is expected to be rather active, especially for a high pump regime where the population of excitons becomes very dense. Once the pump beam is turned off, since the excitons recombine within few μs while excited Er ions decay within few ms, AEE is no more a dominant effect. For this reason, Er^{3+} PL decay traces provide little clue about how to estimate AEE effect.

Cooperative up-conversion (CUC) in SiO_2 doped with Si-nc and Er has been previously studied by several authors and it is known to play a significant role when either the Er concentration or the pumping flux, or both, are high. As stated above, CUC refers to the interaction of two Er ions in the first excited level, resulting in one remaining in the first excited state (after few μs of decay from upper levels) and the no radiative de-excitation of the other to the fundamental level. Therefore the first excited state will be depleted and thus CUC can significantly affect the pumping flux regime needed for population inversion.

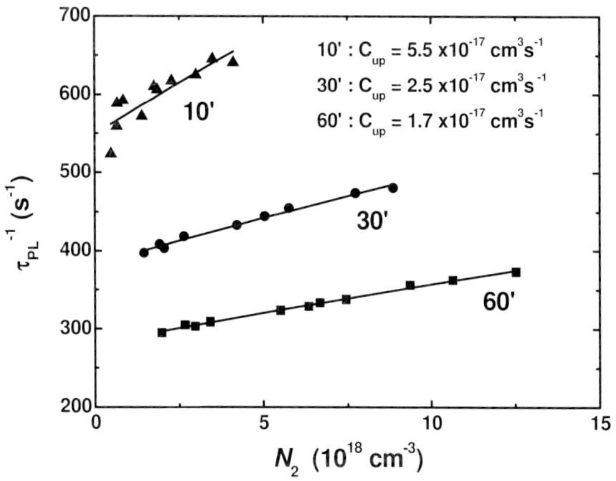

Fig. 11. Inverse of the PL lifetime plotted against the density of excited Er^{3+} ions.

When the pump beam is turned off ($\varphi = 0$), Eq. (5) is simplified to

$$\frac{\mathrm{d}N_2(t)}{\mathrm{d}t} = -\frac{N_2(t)}{\tau_2} - C_{\mathrm{up}} N_2^2(t)\,. \tag{11}$$

As now the decay of N_2 will not follow a single exponential decay law, let's define the PL lifetime τ_{PL} as the integral average:

$$\tau_{\mathrm{PL}}(\varphi) = \int_0^\infty \frac{N_2(t)}{N_2(0)} dt = \frac{\ln\left(1 + N_2(0)\ \tau_2\ C_{\mathrm{up}}\right)}{N_2(0)\ C_{\mathrm{up}}}\,. \tag{12}$$

Note that τ_{PL} is a function of φ since $N_2(0)$ is also φ dependent.

Figure 11 compares the evolution of the PL decay-rate with respect to the density N_2 of inverted ions, to that expected from equation (13). A simplified version of Eq. (13) is deduced from the rate equation of Eq. (10) as $1/\tau_{\mathrm{PL}} \sim 1/\tau_2 + C_{\mathrm{up}}N_2$, a linear dependence as shown in Figure 11. The up-conversion coefficients have been estimated between 2 and 5.5×10^{-17} $\mathrm{cm^3s^{-1}}$, consistent with previous results on similar samples. The predominant de-excitation process in determining the value of τ_{PL} is then $1/\tau_2$, which is in the range of 10^2 - 10^3 $\mathrm{sec^{-1}}$, independent on either φ or N_{Er}.

From Eq. (7), both CUC and AEE induced an 'effective' de-excitation rate for a single excited Er^{3+}, as to say $C_{\mathrm{up}}\,N_2$ and $C_{\mathrm{A}}\,N_{\mathrm{Si}}*$, respectively. Approximately 10^{18} $\mathrm{cm^{-3}}$ Er ions can be usually excited at a

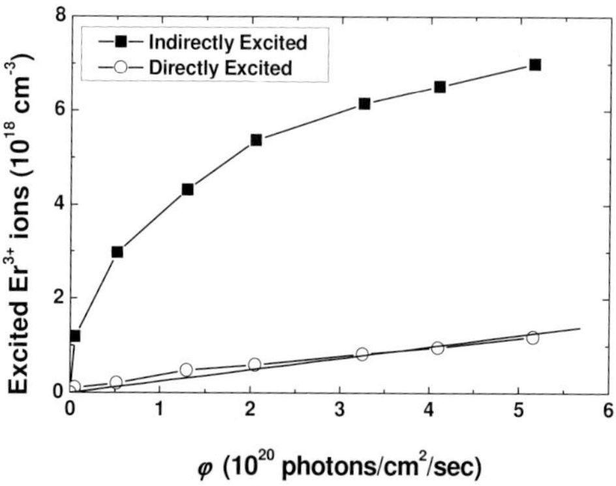

Fig. 12. Contribution of the direct and indirect excitation to the density of excited Er ions when pumping at 488 nm after subtracting the excitation at 476 nm.

moderate pump regime (~10^{19} $cm^{-2}s^{-1}$), implying that CUC rate is ~ 10 s^{-1}. The predominance of Auger effect over CUC ($C_A N_{Si}^* > C_{up} N_2$) would require a population of excited Si-nc more than two orders of magnitude higher than that of 'excited' Er^{3+} ions ($C_{up}/C_A \sim 10^2$), a not realistic occurrence. Thus mainly the Er^{3+} intrinsic decay plays a significant role in determining Er^{3+} de-excitations, while CUC will be rather active, but just as a second order perturbation.

3.6. *Optically active Er ions*

A comparison has been made for PL spectra at 1.54 μm from Er-doped Si-nc layers pumped at 477 nm and 488 nm. The interesting feature is the rather higher Er^{3+} PL intensity under excitation at 488 nm. As σ_{Si} has a similar value at both wavelengths, the additional Er^{3+} PL for the 488 nm pumping is thought to be due to resonantly excited Er^{3+} ions, since the 488 nm line is also resonant with 4*f* levels of the Er^{3+} ions.

The comparison between resonant (488 nm) and non-resonant (477 nm) excitation at different pump powers allows to estimate the amount of Er ions that can be resonantly excited (see Figure 12). For this purpose, the population N_2 after excitation with 488 nm is subtracted by that after 476 nm pumping. By this trick one separates the contributions of indirectly excited Er and the resonantly excited one, which follow a

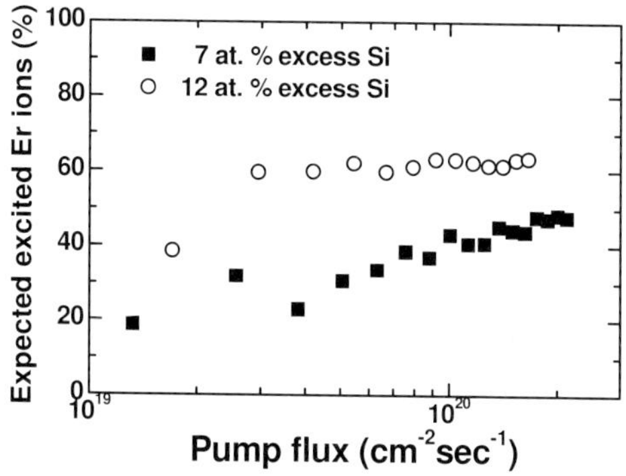

Fig. 13. Expected excited Er^{3+} fraction among Er^{3+} ions coupled to Si-nc as function of pump photon flux.

linear increase with pumping power. By a linear fitting one obtains N_{Er}, the total amount of optically active Er ions, equal to 2.7×10^{20} cm^{-3}, consistent with the amount experimentally measured. Most of the introduced Er ions are thus thought to be optically active in the samples.

From Eq. (8) and (9) at the steady state, the ratio N_2 / N_0 is described only by $R(\varphi) / [R(\varphi) + W(\varphi)]$, that can be experimentally evaluated from time-resolved PL traces. As shown in Fig. 13, the values of N_2 / N_0 are of several tens percent, implying that a significant fraction of Er^{3+} ion coupled to Si-nc should be already excited for such pump flux. The comparison between the result of Figure 12 (N_2/N_{Er}) and Figure 13 (N_2/N_0) brings interesting implications. At a pump flux of 10^{20} $cm^{-2}sec^{-1}$, Figure 12 tells that just 1% of Er^{3+} is emitting, even if from Figure 13 around half of all Er^{3+} excitable ions should be active. This occurrence tells us that only a few percent of the optically active Er^{3+} ions is coupled to Si-nc.

3.7. *Location of Er ions and their accessibility by Si-nc*

When Er^{3+} ions are located at the correct distance (0.5 – 2 nm) from a Si-nc, they can get efficiently excited and be luminescent.[29,38,39] All the ions located beyond that distance are hindered from the energy transfer from Si-nc. Although unable to be accessed by excitons in Si-nc, still these ions can be resonantly excited. The ratio N_0/N_{Er} is expected to be

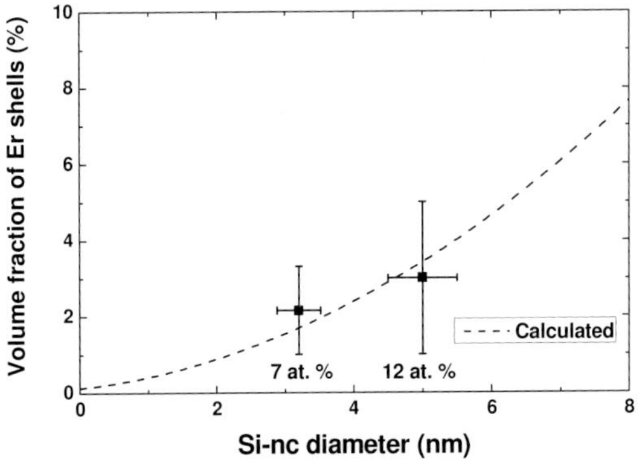

Fig. 14. Simulation curve of the volume fraction of Er3+ 'virtual' interacting shells of Si-nc depending on Si-nc diameter. Symbols are experimental results of the fraction of Er3+ ions coupled to Si-nc among optically active ones.

consistent with the volume fraction of these shells of active interaction around the Si-nc. This volume fraction is dependent on the interaction distance R_0, as well as the Si-nc diameter d and density $N_{\text{nc-Si}}$, and can be simply described as $4\pi/3[(d/2 + R_0)^3 - R_0^3]N_{\text{nc-Si}}$.

A quite good agreement is found between N_0/N_{Er} values experimentally determined and the theoretical prediction using the values of R_0 of 1 nm and $N_{\text{nc-Si}}$ of 3×10^{17} cm^{-3} confirming the validity of the correlation between volume fraction of the interaction shells and coupling ratio, as shown in Figure 14. The difference among the reported interaction distances might just come from a different film preparation. Nonetheless, shorter distances have been reported mostly for amorphous Si-ncs compared to crystalline ones,[29,39] possibly caused by a shorter bounding radius of excitons in amorphous nano-structured silicon. A great uncertainty regards the fraction of doped Er that can be excited via Si-nc, as the reported values range from less than 1%[29] to 10% – 20%[26] up to almost 100%,[25] more probably due to a different interaction distance and Si-nc density of the corresponding films.

3.8. *Device realization*

The most advanced demonstration that highlighted the advantage of Si-nc sensitizing for Er^{3+} ions was a ridge-waveguide amplifier based on

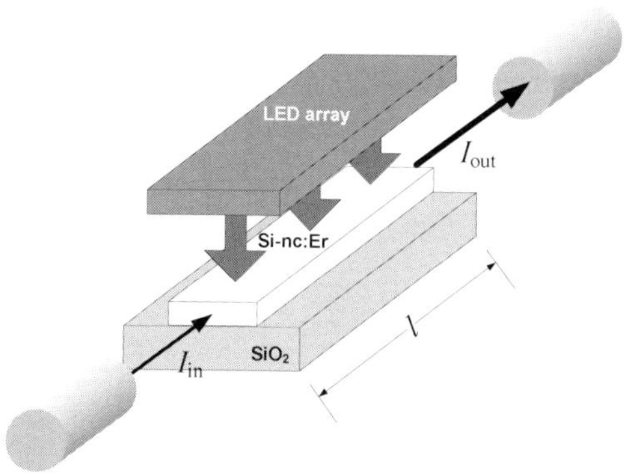

Fig. 15. Waveguided amplifier based on Er-doped Si-nc adopting vertical pump-source.

Er-doped Si-nc thin film, with a visible LED array as vertical pump source.[25] Figure 15 shows the schematic explanation of such waveguide amplifier which adopted vertical broad-band visible pump-source.

Similarly to many other optical amplifiers, the output signal power I_{out} is described as

$$I(P) = cI_0 \times \exp\left[-\left(\alpha + \left(\sigma_{abs}N_1 - \sigma_e N_2\right)\Gamma\right)l\right], \tag{13}$$

where c, I_{in}, Γ, and l are the coupling coefficient, the input signal power, the confinement factor, and the length of a waveguide respectively, while the other notations were previously defined in this chapter.

The net gain per length $G(P)$ in dB/cm units is defined as $\log_{10}(I/I_0)/l$ and is

$$G(P) = \log_{10}\left(\frac{I}{I_0}\right)/l = L_{coup} + L_{WG} + (\log_{10} e) \times \sigma_e N_{Er}\left(2\frac{N_2}{N_{Er}} - 1\right)\Gamma, \tag{14}$$

where L_{Coup} (= $\log_{10}(c)$) is the coupling loss and L_{WG} (= - $\log_{10}$ (e) α l) is the waveguide loss. The overall optical loss is the sum of these two contributions. While both L_{coup} and Γ are technological parameters, the interesting ones in ruling the net-gain are the optical losses, the emission cross-section, the active Er^{3+} ions density, and the fraction of population of excited Er^{3+} ions.

In the Si-nc sensitizing system, despite that optically active Er^{3+} can absorb resonant light signals, the visible pump light selectively excites

Er^{3+} ions that are coupled to Si-nc among all excitable Er^{3+} ions. At high pump flux regime, for conventional silica based EDWAs, N_2 is expected to approach N_{Er} resulting in positive value of $(2N_2/N_{Er} - 1)$. However for Si-nc sensitized WGs, N_2 merely approaches N_0 instead of N_{Er}. Therefore the ultimate net gain will determined by the coupling ratio N_0 / N_{Er}, as described by:

$$G(P) \propto \sigma_e N_{\mathrm{Er}} \left(2\frac{N_0}{N_{\mathrm{Er}}} - 1 \right) \Gamma \tag{15}$$

Even by working under ideal conditions of no optical losses, no positive net gain might be reached unless the Si-nc / Er^{3+} coupling fraction exceed 50%. At the present time, even though few studies have provided evidences of signal enhancement, no positive net gain has been reported yet.

References

1. M. Green, J. Zhao, A. Wang, P. Reece and M. Gal, *Nature* **412**, 805 (2001).
2. W. L. Ng, M. A. Loureno, R. W. Gwilliam, S. Ledain, G. Shao and K. P. Homewood, *Nature* **410**, 192 (2001).
3. L. Rebohle, J. von Borany, D. Borchert, H. Frb, T. Gebel, M. Helm, W. Mller and W. Skorupa, *Electrochem. Solid-State Lett.* **4**, G57 (2001).
4. J. D. L. Torre, A. Souifi, A. Poncet, C. Busseret, M. Lemiti, G. Bremond, G. Guillot, O. Gonzalez, B. Garrido, J. R. Morante and C. Bonafos, *Physica* **E16**, 326 (2003).
5. O. Jambois, H. Rinnert, X. Devaux and M. Vergnat, *J. Appl. Phys.* **98**, 46105 (2005).
6. R. Walters, G. I. Bourianoff and H. A. Atwater, *Nature. Mat.* **4**, 143 (2005).
7. O. Jambois, B. Garrido, P. Pellegrino, J. Carreras, A. Pérez-Rodríguez, J. Montserrat, C. Bonafos, G. BenAssayag and S. Schamm, *Appl. Phys. Lett.* **89**, 253124 (2006).
8. M. Perálvarez, C. García, M. López, B. Garrido, J. Barreto, C. Domínguez and J. A. Rodríguez, *Appl. Phys. Lett.* **89**, 51112 (2006).
9. J. Sun, W. Skorupa, T. Dekorsy, M. Helm, L. Rebohle and T. Gebel, *J. Appl. Phys.* **97**, 123513 (2005).
10. R. Walters, J. Carreras, T. Feng, L. D. Bell and H. A. Atwater, *IEEE J. Sel. Top. Quantum El.* **12**, 1647 (2006).
11. B. Gelloz and N. Koshida, *J. Appl. Phys.* **88**, 4319 (2000).
12. J. G. Simmons, *J. Phys.* **D4**, 613 (1971).

13. K. S. Cho, N.-M. Park, T.-Y. Kim, K.-H. Kim, G. Y. Sung and J. H. Shin, *Appl. Phys. Lett.* **88**, 209904 (2006).
14. M. E. Castagna, S. Coffa, M. Monaco, L. Caristia, A. Messina, R. Mangano and C. Bongiorno, *Physica* **E16**, 547 (2003).
15. O. Jambois, J. Carreras, A. Pérez-Rodríguez, B. Garrido, C. Bonafos, S. Schamm and G. B. Assayag, *Appl. Phys. Lett.* **91**, 211105 (2007).
16. B. Garrido, M. López, O. Gonzálcz, A. Pérez-Rodríguez, J. R. Morante and C. Bonafos, *Appl. Phys. Lett.* **77**, 3143 (2001).
17. L. Rebohle, T. Gebel, H. Frb, H. Reuther and W. Skorupa, *Appl. Surf. Sci.* **184**, 156 (2001).
18. A. Pérez-Rodríguez, O. González-Varona, B. Garrido, P. Pellegrino, J. R. Morante, C. Bonafos, M. Carrada and A. Claverie, *J. Appl. Phys.* **94**, 254 (2003).
19. P. Pellegrino, A. Pérez-Rodriguez, B. Garrido, O. González-Varona, J. R. Morante, S. Marcinkevicius, A. Galeckas and J. Linnros, *Appl. Phys. Lett.* **84**, 25 (2004).
20. T. Gebel, L. Rebohle, J. Sun and W. Skorupa, *Physica* **E16**, 366 (2003).
21. H. Ennen, J. Schneider, G. Pomrenke and A. Axmann, *Appl. Phys. Lett.* **43**, 943 (1983).
22. J. Palm, F. Gan, B. Zheng, J. Michel and L. C. Kimerling, *Phys. Rev.* **B54**, 17603 (1996).
23. Minoru Fujii, Masato Yoshida, Yoshihiko Kanzawa, Shinji Hayashi and Keiichi Yamamoto, *Appl. Phys. Lett.* **71**, 1198 (1997).
24. H.-S. Han, S.-Y. Seo and J. H. Shin, *Appl. Phys. Lett.* **79**, 4568 (2001).
25. J. Lee, J. H. Shin and N. Park, *J. Lightw. Technol.* **23**, 19 (2005).
26. N. Daldosso, D. Navarro-Urrios, M. Melchiorri and L. Pavesi, F. Gourbilleau, M. Carrada and R. Rizk, C. García, P. Pellegrino and B. Garrido, *Appl. Phys. Lett.* **86**, 261103 (2005).
27. P. G. Kik and A. Polman, *J. Appl. Phys.* **91**, 534 (2002).
28. H. Mertens, A. Polman, I. M. P. Aarts, W. M. M. Kessels and M. C. M. van de Sanden, *Appl. Phys. Lett.* **86**, 241109 (2005).
29. B. Garrido, C. García, P. Pellegrino, D. Navarro-Urrios, N. Daldosso, L. Pavesi, F. Gourbilleau and R. Rizk, *Appl. Phys. Lett.* **89**, 163103 (2006).
30. D. Pacifici, G. Franzò, F. Priolo, F. Iacona and L. Dal Negro, *Phys. Rev.* **B67**, 245301 (2003).
31. K. Watanabe, M. Fujii and S. Hayashi, *J. Appl. Phys.* **90**, 4761 (2001).
32. B. Garrido, C. García, S.-Y. Seo, P. Pellegrino, D. Navarro-Urrios, N. Daldosso, L. Pavesi, F. Gourbilleau and R. Rizk, *Phys. Rev.* **B76**, 245308 (2007).
33. C. García, P. Pellegrino, Y. Lebour, B. Garrido, F. Gourbilleau and R. Rizk, *J. Lumin.* **121**, 204 (2006).
34. S.-Y. Seo, O. Jambois, P. Pellegrino, B. Garrido, F. Gourbilleau and R. Rizk (in preparation).
35. C. Garcia, B. Garrido, P. Pellegrino, R. Ferre, J. A. Moreno, J. R. Morante, L. Pavesi and M. Cazzanelli, *Appl. Phys. Lett.* **82**, 1595 (2003).

36. C. J. Oton, W. H. Loh and A. J. Kenyon, *Appl. Phys. Lett.* **89**, 031116 (2006).
37. N. Daldosso, D. Navarro-Urrios, M. Melchiorri, L. Pavesi, C. Sada, F. Gourbilleau and R. Rizk, *Appl. Phys. Lett.* **88**, 161901 (2006).
38. J. H. Shin, W.-H. Lee and H.-S. Han, *Appl. Phys. Lett.* **74**, 1573 (1999).
39. J.-H. Jhe, J. H. Shin, K. J. Kim and D. W. Moon, *Appl. Phys. Lett.* **82**, 4489 (2003).

CHAPTER 6

SILICON NANOCRYSTALS: STRUCTURAL AND OPTICAL PROPERTIES AND DEVICE APPLICATIONS

Fabio Iacona

CNR-IMM, stradale Primosole 50, 95121 Catania, Italy
E-mail: fabio.iacona@imm.cnr.it

Giorgia Franzò, Alessia Irrera, Simona Boninelli,
Maria Miritello and Francesco Priolo

MATIS CNR-INFM & Dipartimento di Fisica e Astronomia dell'Università di Catania, via Santa Sofia 64, 95123 Catania, Italy

This chapter describes the structural and optical properties of Si nanocrystals (Si-nc) and their application for the fabrication of light emitting devices. Transmission electron microscopy (TEM) has been extensively used to characterize the steps leading to the formation of Si-nc in thermally treated SiO_x films; from the quantitative analysis of the TEM data the nc size, their density and the crystalline fraction as a function of the annealing temperature have been obtained. The optical properties of Si-nc have been investigated by means of steady state and time resolved photoluminescence (PL) measurements and correlated with the structural data; the peculiar PL properties exhibited by Si-nc embedded inside an optical microcavity are also presented. A relevant part of the chapter is devoted to light emitting devices based on Si-nc. Si-nc can be efficiently excited by impact with hot electrons, producing a room temperature electroluminescence signal. The perspectives and the performances of these devices can be significantly enhanced if amorphous Si nanoclusters are used, since they are formed at lower temperature than Si-nc; furthermore, the integration of a photonic crystal structure in these devices leads to an increased external efficiency. The relevance and the impact of the above results for the development of silicon nanophotonics are finally discussed.

1. Introduction

Since the discovery of strong visible photoluminescence (PL) at room temperature from porous Si done by Canham in 1990,[1] the possibility to obtain efficient light emission from Si nanostructures has attracted the efforts of a wide scientific community. Very rapidly the focus of the research moved from the brightly-emitting, but brittle and unstable, porous Si to Si nanocrystals (Si-nc) embedded in SiO_2 films. Si-nc exhibit a remarkably higher stability (both from mechanical and optical point of views) with respect to porous Si and are fully compatible with the technology used for Si-based microelectronics with regard to the techniques used for their synthesis (chemical vapor deposition, sputter deposition, ion implantation) and processing (conventional dry and wet etching processes are easily applicable for the patterning of SiO_2 layers containing Si-nc). The efforts done by several research groups have led to a deep comprehension of the complex phenomena ruling the process of light emission from Si-nc and to the optimization, in terms of efficiency and stability, of their properties.[2-11] A major advancement has been represented by the discovery that Si-nc can exhibit optical gain,[12-15] so opening the route towards the development of a Si laser. It is generally recognized that the unusual properties of Si-nc are tightly connected to the very stable passivation ensured by the surrounding SiO_2, mainly by eliminating dangling bonds through the formation of Si=O bonds;[16] furthermore, the SiO_2 matrix in which the nc are embedded constitutes also a low defect density medium in which the efficiency of non-radiative recombination processes is strongly reduced.

Unfortunately, SiO_2 is also a very good insulating material, and this means that carrier injection through a SiO_2 layer containing Si-nc is very difficult. As a consequence, this material, in spite of its very promising optical properties, has been considered for a long time not useful for the realization of efficient Si-based light emitting devices. However, in spite of this drawback, several prototypes of light emitting devices based on Si nanostructures have been proposed. Well before the discovery of the surprising properties of porous Si, the electroluminescence (EL) of small Si clusters embedded in SiO_2 and its possible attribution to quantum confinement effects were reported in the pioneering work of DiMaria *et al.*[17] After the Canham's discovery several devices based on porous Si have been proposed, the most relevant one being probably the integrated bipolar transistor/porous Si LED structure developed by Hirschman *et al.*[18] The first examples of Si-nc devices exhibiting a room temperature EL in the visible region are based on Si-rich oxide layers,[19] nanocrystalline Si layers[20,21] or Si-nc formed by Si ion implantation in SiO_2.[22] The approach commonly used to overcome the difficulty of carrier injection is to confine the Si-nc in ultra-thin layers; in this way an efficient carrier injection can be obtained also for operating voltages of few volts. More recently it has been demonstrated that it is possible to observe intense EL signal at reasonably low voltages also from thicker layers (80 nm), by fully taking advantage of the tuning of size and density which can be obtained when Si-nc are generated by thermal annealing of SiO_x layers.[23] The demonstration that the cross section for the electrical excitation of Si-nc is about two orders of magnitude higher than that one observed under optical pumping has reinforced the idea that they can have important perspectives also for practical applications.[24] Recently a new device structure for an optimized light emission from Si-nc has been proposed.[25] To improve the carrier injection efficiency, a field-effect EL mechanism has been proposed. In this excitation process electrons and holes are sequentially injected in array of Si-nc embedded in the gate oxide of a transistor. The formed excitons recombine radiatively giving origin to the EL signal. This new approach is expected to reduce non-radiative processes limiting EL efficiency, such as Auger recombination, as well as improve the device reliability since it is expected to lead to a less extended oxide wearout with respect to the impact excitation by hot carriers.

Efficient PL and EL emission in the visible region has been also observed from crystalline or amorphous Si clusters embedded in silicon nitride.[26-30] The replacement of the conventional SiO_2 matrix with silicon nitride could lead to a significant improvement of the performances of devices based on Si-nc, since an easier carrier injection due to the lower bandgap of this material is expected. However, the real performances and perspectives of these devices have still to be fully explored; indeed, the origin of the luminescence from Si-rich nitride films remains quite unclear, since it is well known that silicon nitride layers can be luminescent independently of the presence of Si nanoclusters.

Another approach that has been used to efficiently produce photons from Si is based on the introduction of light emitting impurities, such as erbium ions. Er ions emit at 1.54 μm, which represents a strategic wavelength for telecommunication, because it corresponds to a minimum in the loss spectrum of silica optical fibers. It has been demonstrated that Si nanostructures may act as efficient sensitizers for rare earth PL;[31-34] the energy transfer process from the excited nanoclusters to the neighboring Er ions increases by about a factor of 100 the PL efficiency of the system.[32,33] Efficient, stable and room temperature operating electroluminescent devices based on Er-doped Si nanostructures have been proposed.[35,36]

The above described efforts done during the last decade by the scientific community working in the field of silicon photonics have been mainly devoted to the final goal of the development of a Si-based light source able to be used for the monolithic integration of optical and electrical functions on a single Si chip. Indeed, electrical interconnections based on metal lines represent today the most important limitation on the performances of Si-based microelectronic devices.[37] The parasitic capacities generating at the several metal/insulating/metal capacitors present in complex multilevel metallization schemes constitute the major contribution to the delay in the signal propagation. The intrinsic resistivity of the metal lines, as well as the contact resistance at the various metal/metal interfaces constitutes other relevant delay sources. An important reduction of the delay-times has been achieved by replacing the well established metallization schemes based on Al and SiO_2 with new materials, such as Cu-based metal films and low dielectric constant insulating layers, but, as fast as the minimum feature size of the devices will further reduce, the delay due to metal interconnections will represent again an unacceptable bottleneck for device performance. An almost definitive solution to this problem could be represented by the use of optical interconnections for the transfer of information inside a chip.[37-40] To develop this strategy, materials able to generate, guide, amplify, switch, modulate and detect light are needed. The most severe requirement for such materials is represented by the compatibility with the technology used for the fabrication of Si devices. However, because Si is intrinsically unable to efficiently emit light due to its indirect bandgap, it is evident that the main limitation to the above described strategy is the lack of an efficient Si-based light source. For all the reasons listed above (including the compatibility with Si technology), Si-nc probably represent the most important candidate for the realization of efficient electrically-pumped optical sources to be employed in silicon photonics.

In this chapter we will discuss the structural and optical properties of Si-nc. A particular emphasis will be given to the studies performed by transmission electron microscopy (TEM) to elucidate the steps leading to the formation of Si-nc in annealed SiO_x films. The PL properties of Si-nc will be presented, by showing also how they can be strongly modified inside an optical microcavity. A relevant part of the chapter will be devoted to the application of Si nanoclusters to the fabrication of light emitting devices. We will show that very promising performances can be obtained by coupling amorphous nanoclusters with a photonic crystal structure. The relevance of the above results for the development of silicon nanophotonics will be underlined.

2. Formation and Evolution of Si-nc Synthesized by Thermal Annealing of SiO_x Films

Si-nc embedded in SiO_2 can be prepared by using several different techniques, including chemical vapor deposition,[6,10,15] sputter deposition,[41,42] ion implantation[3-5,7-9,12,14] and reactive evaporation.[13,43] The application of deposition techniques for this purpose usually implies the growth of a SiO_x film (i.e. a SiO_2 film containing a Si excess) followed by a thermal annealing to induce a thermodynamically-driven phase separation in the substoichiometric layer, leading to the formation of Si-nc embedded in an almost stoichiometric SiO_2 matrix. This process of growth of Si-nc can be properly followed by using TEM techniques. Figure 1 reports the plan view energy filtered transmission electron microscopy (EFTEM) images of SiO_x films as deposited and annealed at 900 and 1000°C for 1 h in N_2 environment. The SiO_x films have a total Si concentration of 46 at.%, are 80-nm thick and have been synthesized by plasma enhanced chemical vapor deposition (PECVD) by properly adjusting the ratio of the gaseous precursors (ultrapure SiH_4 and N_2O).[10,44,45] The EFTEM technique allows to generate an image by using only electrons that have lost a specific amount of energy due to the interaction with the sample. This allows to obtain a chemical mapping of all the species present in a sample with the very high spatial resolution typical of TEM.[44,45] To map the presence of Si clusters formed inside the SiO_x layer by the annealing processes, we use an energy filter to select an energy window (4-eV wide) centered in correspondence of the Si plasmon loss (16 eV). The high energy shift between the Si and SiO_2 plasmon peaks (about 7 eV) allows to neatly discriminate the Si contribution from the SiO_2 one and to detect Si clusters.[44,45] The image reported in Fig. 1a shows a uniform grey background without any appreciable intensity contrast, demonstrating the absence of any phase separation effect in the as deposited samples. On the other hand, the image reported in Fig. 1b, referring to a sample annealed at 900°C, exhibits the presence of brighter zones, associated to the presence of partially interconnected Si clusters. The phase separation between Si and SiO_2 becomes much more evident by increasing the annealing temperature, and well defined and isolated Si clusters embedded in the oxide matrix are clearly visible in samples annealed at 1000°C (see Fig. 1c).

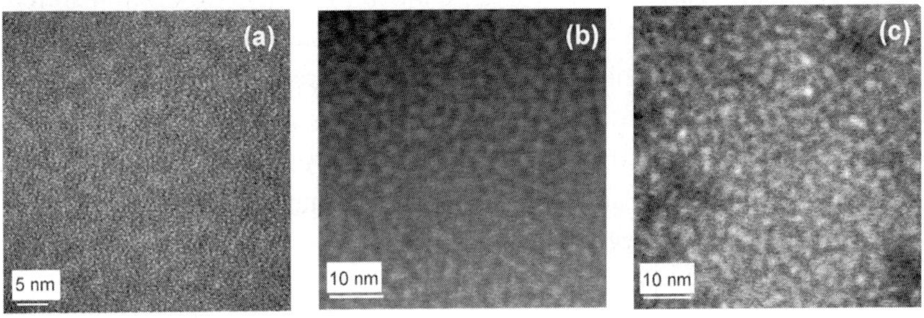

Fig. 1. EFTEM plan view images obtained from (a) as deposited SiO_x films, and SiO_x films annealed at (b) 900°C, and (c) 1000°C. The bright zones are associated to the presence of Si clusters.

To gain a deeper knowledge on the structural properties of this material, we have also employed the more conventional dark field transmission electron microscopy (DFTEM) technique. DFTEM is sensitive to the presence of crystalline planes in the analyzed sample, and it is therefore able to map the system for the presence of crystalline Si clusters. In agreement with literature data,[10] it was not possible to observe any diffraction pattern corresponding to the presence of a crystalline phase for SiO_x samples as deposited and annealed at 900 and 1000°C, demonstrating that all the clusters shown in Figs. 1b and 1c are amorphous.

The direct comparison between EFTEM and DFTEM images is shown in Fig. 2 for samples annealed at temperatures ranging from 1100 to 1250°C. SiO_x films annealed in this temperature range exhibit the presence of a diffraction pattern, shown as an inset in Fig. 2b, mainly consisting of three well distinct rings corresponding to the (111), (220), and (311) planes of crystalline Si, so that it is possible to conclude that they contain a significant fraction of crystalline clusters. DFTEM plan view images, shown in Figures 2b, 2d and 2f, have been obtained by selecting a small portion (about 10%) of the diffraction ring due to the (111) Si planes, while the energy filtered images have been collected again by mapping the presence of Si clusters through the use of an energy window centered in correspondence of the Si plasmon loss. In dark field images, Si-nc appear as bright spots on a dark background. The comparison between images obtained with the two different techniques on the same sample (Figures 2a and 2b for the annealing at 1100°C; Figures 2c and 2d for the annealing at 1150°C; Figures 2e and 2f for the annealing at 1250°C) clearly demonstrates that energy filtered images, due to their capability to show all the Si clusters independently of their crystalline structures, allow a much more complete characterization of this material, if compared with the dark field technique, showing only a small portion of the crystalline population.

The qualitative analysis of the TEM images shown in Figures 1 and 2 allows the comprehension of the steps of the thermal evolution of SiO_x layers leading to the formation of Si-nc. As deposited SiO_x films are homogeneous and fully amorphous materials, without any evidence of phase separation between Si and SiO_2. This is in agreement with the description of SiO_x films given by the random bonding model,[46] which can be reasonably assumed to describe as deposited films, since they have been deposited at low temperature (300°C). The first stages of the phase separation between Si and SiO_2 become visible at 900°C, but well defined and amorphous Si nanoclusters (na) are formed only at 1000°C. At 1100°C na begin to crystallize. We remark that this general behaviour has to be considered valid for all SiO_x compositions; on the other hand, an increase of the clustering and crystallization temperatures we report (1000 and 1100°C, respectively) is expected when dealing with SiO_x films characterized by lower Si concentrations.[10,47]

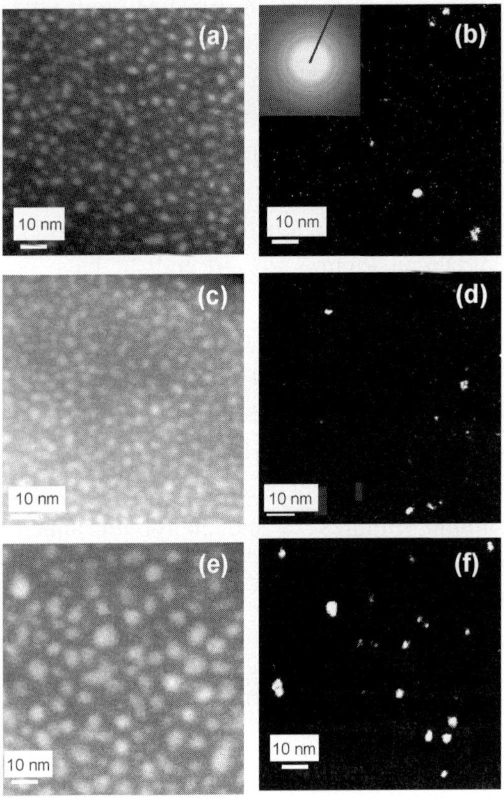

Fig. 2. EFTEM plan view images ((a) 1100°C, (c) 1150°C, (e) 1250°C) and DFTEM plan view images ((b) 1100°C, (d) 1150°C, (f) 1250°C) obtained from SiO_x films annealed for 1 h. The inset of (b) shows the electron diffraction pattern of the sample annealed at 1100°C.

Very detailed information on the structure of the annealed SiO_x samples can be obtained by the quantitative analysis of the TEM images above presented. The first step is the determination of the size of the Si nanoclusters. The mean radius of the Si clusters, as obtained by analyzing EF and DF images according to the procedure described in Refs. 44 and 45, is reported in Fig. 3a as a function of the annealing temperature. The mean radii detected fall in the range 1-3 nm. Both techniques demonstrate that the mean radius increases by increasing the annealing temperature, but they do not give the same mean size for the same sample. In particular, EFTEM images give systematically larger mean radii than those detected in the dark field images. Few possible explanations for this difference are presented elsewhere.[44,45]

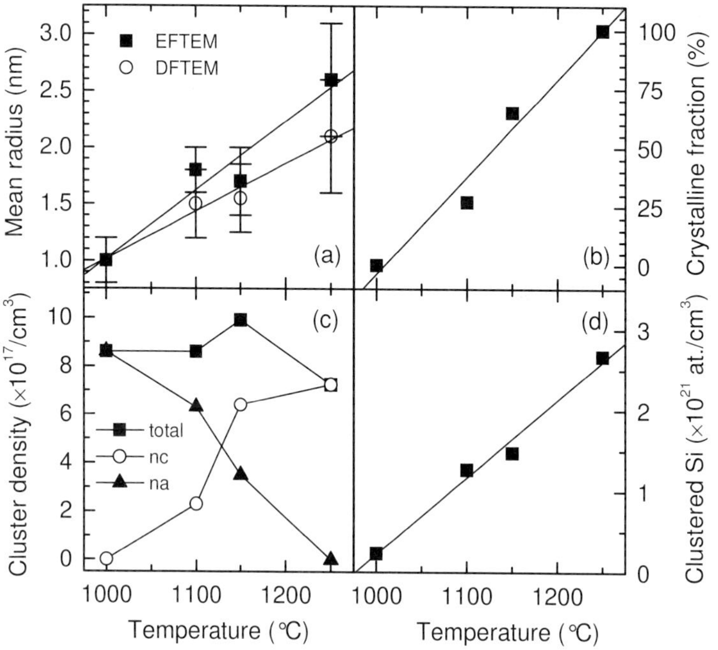

Fig. 3. (a) Si nanocluster mean radius, (b) crystalline fraction, (c) density of amorphous (na) and crystalline (nc) nanoclusters and their sum, (d) concentration of clustered Si atoms, as a function of the annealing temperature. The data are obtained from EFTEM and DFTEM analyses. The lines are drawn to guide the eye.

For samples in which the amorphous and crystalline phases coexist, we have estimated the crystalline fraction present at a given temperature. In order to do this, the ratio between the number of nc (as detected by the dark field analysis) and the total number of clusters (as detected by EFTEM) has been evaluated at the different temperatures. The cluster count has been performed in the same sample regions for both techniques; furthermore, to avoid shadowing effects between clusters placed at different depths, we have selected for the analysis very thin regions. The data have been reported in Fig. 3b, and a clear trend, showing the progressive increase of the crystalline fraction by increasing the temperature, has been found. Under the hypothesis that at 1250°C all clusters are crystalline,[44,45] the estimation of the nc fraction at lower temperatures leads to values of about 30% at 1100°C (the temperature at which the crystallization process starts) and 60% at 1150°C, clearly demonstrating that in this range, the temperature plays a role not only in the cluster growth, but also in extensively promoting the amorphous to crystal transition.[41]

Plan view EFTEM images allow also a very reliable estimation of the number of nanoclusters per unit volume at the different temperatures. In order to gain this information, however, it is necessary to estimate the thickness sampled by the EFTEM experiment, in order to determine the analyzed volume. To measure the sampled thickness for each analyzed EFTEM plan view image we have collected, under the same experimental conditions and in the same sample region, also an electron energy loss spectrum. The analyzed depth is proportional to the logarithm of the ratio between the integral intensity of the whole energy loss spectrum and the integral intensity of the zero-loss peak.[44,45] This method is able to give us a thickness in units of the electron mean free path; this number has been converted in depth units by using an experimentally determined value for the electron mean free path of 160 nm for 200 kV electrons,[44,45] which is in good agreement with other literature reports.[48]

The values for the Si nanocluster density we have obtained by using the above described method are reported in Fig. 3c as a function of the annealing temperature. The nanocluster density is almost constant in the 1000-1150°C range (9×10^{17} cm^{-3}), while it decreases (about 7×10^{17}

cm^{-3}) by increasing the annealing temperature up to 1250°C. This result, coupled with the continuous increase in cluster size with temperature reported in Fig. 3a, suggests that nanocluster growth is not simply due to the inclusion of Si atoms diffusing from the oxide matrix, but also Ostwald ripening effects, leading to the disappearance of small clusters, are operating. By taking into account the data on crystalline fraction reported in Fig. 3b it is possible to calculate the concentration of nc and of na as a function of the annealing temperature. The obtained data are reported in Fig. 3c and demonstrate that the temperature progressively induces the transformation of 9×10^{17} na/cm^3 in 7×10^{17} nc/cm^3, with the loss of about 20% of the clusters present at 1000°C, due to the above described occurrence of Ostwald ripening phenomena.

Finally, from the above reported data on the nanocluster density and mean radius the number of clustered Si atoms per unit volume at the different temperatures can be easily calculated. The obtained data are reported in Fig. 3d and demonstrate that the number of Si atoms forming detectable clusters remarkably increases by increasing the annealing temperature. It is noteworthy that, also at 1250°C, a relevant fraction of excess Si atoms not detectable by EFTEM exists. Indeed, the value of 3×10^{21} cm^{-3} has to be compared with the excess Si content of about 1×10^{22} cm^{-3}. On the basis of the present data, it is not possible to unambiguously determine the actual state of the missing Si atoms, and, in particular, if they are embedded in clusters too small to be detected by TEM (Si clusters having a radius lower than 1 nm can be hardly detected also by using EFTEM), or if they are still dissolved in the SiO_x matrix. The existence of this "invisible" fraction of Si atoms, which has been confirmed by using an analytical methodology based on the combination between EFTEM imaging and electron energy loss spectroscopy,[49] has to be taken into account when the density data of Fig. 3c are discussed. Indeed, the existence of a wide temperature range (1000-1150°C) in which the cluster density seems to be almost constant, is probably the result of a balance between the "appearance" of new clusters (we cannot distinguish if this is due to the growth of smaller ones, or to the nucleation of new ones) and the "disappearance" of small clusters due to ripening effects. On the other hand, at 1250°C the Si atoms dissolved in the matrix or forming undetectable clusters are strongly diminished, and therefore ripening effects become more marked.

3. Optical Properties of Si-nc

Si-nc prepared by thermal annealing of SiO_x layers emit light at room temperature in the range 700-1100 nm. As an example Fig. 4 reports the normalized PL spectra of Si-nc obtained by annealing at 1250°C for 1 h in N_2 ambient SiO_x films having different total Si contents (from 35 to 44 at.%), synthesized by PECVD by changing the ratio of the gaseous precursors (SiH_4 and N_2O). PL measurements were performed by pumping with the 488 nm line of an Ar^+ laser. Spectra were measured at room temperature, with a pump power of 10 mW. The PL signal shows a marked blue shift with decreasing Si content as a result of the smaller size of the Si-nc.[10] Indeed, the average nc radius determined by TEM increases from 1.1 to 2.1 nm by increasing the Si content from 37 at.% to 44 at.%. Note that a similar tuning of the wavelength of the emitted light can be also obtained by changing the annealing temperature for a fixed Si concentration in the SiO_x films.[10]

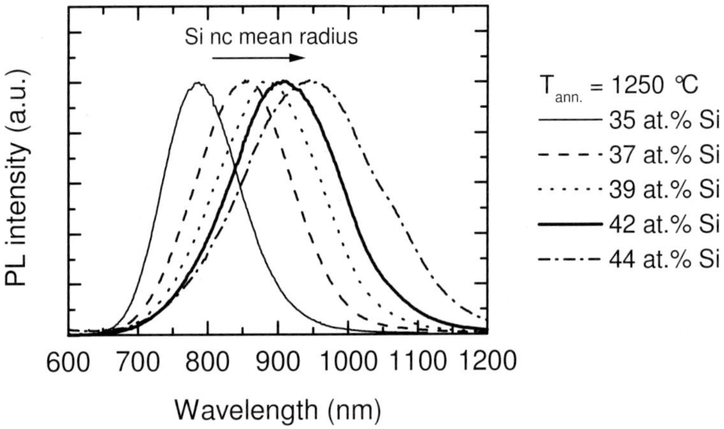

Fig. 4. Normalized PL spectra of SiO_x thin films with different Si concentrations annealed at 1250°C for 1 h. Spectra were measured at room temperature, with a laser pump power of 10 mW.

The data shown above seem to be in agreement with the carrier quantum confinement theory, which predicts the progressive blueshift of the PL peaks with decreasing the crystal size, due to the enlargement of the band gap of the Si-nc with respect to bulk crystalline Si.[50] However, by comparing the experimental PL data with theoretical calculations of the Si-nc bandgap taken from literature,[50] we find that our data are remarkably shifted towards lower energies. Indeed, our results seem to be more consistent with a model in which light emission is not due to band-to-band recombination processes within the nc, but the recombination occurs on intermediate states introducing energy levels inside the band gap and corresponding to SiO_2/Si interfacial states.[8,51] The nature of these states is such that their energy level is not fixed, like the case of luminescence from defects, (otherwise we could not observe any dependence of the emitted wavelength on Si concentration and temperature) but depends on the nc size, i.e. smaller grains, having a larger band gap, can also have energy levels at higher energy with respect to larger grains, characterized by a narrower band gap. This effect has been quantitatively explained with the trapping of an electron (or even an exciton) by Si=O bonds producing localized states in the bandgap of Si-nc with diameters smaller than 3 nm.[16]

The decay time of the PL signal is of particular interest since it reflects the confinement properties of the Si-nc.[52] Figure 5 shows the room temperature decay-time of the PL signal at two fixed detection wavelengths, 700 nm and 950 nm, for different SiO_x samples after shutting off a 10 mW 488 nm Ar^+ laser beam chopped at a frequency of 55 Hz. From the analysis of Fig. 5a we note that the decay time at 700 nm increases with decreasing Si content and it is characterized by a stretched exponential shape which becomes more and more similar to a single exponential with decreasing Si content. Stretched exponential functions have been widely observed in the literature in the decay time of both porous Si and Si-nc.[9,53] In a stretched exponential the decay line shape is given by:

$$I(t) = I_0 \exp [-(t/\tau)^{\beta}]. \quad (1)$$

In Eq. (1), $I(t)$ and I_0 are the intensity as a function of time and at $t = 0$, τ is the decay time and β is a dispersion factor ≤ 1. The smaller is β the more stretched is the exponential. The results of the fits to the data in Fig. 5a with Eq. (1) demonstrates that the factor β decreases from 0.85 to 0.63 and τ from 65 μs to 10 μs by increasing the Si content from 35 to 44 at.%. On the other hand, extremely interesting is the behavior of the 950 nm signal decay time, shown in Fig. 5b, which is characterized by almost single exponentials with the same lifetime of 175 μs for all the Si contents.

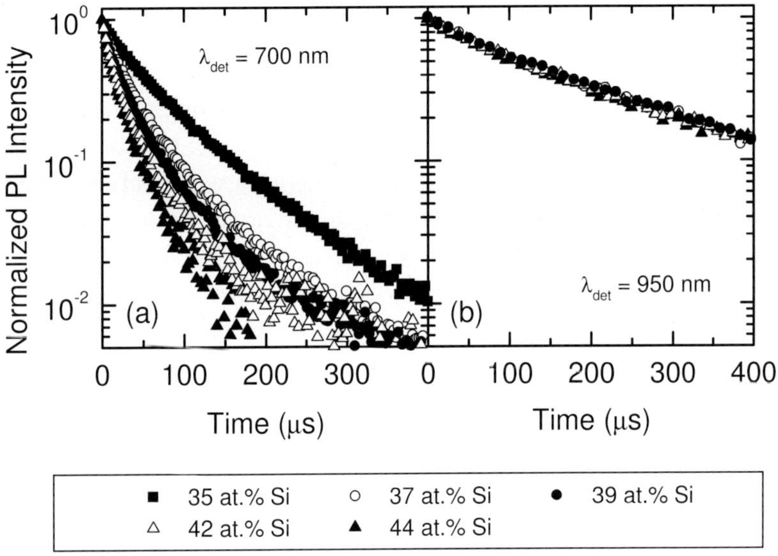

Fig. 5. Measurements of the time-decay of the PL signal (a) at 700 nm and (b) at 950 nm for SiO_x films with different Si contents annealed at 1250°C for 1 h. Data were taken at room temperature and with a laser pump power of 10 mW.

A decrease in β has been associated with a redistribution of the energy within the sample with a transfer from smaller nc (having larger gaps) to larger nc (having smaller gaps).[9] This picture is consistent with our data, showing smaller β and τ values in Si-rich samples in which the nc concentration is higher and hence the energy transfer is more probable. It should be stressed that, since all measurements are performed at the same wavelength they reveal the properties of the same class of nc (i.e. having the same size) embedded within different samples. The markedly different behavior observed demonstrates that the environment of the nc plays a quite important role in determining its decay time. The more the nc are isolated (larger β) the larger is the decay time τ (since energy transfer becomes less probable). Moreover, larger nc (those emitting at 950 nm, see Fig. 5b) cannot transfer their energy to the surrounding nc since their energy is not sufficient due to their smaller band gaps. Therefore they act as "isolated" nc in all systems in the sense that once excited they will re-emit the energy only radiatively. This explains the identical lifetime and the almost single exponential behavior shown in Fig. 5b.

3.1. *Si-nc inside an optical microcavity*

The main problems which prevent practical applications of Si-nc in nanophotonics are: (i) the difficulty in obtaining the electrical excitation of the Si-nc, since they are embedded in an insulating host (SiO_2); (ii) the absence of spectral purity in the emission; (iii) the absence of directionality in the emission; (iv) the luminescence yield which needs further enhancement. We will show in Sec. 4 that the efficient electrical excitation of Si-nc is possible. On the other hand, points (ii)-(iv) can be overcome by using photonic crystals (which will be discussed in Sec. 4.1) or microcavities.

When a photon is confined in a microcavity in resonance with the emission of the active medium, the light emission becomes spectrally sharp and it is strongly enhanced in the direction of confinement. Optical microcavities have been demonstrated for porous Si[54] and for Er doped SiO_2.[55] Si/SiO_2 Fabry-Perot microcavities in resonance with the Si-nc emission band consist of two distributed Si/SiO_2 Bragg reflectors and a central SiO_x region, with $x = 1.7$. They have been grown by sequential PECVD processes on top of a Si substrate.[56] The thicknesses of the layers forming the mirror and of the active region were varied in order to match $\lambda_o/4n$ and $\lambda_o/2n$, respectively (being λ_o the intended value of the resonance and n the refractive index of the medium). To obtain a resonance at $\lambda_o = 830$ nm the thicknesses were 57 nm for Si, 141 nm for SiO_2 and 241 nm for SiO_x. The number of Si/SiO_2 pairs was 5 pairs for the bottom mirror and 4 pairs for the top mirror. After deposition, a thermal annealing at 1250°C for 1 h induces the separation of the Si and

SiO_2 phases in the active SiO_x region and Si-nc with a mean radius of 1.1 nm were formed, as evidenced by TEM. A cross sectional TEM image of a microcavity is shown in Fig. 6a. Both distributed Bragg reflectors and the Si-nc containing active region are clearly visible. The presence of Si-nc can be appreciated as bright spots in the dark field image showing an expanded view of the active layer. The stop band of this microcavity is at wavelengths between 600 and 1200 nm (as deduced from reflectivity measurements, not shown) and the quality factor ($Q = \lambda_o/\Delta\lambda$ being $\Delta\lambda$ the full width at half maximum of the resonance dip in the reflectivity) is $Q = 500$. Figure 6b shows the comparison between the PL spectrum of the microcavity and the PL spectrum of the same Si-nc outside the microcavity. It has to be noted that the PL intensity within the microcavity is strongly enhanced (the spectrum of Si-nc is multiplied by a factor of 40 in the figure). Secondly, the spectrum of the Si-nc inside the cavity is extremely sharp; a full width at half maximum lower than 1.5 nm, to be compared with the value of 180 nm of the original emission, is noticed.

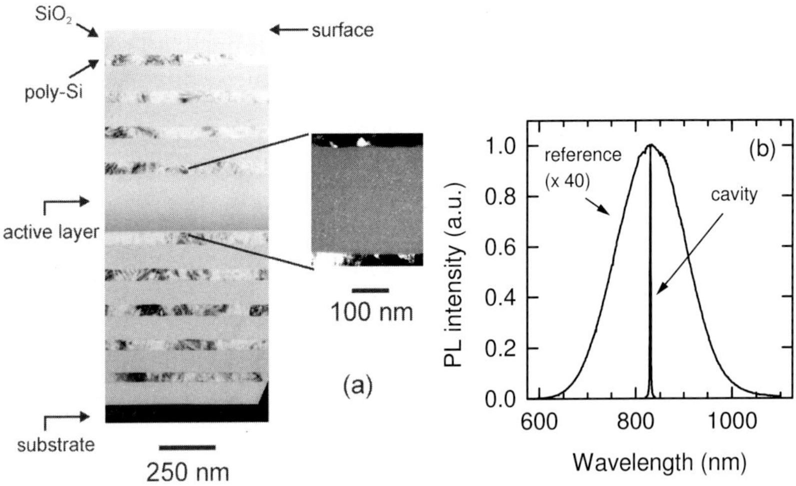

Fig. 6. (a) Cross sectional bright field TEM image of a Si/SiO_2 microcavity with Si-nc in the active region. Si-nc are clearly evidenced as bright spots in the dark field image showing a detail of the central region of the cavity. (b) Comparison between the room temperature PL spectra of Si-nc not embedded in a microcavity and the Si-nc-based microcavity shown in panel (a). The spectrum of the cavity is about 40 times more intense than outside the cavity.

The directionality properties of the microcavity have been also studied; the integrated PL intensity as a function of detection angle with respect to the normal to the sample has been found to strongly reduce with increasing detection angle, evidencing a marked directionality. Indeed, at 30° out of the normal direction the integrated intensity falls below 20% of its maximum value.

4. Light Emitting Devices Based on Si-Nanoclusters

Si-nc embedded in SiO_2 prepared by thermal annealing of SiO_x films synthesized by PECVD can be successfully used to fabricate light emitting MOS devices. The active region of a typical device is shown in the TEM cross-sectional micrograph of Fig. 7a: from the top, a n-type polysilicon film (resistivity of 1 mΩ·cm), the SiO_x active region (80-nm thick) and a p-type Si substrate (resistivity of 10 mΩ·cm) are distinguishable. The Si-nc contained in the SiO_x layer are clearly shown as white dots on the dark SiO_2 background in the DFTEM cross section shown in Fig. 7b. A top view of the device is shown in the scanning electron microscopy (SEM) micrograph of Fig. 7c. Aluminium-based contacts, defined as rings, were made to the polysilicon film and to the Si substrate. A metal-free circular area for the exit of the light is also visible.

SiO_x films are conductive due to the excess Si they contain and allow an electrical current to flow from the highly doped p-type substrate to the n-type polysilicon top electrode once the devices are biased. The main mechanisms of conduction are tunnelling (either direct or Fowler-Nordheim[23]), or Poole-Frenkel.[57] Si-nc excitation mainly occurs through impact excitation by hot electrons.[23] A preliminary test of the characteristics of the emission of these devices has been performed by emission microscopy (EMMI) measurements. The EMMI apparatus consists of an optical microscope and a CCD camera sensitive to visible and infrared radiation (from 0.3 to 1.2 μm), surrounded by a shielding box. The EMMI image of a device with a Si content of 46 at.% and annealed at 1100°C is reported in Fig. 8a. The image reveals an intense emission, very stable and almost homogeneous over the whole active area. The bright emission we observe starts at a voltage of −15 V, with a current density of the order of about 90 μA/cm^2.

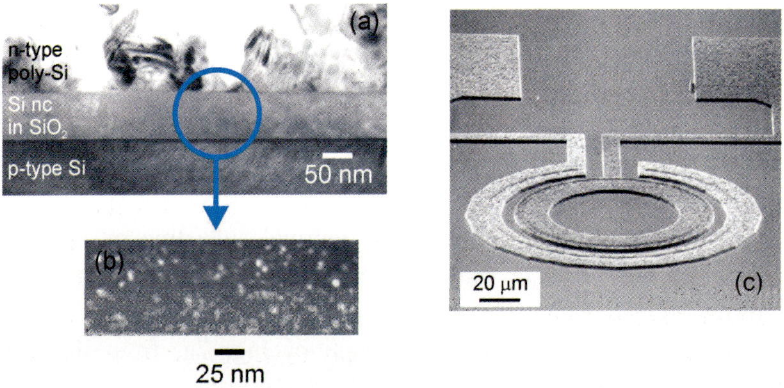

Fig. 7. (a) Cross-sectional bright field TEM view of the active area of a device based on Si-nc. (b) Cross-sectional dark field TEM view of the active layer containing Si-nc. (c) SEM image of the device.

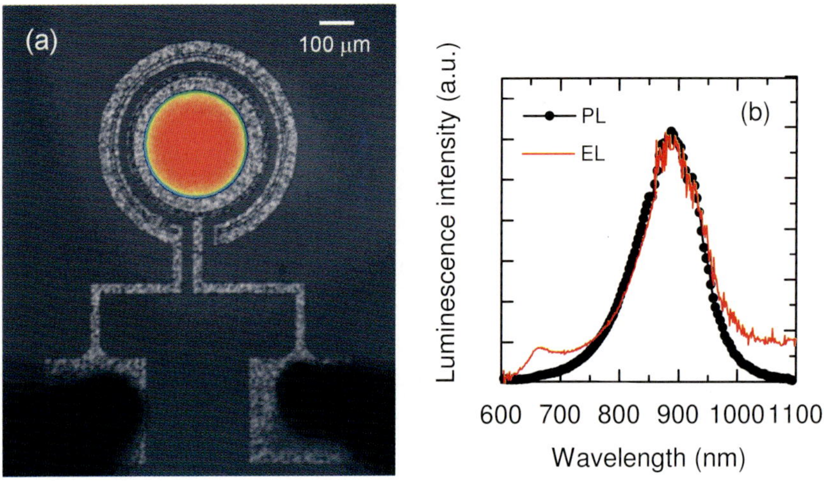

Fig. 8. (a) EMMI image of a device with 46 at.% Si annealed at 1100°C for 1 h. The colors indicate different intensities of the emitted light, red being the highest. (b) Comparison between the PL and the EL spectra of a device with a Si content of 42 at.% in the SiO_x layer. The EL spectrum was measured with a voltage of −48 V and a current density of 4 mA/cm^2.

A spectral analysis of the light emitted from the devices has been performed by EL measurements. Fig. 8b shows the room temperature EL spectrum measured in a device with a Si content of 42 at.% in the SiO_x layer. The spectrum was measured by biasing the device with a square pulse at a frequency of 55 Hz with a voltage of −48 V; under these conditions the current density passing through the device is 4 mA/cm^2. The EL spectrum presents two peaks: a weak one at 660 nm, which can be attributed to the presence of radiative defects in the oxide matrix and a most intense one at 890 nm. It is interesting to note that the main EL peak is very similar both in position and shape to the PL peak measured in the same sample and shown for comparison in Fig. 8b. The PL spectrum was taken by illuminating the device with the 488 nm line of an Ar^+ laser with a pump power of 10 mW. It is therefore straightforward to attribute the EL signal to electron-hole pair recombination in the Si-nc dispersed in the oxide layer. Devices based on Si-nc are very stable and can work continuously for several days without deterioration. The estimated internal efficiency is about 1%.

Amorphous Si nanoclusters may constitute an interesting alternative to Si-nc for the fabrication of light emitting devices to be used for the monolithic integration of optical and electrical functions in Si VLSI technology. In fact, they exhibit an intense room temperature EL in the same range of wavelengths already found for their crystalline counterpart, with the advantage to be formed at a temperature of 900°C, while, as above discussed, at least 1100°C are needed for the formation of Si-nc.[57,58]

The de-excitation properties of the devices based on amorphous and crystalline Si nanoclusters have been studied by means of time-resolved EL measurements. In Fig. 9 the comparison between the decay time of the EL signal measured at 950 nm for a device based on Si-nc (obtained by annealing at 1100°C a SiO_x layer containing 46 at.% of Si) and a device based on amorphous Si-nanoclusters (obtained by annealing at 900°C a SiO_x layer containing 46 at.% of Si) is reported. The data have been collected by biasing the devices with a square pulse at a frequency of 55 Hz, with a current density passing through both devices of 3 A/cm^2. When the excitation is switched off we observe the decay of the EL signal as a function of time. The lifetime of the EL signals has been extracted by fitting the experimental curves reported in Fig. 9 with Eq. (1). The lifetime of the EL signal is 57 μs for Si-nc and 600 ns for amorphous Si nanoclusters. It is possible to explain the faster EL lifetime of the amorphous clusters with the presence of non-radiative recombination processes, strongly related with their structural properties.

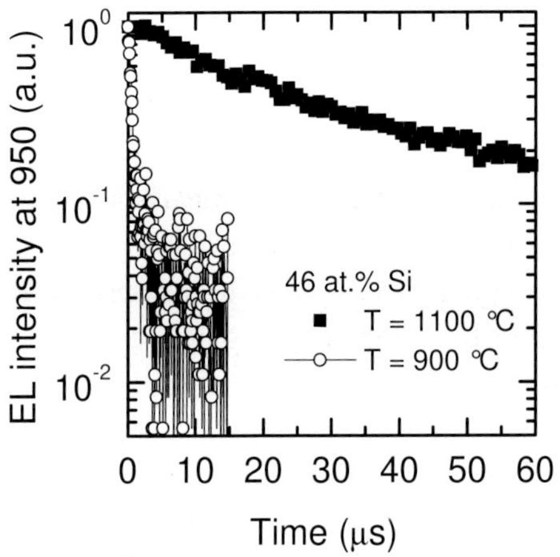

Fig. 9. Comparison between the decay time of the EL signal measured at 950 nm for a device based on Si-nc (obtained by thermal annealing at 1100°C of a SiO_x layer) and a device based on amorphous Si nanoclusters (obtained by thermal annealing at 900°C of a SiO_x layer). The Si concentration in the SiO_x layers is 46 at.% for both cases.

The comparison of the EL properties of devices based on different kinds of Si nanostructures can be found in Fig. 10a, where we report the EL intensity at 850 nm as a function of the applied voltage under forward bias conditions. In particular, devices based on SiO_x layers with 39, 42 and 46 at.% of Si annealed at 1100 and 1250°C (based on Si-nc) and with 46 at.% Si annealed at 900°C (based on amorphous nanoclusters) have been investigated. The figure demonstrates that for each device the EL intensity strongly increases by increasing the applied voltage; it is also possible to note that each device exhibits a different voltage threshold for light emission, and that this threshold strongly increases by decreasing the Si concentration in the SiO_x film. Moreover, it is noteworthy that amorphous Si nanoclusters exhibit an EL intensity very similar to that one observed in a device with Si-nc by applying much lower voltages, although the two systems have been obtained from SiO_x layers with the same Si concentration (46 at.%).

The EL properties of SiO_x films have been investigated also by studying the emission at 850 nm as a function of the current density passing through the device under forward bias conditions (see Fig. 10b). It is instructive to compare this figure with Fig. 10a; while the various devices exhibit different operating voltages, by observing the behaviour of the EL intensity as a function of the current density it is possible to conclude that all the devices based on Si-nc show a common trend, i.e. an almost linear increase of the EL intensity with current density. On the other hand, the devices based on amorphous Si-nanoclusters are characterized by a very different behavior, since for the same current density they exhibit an EL intensity about one order of magnitude lower than those based on Si-nc.

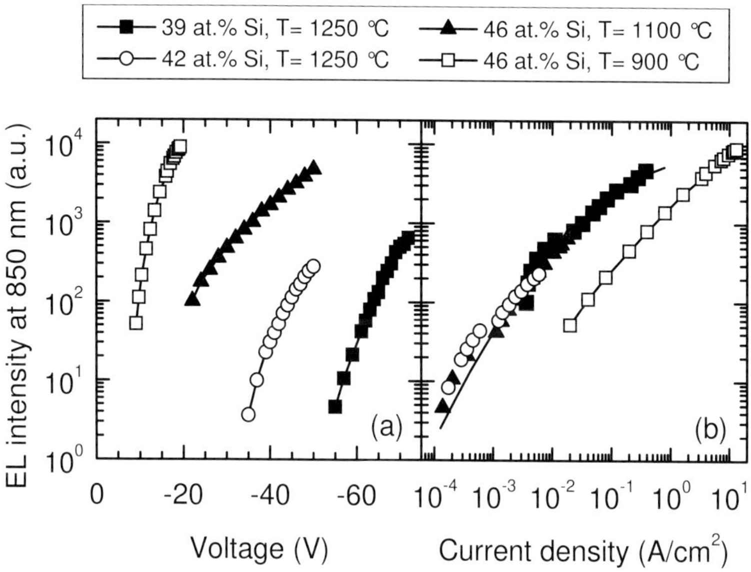

Fig. 10. EL intensity at 850 nm as a function of (a) the applied voltage under forward bias conditions and (b) of current density passing through the device under forward bias conditions for devices based on Si-nc (formed by using different combinations of annealing temperature and Si excess), and a device based on amorphous Si nanoclusters (obtained by annealing at 900°C a SiO_x film containing 46 at.% of Si).

The analysis of Fig. 10 allows us to draw some conclusions about the properties of electroluminescent devices based on amorphous and crystalline Si nanoclusters:

(i) the operating voltage strongly depends on the characteristics of the active layer; as a general trend, it decreases by increasing the Si concentration and by decreasing the annealing temperature of the SiO_x layer. These effects are strongly related with the structural properties of the active layer. In the case of nanocrystalline layers, the best performances have been obtained by annealing at 1100°C a SiO_x film containing 46 at.% of Si. Under these conditions, Si-nc having the proper size (from 1 to 2 nm) and density (about 1×10^{18} cm^{-3}) for light emission are formed. Note that in devices with a reduced thickness (25 nm) of the active layer operating voltages as low as −4 V have been obtained.[24]

(ii) The current density flowing through the devices is the key parameter determining the EL intensity of devices based on Si-nc. This demonstrates that, for the range of Si concentrations and annealing temperatures we have explored, the excitation and de-excitation mechanisms are essentially the same. On the other hand, devices based on amorphous nanoclusters do not exhibit the same brightness as a function of the current density and seem to be less efficient.

(iii) The main reason why EL intensity in amorphous nanoclusters is weaker with respect to Si-nc for the same current density is the faster lifetime. However, this explanation is not sufficient. In fact, the lifetime of amorphous nanoclusters is two orders of magnitude smaller than that one of nc, while at a constant current density the EL of amorphous nanograins is only one order of magnitude smaller. This suggests that indeed part of the decrease in lifetime for the amorphous clusters should be due to a much shorter radiative lifetime. Since amorphous nanoclusters are in average smaller than nc this assumption is very likely.

(iv) In spite of the above considerations, the efficiencies of devices based on amorphous and crystalline nanoclusters are similar; the high current density needed to observe intense EL signals is balanced by the low operating voltage and these unusual working conditions have no detrimental effect on the device reliability.[57,58] This means that amorphous nanoclusters may have strong potentialities in silicon photonics, especially if this material can be further optimized by reducing the influence of non radiative processes.

4.1. *Enhancement of the efficiency of light emitting devices based on Si nanoclusters by coupling with photonic crystals*

The external efficiency of devices based on Si nanostructures can be considerably enhanced by the integration with a two dimensional photonic crystal, thus paving the way to a whole new class of practical Si light emitters. A proper design of the photonic crystal is able to extract a large amount of light, which would be otherwise wasted, thus enhancing the overall efficiency by over a factor of four. In fact, in a typical Si LED, as a result of the high refractive index of Si (n = 3.5), most of the emitted light remains trapped inside the device by total internal reflection, thus severely limiting the overall performance. The realization of a two-dimensional photonic crystal on the device surface was proposed to overcome this limitation, and proved its effectiveness within an optically pumped III-V quantum well embedded in a multilayer microcavity.[59]

The basic structure of the device is again the simple MOS-like one previously described. The active layer is a 70-nm thick SiO_x film with a Si content of 44 at.%, prepared by PECVD and annealed in N_2 at 900°C for 1 h to induced a phase separation and the formation of amorphous Si nanoclusters. We underline again that the use of amorphous nanoclusters as active centers allowed to strongly reduce the overall thermal budget for device fabrication (as above discussed, temperatures as high as 1100°C are needed for Si-nc synthesis) without significantly affecting the device performances. The main peculiarity of this device concerns the 300 nm n-type polysilicon layer, which acts as the top electrode. With the aim of enhancing vertical light extraction, this polysilicon cap has been patterned by using conventional lithography to produce a triangular two-dimensional photonic crystal.[60] The photonic crystal is composed of holes having a diameter of 400 nm and has a lattice constant of 900 nm. The depth of the holes is only 200 nm, hence leaving an uninterrupted 100-nm thick polysilicon layer for the electrical contact.

The SEM image reported in Fig. 11a shows the top view of the photonic crystal, demonstrating the good uniformity of the hole shape and spacing. A detail of a single hole is shown in the TEM cross section reported in Fig. 11b. The lattice constant of the photonic crystal has been tuned in order to introduce a fourth-order diffraction escape path at 860 nm (i.e. within the emission band of Si nanoclusters) for those light modes which are index-guided in the top Si layer. We have chosen on purpose to use conventional lithography at the expenses of exploiting high-order diffraction, in order to prove the full compatibility of the processing with standard Si VLSI technology. For comparison, reference devices were fabricated by designing a random hole pattern on top of the polysilicon cap (random devices) or by leaving the polysilicon layer totally unetched (unpatterned devices).

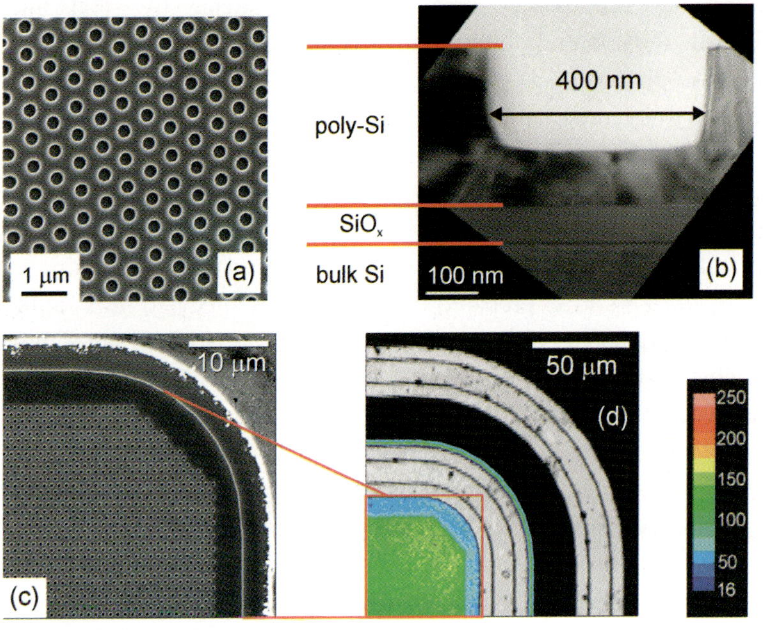

Fig. 11. (a) SEM image of the photonic crystal fabricated on top of the device. (b) Cross sectional TEM view of a single hole of the photonic crystal. (c) SEM image of the upper right corner of the device. (d) EMMI image of the same corner of the biased device. The intensity scale is in false colors.

A very interesting feature of photonic crystal devices can be noticed by observing one of the four corners, both by SEM (Fig. 11c) and by EMMI (Fig. 11d). The EMMI image was taken at room temperature by biasing the device at -12 V, causing a current flow of 0.55 A/cm^2. The EL intensity is represented in false colours. It is evident the presence of a small edge region (coloured in blue in Fig. 11d), where the EL intensity drops by approximately a factor of four. It is quite impressive the strong spatial correlation found between the step-like intensity variation and the end of the photonic crystal pattern, as identified in the SEM image of the same corner. This is a clear evidence of the effectiveness of the photonic crystal structure in enhancing light extraction from this device.

In order to better explore this phenomenon we have performed EL measurements on photonic crystal, random and unpatterned devices. The EL data reported in Fig. 12a were taken under a –20 V forward bias and at a current density of 22 A/cm^2. The unpatterned device shows an EL emission characterised by two peaks at around 750 nm and 1000 nm. As previously discussed, the EL from amorphous Si nanoclusters is similar to the nc counterpart. The peculiar shape of the spectrum shown in Fig. 12a cannot be considered an intrinsic characteristic of the amorphous clusters but the result of interferences occurring within the polysilicon layer. Indeed, though the top layer is almost transparent to the emitted radiation, some of the emitted light will be index-guided in it and is not extracted. Indeed, the same peaks were also observed in PL experiments, but disappear as soon as the top polysilicon layer is removed, leaving a single band centered at about 800 nm. The EL spectrum of the random device is very similar to the unpatterned one, indicating that a random hole distribution has almost no effect on the extraction of trapped light. In contrast, the EL of the photonic crystal device is strongly enhanced with respect to the previous two. In particular, at about 860 nm, in resonance with the designed diffraction spot of the two dimensional photonic pattern, a sharp emission peak is observed, demonstrating an enhanced coupling to vertical radiation of the guided modes. The enhancement achieved accounts for more than a factor of four.

The coherent scattering of guided modes, with a well defined in-plane momentum, should produce a vertical extraction. We hence characterised the EL intensity of the photonic crystal device as a function of observation angle. During all measurements, a small aperture was placed in front of the device, selecting an emission cone of 4° of semi-aperture. Wavelengths far from the photonic crystal diffraction peak smoothly decrease in intensity with increasing angle. In contrast, the peak at 860 nm suddenly disappears after only 10° of inclination, demonstrating its coherent scattering character related to the photonic-crystal diffraction mechanism. This feature is particularly evident in Fig. 12b, which is the experimental polar distribution of the emitted radiation at 860 nm taken by tilting the device along the ΓM direction. It is interesting to note that the radiation pattern of the photonic crystal device is extremely directional, once more demonstrating the presence of diffraction effects.

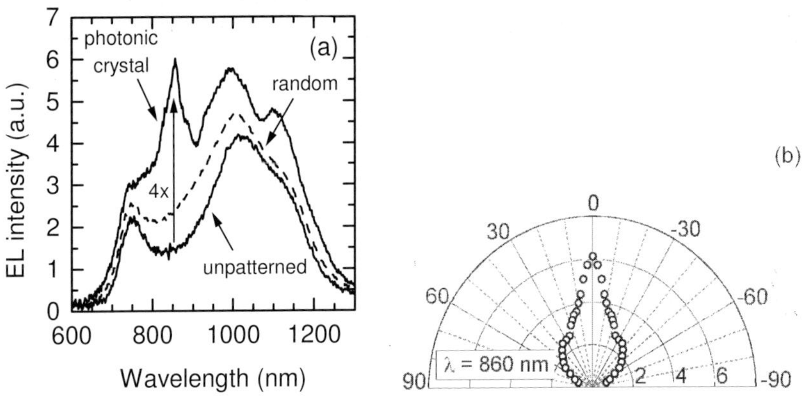

Fig. 12. (a) EL spectra for the unpatterned, random, and photonic crystal devices. An EL intensity enhancement accounting for a factor of 4 at 860 nm is demonstrated in the photonic crystal device. (b) Polar distribution of the emitted radiation at a wavelength of 860 nm. Tilt is along ΓM direction.

5. Conclusions

In this chapter we have presented an overview of the properties and applications of Si-nc. We have shown that the detailed analysis of the thermally-induced evolution of the structure of the SiO_x films used as precursors for Si-nc formation constitutes an invaluable tool for the comprehension and the optimization of the optical properties of this material. This knowledge represents the key factor for the fabrication of electroluminescent devices based on Si-nc or amorphous Si nanoclusters characterized by working conditions close to those required for practical applications in terms of operating voltage, stability and reliability. We have also proposed some examples of the integration of Si-nc with optical microcavities and photonic crystals; these applications have allowed significant improvements in the optical efficiency and in the directionality and a linewidth narrowing of the luminescence signals. It is important to remark that Si-nc formation and their processing, and the fabrication of microcavities, photonic crystals and devices have been realized by using methodologies fully compatible with current Si technology. The above findings constitute significant advancements towards the development of silicon nanophotonics and, in particular, towards the fabrication of new generations of silicon devices in which photonic and electronic functions are integrated together by fully exploiting the unusual properties of Si-nc.

Acknowledgements

The contributions given by C. Spinella, C. Bongiorno, D. Pacifici, C. D. Presti, I. Crupi, V. Vinciguerra, E. C. Moreira, D. Sanfilippo, G. Di Stefano, A. Piana and P. G. Fallica to some of the experiments reported in this chapter are gratefully acknowledged.

References

1. L. T. Canham, *Appl. Phys. Lett.* **57**, 1046 (1990).
2. Y. Kanemitsu, T. Ogawa, K. Shiraishi and K. Takeda, *Phys. Rev.* **B48**, 4883 (1993).
3. T. Shimizu-Iwayama, K. Fujita, S. Nakao, K. Saitoh, T. Fujita and N. Itoh, *J. Appl. Phys.* **75**, 7779 (1994).
4. J. G. Zhu, C. W. White, J. D. Budai, S. P. Withrow and Y. Chen, *J. Appl. Phys.* **78**, 4386 (1995).
5. K. S. Min, K. V. Shcheglov, C. M. Yang, H. A. Atwater, M. L. Brongersma and A. Polman, *Appl. Phys. Lett.* **69**, 2033 (1996).
6. T. Inokuma, Y. Wakayama, T. Muramoto, R. Aoki, Y. Kurata and S. Hasegawa, *J. Appl. Phys.* **83**, 2228 (1998).

7. M. L. Brongersma, A. Polman, K. S. Min, E. Boer, T. Tambo and H. A. Atwater, *Appl. Phys. Lett.* **72**, 2577 (1998).
8. T. Shimizu-Iwayama, N. Kurumado, D. E. Hole and P. D. Townsend, *J. Appl. Phys.* **83**, 6018 (1998).
9. J. Linnros, N. Lalic, A. Galeckas and V. Grivickas, *J. Appl. Phys.* **86**, 6128 (1999).
10. F. Iacona, G. Franzò and C. Spinella, *J. Appl. Phys.* **87**, 1295 (2000).
11. J. Valenta, R. Juhasz and J. Linnros, *Appl. Phys. Lett.* **80**, 1070 (2002).
12. L. Pavesi, L. Dal Negro, C. Mazzoleni, G. Franzò and F. Priolo, *Nature* **408**, 440 (2000).
13. L. Khriachtchev, M. Rasanen, S. Novikov and J. Sinkkonen, *Appl. Phys. Lett.* **79**, 1249 (2001).
14. K. Luterova, I. Pelant, I. Mikulskas, R. Tomasiunas, D. Muller, J.-J. Grob, J.-L. Rehspringer and B. Honerlage, *J. Appl. Phys.* **91**, 2896 (2002).
15. L. Dal Negro, M. Cazzanelli, L. Pavesi, S. Ossicini, D. Pacifici, G. Franzò, F. Priolo and F. Iacona, *Appl. Phys. Lett.* **82**, 4636 (2003).
16. M. V. Wolkin, J. Jorne, P. M. Fauchet, G. Allan and C. Delerue, *Phys. Rev. Lett.* **82**, 197 (1999).
17. D. J. DiMaria, J. R. Kirtley, E. J. Pakulis, D. W. Dong, T. S. Kuan, F. L. Pesavento, T. N. Theis, J. A. Cutro and S. D. Brorson, *J. Appl. Phys.* **56**, 401 (1984).
18. K. D. Hirschman, L. Tsybeskov, S. P. Duttagupta and P. M. Fauchet, *Nature* **384**, 338 (1996).
19. G. G. Qin, A. P. Li, B. R. Zhang and B.-C. Li, *J. Appl. Phys.* **78**, 2006 (1995).
20. S. Fujita and N. Sugiyama, *Appl. Phys. Lett.* **74**, 308 (1999).
21. P. Photopoulos and A. G. Nassiopoulou, *Appl. Phys. Lett.* **77**, 1816 (2000).
22. N. Lalic and J. Linnros, *J. Lumin.* **80**, 263 (1999).
23. G. Franzò, A. Irrera, E. C. Moreira, M. Miritello, F. Iacona, D. Sanfilippo, G. Di Stefano, P. G. Fallica and F. Priolo, *Appl. Phys. A: Mater. Sci. Process.* **74**, 1 (2002).
24. A. Irrera, D. Pacifici, M. Miritello, G. Franzò, F. Priolo, F. Iacona, D. Sanfilippo, G. Di Stefano and P. G. Fallica, *Appl. Phys. Lett.* **81**, 1866 (2002).
25. R. J. Walters, G. I. Bourianoff and H. A. Atwater, *Nature Mater.* **4**, 143 (2005).
26. N.-M. Park, C.-J. Choi, T.-Y. Seong and S.-J. Park, *Phys. Rev. Lett.* **86**, 1355 (2001).
27. L. Dal Negro, J. H. Yi, L. C. Kimerling, S. Hamel, A. Williamson and G. Galli, *Appl. Phys. Lett.* **88**, 183103 (2006).
28. N.-M. Park, T.-S. Kim and S.-J. Park, *Appl. Phys. Lett.* **78**, 2575 (2001).
29. K. S. Cho, N.-M. Park, T.-Y. Kim, K.-H. Kim, G. Y. Sung and J. H. Shin, *Appl. Phys. Lett.* **86**, 071909 (2005).
30. L.-Y. Chen, W.-H. Chen and F. C.-N. Hong, *Appl. Phys. Lett.* **86**, 193506 (2005).
31. A. J. Kenyon, P. F. Trwoga, M. Federighi and C. W. Pitt, *J. Phys.: Condens. Matter* **6**, L319 (1994).

32. M. Fujii, M. Yoshida, Y. Kanzawa, S. Hayashi and K. Yamamoto, *Appl. Phys. Lett.* **71**, 1198 (1997).
33. G. Franzò, V. Vinciguerra and F. Priolo, *Appl. Phys. A: Mater. Sci. Process.* **69**, 3 (1999).
34. P. G. Kik, M. L. Brongersma and A. Polman, *Appl. Phys. Lett.* **76**, 2325 (2000).
35. F. Iacona, D. Pacifici, A. Irrera, M. Miritello, G. Franzò, F. Priolo, D. Sanfilippo, G. Di Stefano and P. G. Fallica, *Appl. Phys. Lett.* **81**, 3242 (2002).
36. A. Nazarov, J. M. Sun, W. Skorupa, R. A. Yankov, I. N. Osiyuk, I. P. Tjagulskii, V. S. Lysenko and T. Gebel, *Appl. Phys. Lett.* **86**, 151914 (2005).
37. L. C. Kimerling, *Appl. Surf. Sci.* **159-160**, 8 (2000).
38. A. Polman, *Nature Mater.* **1**, 10 (2002).
39. M. Paniccia and S. Koehl, *IEEE Spectr.* **42**, 38 (2005).
40. B. Jalali, M. Paniccia and G. Reed, *IEEE Microwave Mag.* **7**, 58 (2006).
41. M. Zacharias, J. Blasing, P. Veit, L. Tsybeskov, K. Hirschman and P. M. Fauchet, *Appl. Phys. Lett.* **74**, 2614 (1999).
42. F. Gourbilleau, X. Portier, C. Ternon, P. Voivenel, R. Madelon and R. Rizk, *Appl. Phys. Lett.* **78**, 3058 (2001).
43. M. Zacharias, J. Heitmann, R. Scholz, U. Kahler, M. Schmidt and J. Blasing, *Appl. Phys. Lett.* **80**, 661 (2000).
44. F. Iacona, C. Bongiorno, C. Spinella, S. Boninelli and F. Priolo, *J. Appl. Phys.* **95**, 3723 (2004).
45. S. Boninelli, F. Iacona, G. Franzò, C. Bongiorno, C. Spinella and F. Priolo, *J. Phys.: Condens. Matter* **19**, 225003 (2007).
46. H. R. Philipp, *J. Non-Cryst. Solids* **8–10**, 627 (1972).
47. F. Iacona, S. Lombardo and S. U. Campisano, *J. Vac. Sci. Technol.* **B14**, 2693 (1996).
48. C. Bonafos, B. Garrido, M. Lopez, A. Perez-Rodriguez, J. R. Morante, Y. Kihn, G. Ben Assayag and A. Claverie, *Appl. Phys. Lett.* **76**, 3962 (2000).
49. C. Spinella, C. Bongiorno, G. Nicotra, E. Rimini, A. Muscarà and S. Coffa, *Appl. Phys. Lett.* **87**, 044102 (2005).
50. C. Delerue, G. Allan and M. Lannoo, *Phys. Rev.* **B48**, 11024 (1993).
51. L. N. Dinh, L. L. Chase, M. Balooch, W. J. Siekhaus and F. Wooten, *Phys. Rev.* **B54**, 5029 (1996).
52. F. Priolo, G. Franzò, D. Pacifici, V. Vinciguerra, F. Iacona and A. Irrera, *J. Appl. Phys.* **89**, 264 (2001).
53. L. Pavesi, *J. Appl. Phys.* **80**, 216 (1996).
54. L. Pavesi, C. Mazzoleni, A. Tredicucci and V. Pellegrini, *Appl. Phys. Lett.* **67**, 3280 (1995).
55. E. F. Schubert, A. M. Vredenberg, N. E. J. Hunt, Y. H. Wong, P. C. Becker, J. M. Poate, D. C. Jacobson, L. C. Feldman and G. J. Zydzik, *Appl. Phys. Lett.* **61**, 1381 (1992).
56. F. Iacona, G. Franzò, E. C. Moreira and F. Priolo, *J. Appl. Phys.* **89**, 8354 (2001).

57. A. Irrera, F. Iacona, I. Crupi, C. D. Presti, G. Franzò, C. Bongiorno, D. Sanfilippo, G. Di Stefano, A. Piana, P. G. Fallica, A. Canino and F. Priolo, *Nanotechnology* **17**, 1428 (2006).
58. F. Iacona, A. Irrera, G. Franzò, D. Pacifici, I. Crupi, M. Miritello, C. D. Presti and F. Priolo, *IEEE J. Sel. Top. Quantum Electron.* **12**, 1596 (2006).
59. A. A. Erchak, D. J. Ripin, S. Fan, P. Rakich, J. D. Joannopoulos, E. P. Ippen, G. S. Petrich and L. A. Kolodziejski, *Appl. Phys. Lett.* **78**, 563 (2001).
60. C. D. Presti, A. Irrera, G. Franzò, I. Crupi, F. Priolo, F. Iacona, G. Di Stefano, A. Piana, D. Sanfilippo and P. G. Fallica, *Appl. Phys. Lett.* **88**, 033501 (2006).

CHAPTER 7

OPTICAL SPECTROSCOPY OF INDIVIDUAL SILICON NANOCRYSTALS

Jan Valenta

Department of Chemical Physics & Optics, Faculty of Mathematics & Physics, Charles University, Ke Karlovu 3, CZ-121 16 Prague 2, Czech Republic

Jan Linnros

Department of Microelectronics & Applied Physics, Royal Institute of Technology, Electrum 229, S-164 21 Kista-Stockholm, Sweden

The available data on photoluminescence (PL) spectroscopy of single Si nanocrystals (Si-nc) is reviewed for two types of samples: (i) Regular matrices of Si pillars produced by electron-beam lithography, reactive ion etching and oxidation, (ii) grains of porous Si deposited onto a substrate from a diluted colloidal suspension. A wide-field imaging micro-spectroscope with detection by a CCD camera is used preferably to detect spectra, while a confocal microscope with single-photon-counting detection is applied for detection of PL fluctuations, so called ON-OFF blinking. Cryogenic-temperature PL spectroscopy of Si-nc reveals atomic-like narrow lines that document an unexpectedly large contribution of zero-phonon transitions and also some low-frequency phonon-replicas. The blinking photon-statistics indicates that the transition between bright and dark states of a Si-nc has character of a diffusion-controlled electron-transfer reaction where quenching occurs by Auger recombination. Finally we show that all published results indicate a common PL mechanism of Si-nc, largely insensitive to fabrication methods.

1. Introduction

Single nanocrystal spectroscopy (SNS), based on techniques developed for single molecule studies in the 1990s, proved to be very efficient in revealing fundamental properties of semiconductor quantum dots (QD) of direct band gap semiconductors.[1,2] SNS enables not only to overcome the inhomogeneous broadening of ensemble spectra but it also leads to the discovery of new phenomena like emission intermittency (*ON-OFF* blinking), spectral diffusion and polarization, Stark effect, *etc.*[3,4] The application of SNS to single Si nanocrystals (Si-ncs) is, however, still not fully explored because of two main problems:

- *Very low emission rate*, which is a consequence of the indirect band-gap structure (conserved even in small Si-nc[5]). The radiative lifetime is very long[a] (typically > 0.1 ms at room temperature (RT) depending on wavelength).[5,6]
- *Difficult fabrication of structures* in which Si-ncs are at same time *well defined, efficiently emitting, and sufficiently diluted* in order to enable detection of photoluminescence (PL) from a single Si-nc.

In the literature there are reports on SNS of Si-ncs from only five research groups to the authors knowledge: The groups of Buratto,[7,8,9] Cichos,[10,11,12] and Valenta[13] investigated grains of porous silicon (PSi) deposited on substrates from diluted colloidal suspensions, the Korgel group[14] reported on properties of single Si-ncs prepared by arrested precipitation in a liquid, and the Linnros' group[15,16,17,18,19] investigated Si-ncs formed on top of oxidized nanopillars made by electron-beam lithography and reactive ion etching. Valenta *et al.* reported also on detection of electroluminescence from single Si-nc in a p-i-n light-emitting diode structure containing a thin active layer of SiO_2 with Si-nc[20,21] but we do not include that work in this review.

In this work we give a review of our SNS experiments (with references to other published works) on Si-nc samples prepared either by lithography or from PSi suspensions. The results indicate that there is a common PL mechanism in Si-ncs independent on the nanostucture fabrication technique.

---

[a] This is valid for the so called S-band which is usually the main band observed in PL spectra of Si nanostructures. It is located in the yellow-orange-red part of the optical spectrum and its label comes from the abbreviation of "slow" referring to its slow PL decay. The S-band is by far the most studied PL band of Si-ncs and also in this review we discuss only the behaviour of the S-band emission.

2. Sample Preparation Techniques

Most of the SNS experiments are based on well-developed and widely available techniques of "classical" microscopy. The spatial resolution d of these instruments is diffraction limited[b], i.e. the resolution or size of the diffraction-limited spot is given by $d = 1.22\lambda/2NA$, where λ is the wavelength of light and NA the numerical aperture of an optical system (objective lens). It means that an optically addressed spot has a diameter of about half of the wavelength (or larger) – several hundreds of nanometers, while a studied single nanocrystal has a size of only a few nanometers. In spite of such huge discrepancy of scales, it may still be possible to detect the PL signal from a single Si-nc providing the following conditions of a prepared sample are fulfilled (see Fig. 1):

- The spatial separation of Si-ncs contributing to the detected signal must be larger than the resolution limit of the imaging system.
- PL signals or scattering from the surrounding matrix and substrate must be minimized as well as all other sources of background signal (luminescence of filters etc.).

Below we describe two successful approaches to fabricate diluted and clean samples of Si-nc (with parameters used by our groups).

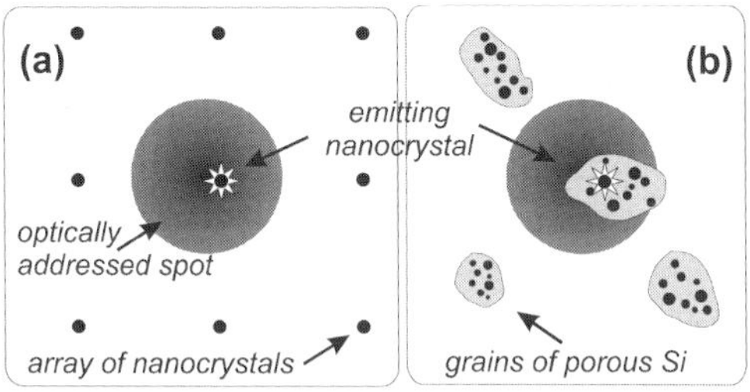

Fig. 1. Two possible scenarios for detection of a single Si-nc in optical far-field: (a) Nanocrystals are organized in a regular array by lithography or (b) PSi grains (Si-nc clusters) are dispersed on a substrate and only a fraction of the present Si-ncs gives a detectable signal. (Objects proportions are not to scale).

---

[b] Special techniques with subdiffraction resolution, e.g. scanning near-field optical microscopy, may be also applied to SNS experiments. However, they have not been successfully applied to Si-nc, to our knowledge. Therefore we are not going to discuss them in this review.

2.1. *Arrays of Si-ncs made by electron-beam lithography*

Electron beam lithography was used to form resist dots with diameters of about 100 nm on an N-type (<100>, 20-40 Ω cm) Si wafer having a 25 nm thermal oxide layer. Reactive ion etching (RIE) using CHF_3/O_2-based chemistry was then performed to etch through the top SiO_2 layer followed by chlorine based RIE for Si etching. The resulting 200 nm tall

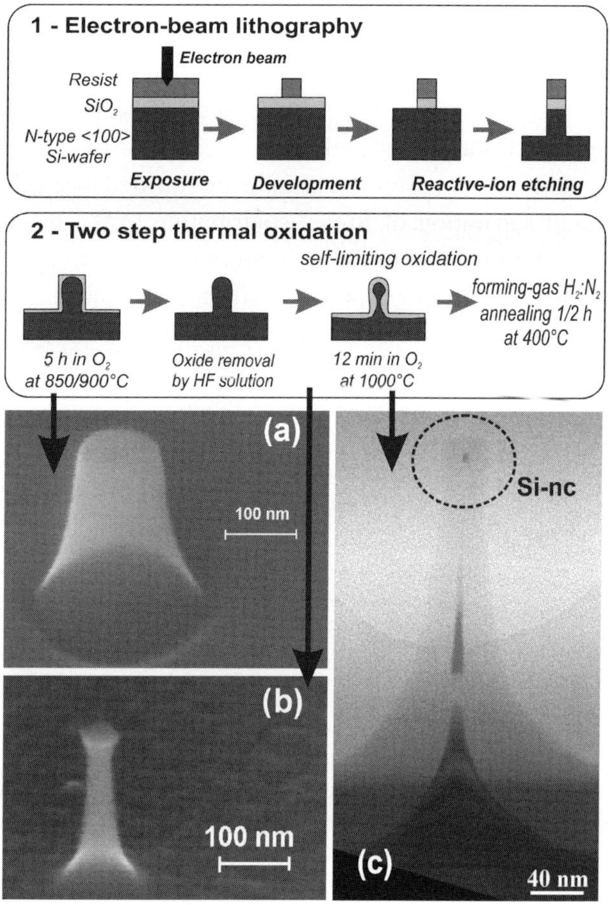

Fig. 2. Fabrication procedure of Si nanopillars by electron beam lithography, reactive ion etching and dry oxidation.[21,22] The lower panels represent SEM images (45° tilt view) after initial etching (a) and after the first oxidation (b), and the high-resolution TEM view through one finished nanopillar where a remaining nanocrystal with d ~4 nm is seen at the top of the pillar (c).

pillars were subsequently thermally oxidized for 5 h in O_2 gas at 850ºC or 900ºC. The different temperatures give slightly different consumption of Si, which combined with the range of different initial pillar diameters resulted in Si cores of different sizes ranging down to the few-ten nanometer regime. The oxide was then removed by buffered wet etching. A second oxidation followed at 1000ºC for 12 minutes. Finally, the samples were annealed for 30 min at 400ºC in a 1:9 mixture of H_2:N_2 gas to passivate surface states in order to enhance the PL. The preparation technology is schematically illustrated in Fig. 2, where we present also two SEM images of the structures after initial patterning and after the first oxidation. High-resolution TEM images (Fig. 2c) were obtained by separating a row of nanopillars with focused-ion-beam processing and subsequent manipulation by micro-tweezers.[22] The crucial point for achieving detectable PL is to find an optimal combination of the initial size of crystals (in our case it was 100 or 130 nm) and the oxidation parameters. Both wider and narrower pillars have no detectable PL as their Si core is either too large or it is completely consumed. Note that a phenomenon of self-limiting oxidation[23] plays a very important role in the formation of a Si-nc at the top of a pillar. The rate of oxidation is significantly retarded on a few nanometer scale when the surface has a large curvature and stress builds up at the oxide-silicon interface. This phenomenon occurs only at temperatures at or below 900ºC where oxide viscous flow does not occur. For pillar geometry, the largest curvature is at the top and an isolated Si-nc may then be created there if size, geometry and oxidation parameters are well tuned.

2.2. *Colloidal suspensions of porous silicon grains*

Porous silicon (PSi) was prepared by electrochemical etching of p-type silicon wafers (<100>, ~0.1 Ω cm) in a solution containing hydrofluoric acid HF (50%), pure ethanol (for UV) and hydrogen peroxide H_2O_2 (3%). Platinum was used as a second electrode and a continuous stirring of the bath was applied. The bath composition was HF:ethanol:H_2O_2 = 1:2.46:0.54 and etching time was 2 hours. The freshly prepared PSi layer on the Si wafer was immediately dipped into pure H_2O_2 (3%) for 5 minutes in order to perform a post etching procedure. The effect of H_2O_2

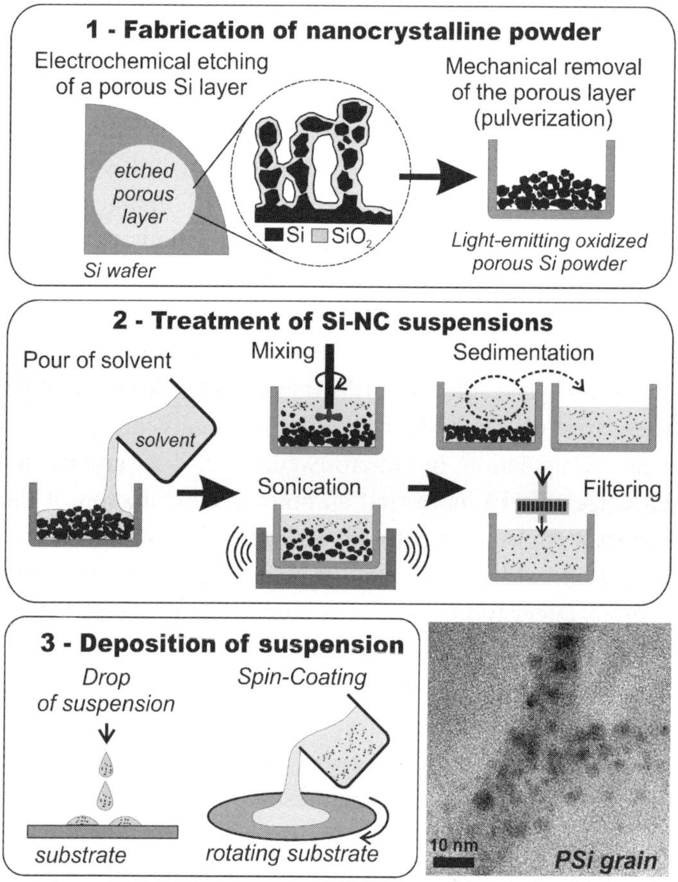

Fig. 3. Schematic representation of fabrication procedure of PSi colloidal suspensions and their deposition on substrates.[13,24] The bottom-right panel shows the HR-TEM image of a single PSi grain containing many Si-ncs in a SiO_2 matrix.

consisted in additional oxidative activity on the PSi surface making the size of crystallite cores to shrink. Small mean sizes of Si-ncs and, consequently, dominant short-wavelength PL band around 600 nm was obtained in this way. Also, a relatively low etching current density (2.3 mA/cm^2) contributed to this effect. PSi powder was then obtained by mechanical pulverization of the PSi film from the Si substrate. Colloidal suspensions were prepared by pouring ethanol or iso-propanol onto the PSi powder and by mixing in an ultrasonic bath.

The original powder contained many large PSi grains of several μm or even tens of μm and sonication is inefficient to break the largest grains to sizes below a μm.[24] Further size selection was therefore necessary. We applied filtering of the supernatant part of the sedimented colloidal suspension using membranes with pores of 100 nm (Millex Millipore). This procedure gives low-concentrated optically clear suspensions that may be further diluted and deposited (by means of spin-coating or simple dropping) on cleaned substrates (Si wafers, glass or fused silica). The preparation procedure is sketched in Fig. 3. The PSi grain concentration in the suspension is not known (sedimentation and filtering removes most of the original PSi powder) so the proper dissolution must be found empirically to optimize the density of emitting objects observed in a microscope. The PSi particles contained in the suspension were characterized with high-resolution TEM. A drop of the suspension was deposited on a grid with carbon membrane and imaged in the JEOL JEM-3010 HR-TEM microscope. These observations show that PSi grains consist of many Si-ncs (sometimes almost interconnected and kept together most probably by amorphous SiO_2) with diameters ranging from about 2 to 5 nm (see the bottom-right panel of Fig. 3). Single isolated Si-ncs are not found.

In the following text we use abbreviations for the two types of Si-nc: NPSi = nano-pillar Si, PSiG = porous Si grains.

3. Experimental Set-Ups for Single Nanocrystal Spectrocopy

PL images and spectra of Si-nc samples were studied using imaging micro-spectroscopy while blinking statistics was measured using a confocal microscope with single-photon-counting (SPC) detection.

3.1. *Imaging micro-spectroscopy*

The set-up is based on an imaging spectrograph (a spectrograph with corrected optical aberrations for good 2D imaging in the output focal plane) connected to an optical microscope (an inverted or up-right construction - Fig. 4). Light from the sample was collected by an objective, imaged onto the entrance slit of a spectrometer and detected by a LN-cooled CCD camera. PL was excited with the blue (442 nm) or

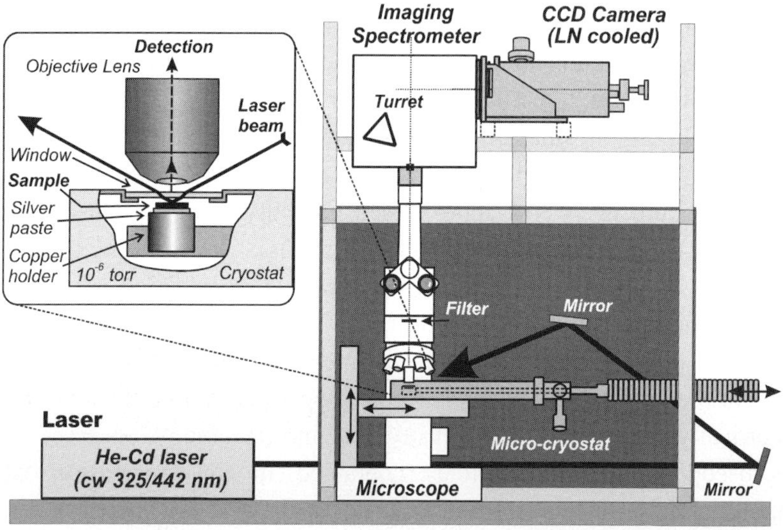

Fig. 4. The imaging spectroscopy set-up constructed at KTH in Stockholm.[21,25] For detailed description see text. The inset in the upper-left corner shows the placement of a sample in the cryostat.

UV-line (325 nm) of a cw He-Cd laser. The laser beam was directed towards the sample through the gap between the objective and the sample surface at grazing incidence. PSi colloids were deposited on a total-internal-reflection quartz prism (not shown in Fig. 4) and excited by an evanescent-field in order to substantially reduce the background.[13] For low-temperature measurements NPSi samples were placed on a cold-finger of a cryostat and imaged through its window by an objective equipped with a variable-thickness window correction.[25]

The imaging-spectroscopy experimental procedure was as follows (Fig. 5): For each sample, first, the images of reflection and PL were obtained using a mirror inside the spectrometer (entrance slit opened to a maximum). Then, an area of interest was placed in the centre of the image, entrance slit closed to the desired width (resolution) and the mirror was switched to a diffraction grating (mirror and two gratings are mounted on the same turret) in order to record a spectrum. PL spectra may be extracted as an intensity profile of the respective part of the spectral image. All spectra were corrected for spectral sensitivity of the detection system. The acquisition time of PL spectra is typically 30 min.

Spectra of several objects may be obtained simultaneously if their images lay in the region restricted by the entrance slit. The imaging spectroscopy system with CCD detection is relatively slow and cannot be used for detection of fast emission changes, for this purpose the following set-up is better suitable.

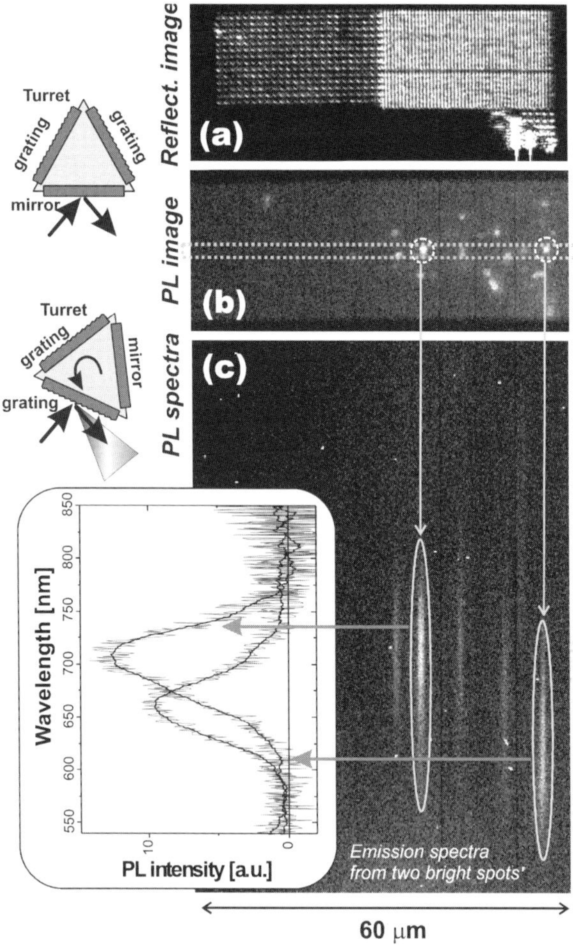

Fig. 5. The procedure of imaging spectroscopy measurement is illustrated on a sample containing arrays of Si nanopillars with spacing of 0.5 and 1 μm.[16] The reflection (a) and PL (b) images are taken using a mirror and an open input slit, while the spectra (c) are taken with a narrow input slit and dispersion of light on a grating. The spectrum is extracted as an intensity cross-section of the spectral image (c).

3.2. *Laser scanning confocal microscopy*

Fluctuations of the PL emission was detected with the SPC detection system connected to an inverted confocal microscope (in the epi-fluorescence configuration, i.e. with excitation and detection through the same objective), Fig. 6. The sample deposited on a cleaned quartz cover slide was excited with a blue diode-laser (444 nm) driven in the continuous-regime. The signal is filtered, focused on a confocal hole and detected by a pair of avalanche photodiode photon counting modules (APD-PC). The arrival time of every detection event is recorded and treated numerically after experiments (events are integrated within chosen time-intervals (bins), and analyzed statistically).

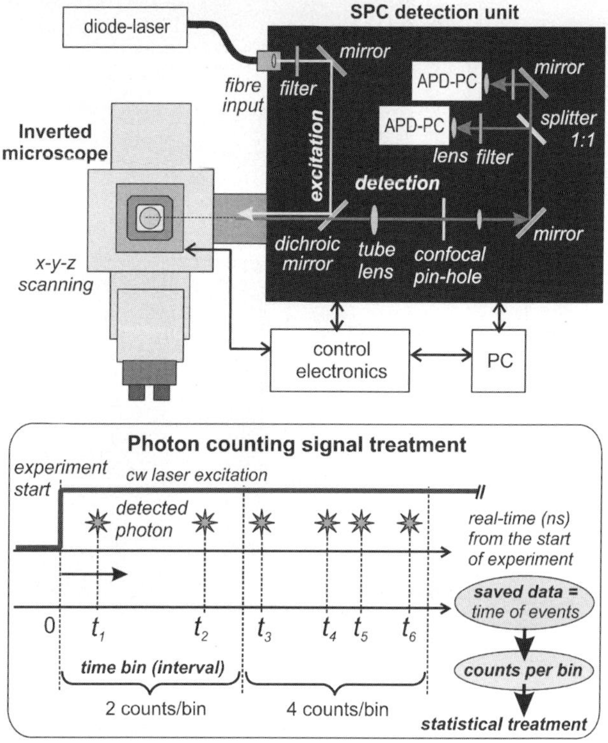

Fig. 6. The laser scanning confocal microscopy set-up with SPC detection unit. The bottom panel illustrates registration of arrival time of all detection events and post-experiment integration of signal counts within selected time-bins.

The pair of APD modules can also be used in a start-stop wiring to measure the distribution of intervals between detection events and obtain the second order autocorrelation function $g^{(2)}(\tau)$. But the technique of recording arrival times of detection events is more flexible, because the maximum information is recorded and various statistical treatments may be applied post-experiment, including calculation of $g^{(2)}(\tau)$.[26]

4. Experimental Results

4.1. *Photoluminescence spectra of individual Si-nc at RT*

PL spectra of individual Si-ncs can be measured at RT for intensively luminescing nanocrystals. In Fig. 7a, we plot PL spectra of three NPSi dots. The detection time was 30 min at the excitation intensity of 0.5 W/cm^2 (the spectral resolution is ~10 nm). The PL spectrum of a single

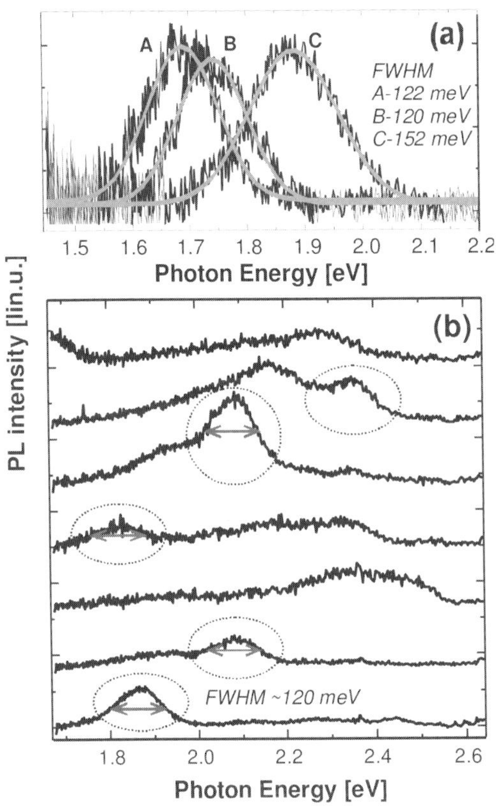

Fig. 7. (a) PL spectra of three different NPSi single Si-nc.[15,13] The spectral bands are fitted with a single Gaussan whose FWHM is indicated. (b) PL spectra of single PSi grains. Typical peak widths of 120 meV are indicated by dashed circles and arrows. Detection time of all spectra is 30 min.

Si-nc is characterized by a single band which peak position varies from dot-to-dot, most likely as a result of a variation in the amount of quantum confinement due to the size dispersion. The PL band can be fitted by a single Gaussian peak lying in the range 1.58 - 1.88 eV (plotted as bold gray lines in Fig. 7a. Full-width at half-maximum (FWHM) is 122, 120, and 152 meV for spectra of dots A, B, and C, respectively, significantly narrower than the usual ensemble PL spectrum of Si-ncs.

The PL spectra of single emitting spots in PSiG samples (Fig. 7b) are more complicated. There may be several bands of different width, but the most abundant one is again a Gaussian band with a FWHM around 120 meV. The spectral structure of an individual object is, however, not constant. In Fig. 8 we plot repeated measurements of PL spectra from three objects (detection time is 30 min for each spectrum). One can see an apparent blinking as well as spectral diffusion – drift of the PL bands. The difference between PL spectra of single dots in NPSi and PSiG samples is likely due to a contribution of more than one Si-nc to the detected signal in the case of PSi grains. Here we do not observe an isolated Si-nc, but a cluster of Si-nc in which one or more Si-ncs gives detectable PL signal.

There are mainly two possible reasons for the wide PL spectrum: dominating phonon-assisted transitions and spectral diffusion during

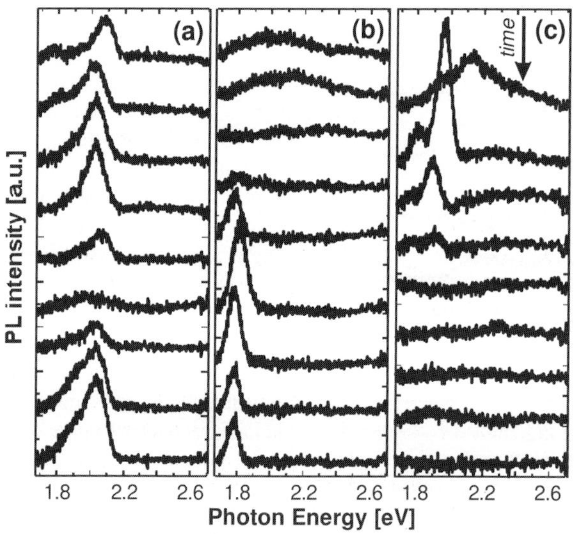

Fig. 8. Sequence of nine 30 min acquisitions of PL spectra from three different emitting spots.[13] The time sequence starts at the top of the panels. The three panels have not the same intensity scale.

long detection acquisitions. More information on spectra of single Si-ncs can be obtained only at low temperatures.

4.2. *Low-temperature PL of individual Si-nc*

PL spectra of the NPSi samples in a cryostat were detected down to 30 K. At temperatures below 30 K, however, we were unable to detect any consistent PL, most probably due to the important lifetime increase observed for Si-ncs at very low temperatures.[27] This has been explained in terms of a singlet-triplet splitting with a lower lying "dark" triplet state. As the emission rate is lowered by a factor ~50, it falls below the detection capability of the detection system. Upon decreasing temperature, the PL band continuously narrows down and a side-band shifted 60 meV from the main peak may be observed, but not for every Si-nc (Figs. 9c and 9d). At 35 K, the main peak is only about 2 meV wide, indeed less than k_BT confirming the atomic-like emission of a quantum dot. At this temperature also a 6 meV satellite is resolved (see Fig. 9).

The origin of the 60 meV side-band may be interpreted as the TO phonon replica, taking into account the TO phonon energy in bulk Si (56 meV at the X-point, 64 meV at the Γ-point). The TO-phonon-assisted

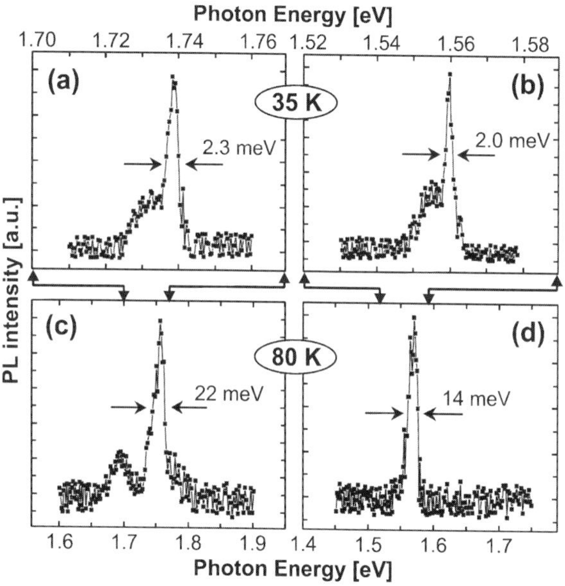

Fig. 9. Low-temperature PL spectra of two single Si-ncs (panels a-c and b-d are from the same Si-ncs) detected at 35 K (upper panels a,b) and 80 K (lower panels c,d).[17] Note the different range of spectra.

transitions have been found to be dominant in resonantly excited PL spectra of PSi by Kovalev *et al.*[28] (excitation energy at lower end of the luminescent band largely suppressing the inhomogenous broadening). It is illustrative to compare resonant PL spectra and spectra of single Si-ncs (see Fig. 10). Because the single Si-nc PL spectra are excited non-resonantly (exciting photons have energy 3.81 eV (325 nm), causing direct Γ-Γ transitions) we can observe only momentum-conserving phonons participating in the emission transitions, while in the resonant experiment there are two-phonon features (mainly 2TO, TO+TA) combining phonons participating in absorption and in emission processes. We note that the possible TA phonon replica (19 meV) is not clearly resolved in single Si-nc spectra.

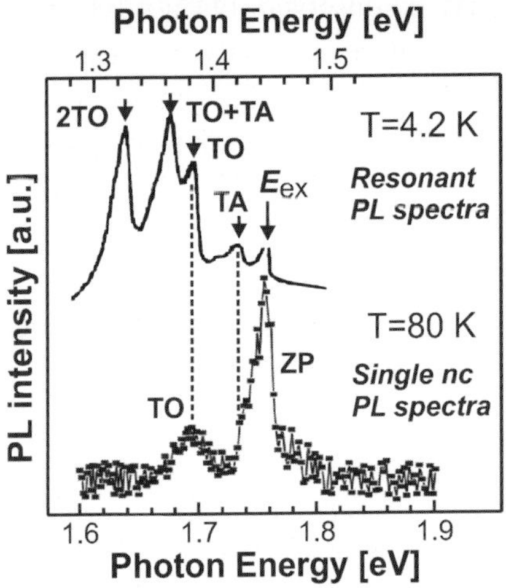

Fig. 10. Comparison of spectral structures in the resonantly excited PL of PSi (upper curve, T=4.2 K) and in PL spectra of a single Si-nc (Fig. 9c). Adapted from Refs. 17 and 28.

The main peak at higher photon energy of single Si-nc spectra (Fig. 10) is then interpreted as a zero-phonon (ZP) transition. ZP optical transition was reported earlier for Si-nc[28] and predicted by theory[29] as a breakdown of *k*-conservation at small dimensions. The ratio of ZP transitions to the phonon-assisted ones would then increase with emission energy (reduced size). For single Si-nc data we cannot find such a trend and, instead, we found a dot-to-dot variation of the phonon-assisted process, suggesting local differences which can only be averaged in ensemble measurements. It is important to note, though, that the data

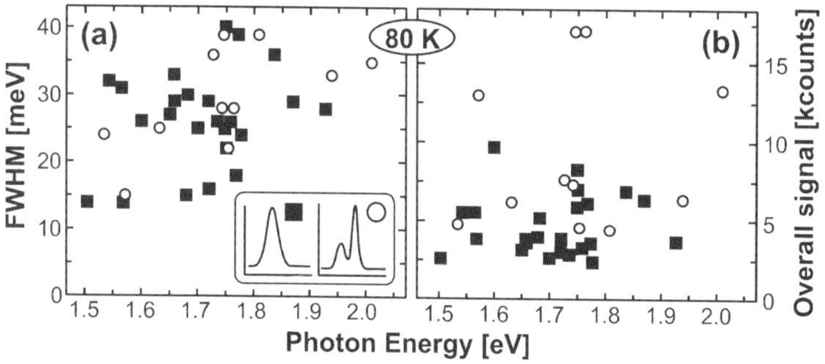

Fig. 11. Statistical summary of spectral characteristics of single Si-ncs emitting at different photon energy at 80 K.[25] (a) The width of the main peak, (b) the overall PL signal integrated for 30 min. The Si-ncs with single and double-peak spectra are indicated by black squares and white circles, respectively. No correlation between spectral characteristics is found.

of Kovalev *et al.*[28] were obtained at temperature of a few K, much lower than the single Si-nc data. In addition, they report on a large change of resonantly excited PL lines intensity ratio for oxidized samples suggesting a dependence on the Si-nc local environment. Theoretical calculations[30] made within the tight-bonding model showed that the existence of strained interface regions in the oxidized nanocrystals leads to the localization of carriers and an enormous increase of the ZP line transition probability.

We would like to point out that only a fraction (about 1/3) of the dots exhibited double peak ZP-TO spectral structure. Most of the PL spectra of single Si-nc are formed by a single peak (except the 6 meV side-band at very low-T). This indicates that only one recombination channel dominates (but it is not evident whether the peak is ZP or TO). We can only speculate about this finding and suggest that in some Si-ncs local differences in geometry and surface quality may enhance or decrease the probability of phonon-assisted emission.

In Fig. 11a we present a statistical breakdown of the linewidth for different single Si-ncs (at 80 K). Note that at this temperature all linewidths are larger than $k_B T$ (6.9 meV). In general, the homogeneous width of a quantum dot is given by the inverse of the dephasing time, which consists of the radiative lifetime and various scattering times (interaction of the exciton with phonons, interface states, defects, etc.).[31]

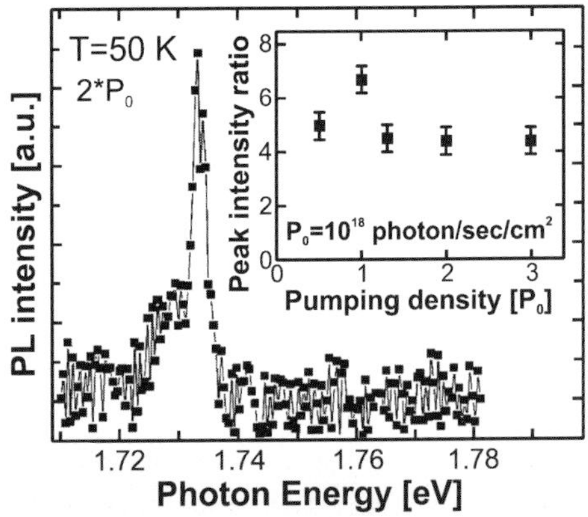

Fig. 12. PL spectrum of a Si-nc at 50 K under pumping of 2×10^{18} photon/sec/cm^2.[25] The inset illustrates the lack of intensity dependence of the intensity ratio of the main-peak and the side-band.

Thus, the linewidth is a unique parameter of an individual quantum dot and depends on its interface with surrounding matrix, dot geometry and purity. Indeed, what we find is a scatter of this parameter without any distinct dependence on the dot size. In Fig. 11b thc overall signal intensity is plotted versus the emission photon energy for a number of dots probed at 80 K. It is seen that within the same spectra integration time (30 min.) the total light output varies significantly from dot-to-dot. This is mainly due to the blinking phenomenon that will be discussed in the next paragraph. But again, the blinking characteristics vary extremely from dot-to-dot and no correlation with spectral characteristics is found.

For the interpretation of the 6-meV side-band resolved at 35 K there is an important observation that the intensity ratio of the side-band and the main peak is not dependent on excitation intensity (see Fig. 12). Therefore a non-linear origin, e.g. biexciton, is excluded (Note that the pumping intensity 1×10^{18} photons/sec/cm^2 – i.e. about 0.6 W/cm^2 for 3.81 eV photon energy - corresponds approximately to 1 exciton average occupancy in the nanocrystal using cross sections from Ref. 5 and a ~10^{-4} s exciton lifetime). We attribute the side-band to a low-frequency phonon-assisted transition. One may invoke torsional or spheroidal modes as calculated by Takagahara.[32] For a 4-nm diameter Si nanocrystals, he calculated a set of discrete values for the acoustic phonon energy spectrum starting from ~5 meV. By Raman spectroscopy[33] the presence of confined acoustic modes with energy of a few meV was, indeed, experimentally observed. The lack of nanocrystal size dependence for the acoustic phonon energy in the present experiment remains however to be explained.

In Fig. 13a, the Lorentzian FWHM of the PL band is plotted versus temperature for a Si-nc that shows no TO-phonon replica. It can be seen that a fit based solely on Bose statistics for a low-frequency phonon mode ($\hbar\omega$ = 6 meV) with the only fitting parameter – the proportionality coefficient – reproduces correctly the main trend in the linewidth temperature dependence. We conclude that exciton interaction with low-frequency phonons is consistent with the observed linewidth temperature evolution. At RT the linewidth of the Si-ncs without TO-replica reaches only about 100 meV, while those exhibiting TO-replica have a somewhat broader emission line ~150 meV (see Fig. 13b).

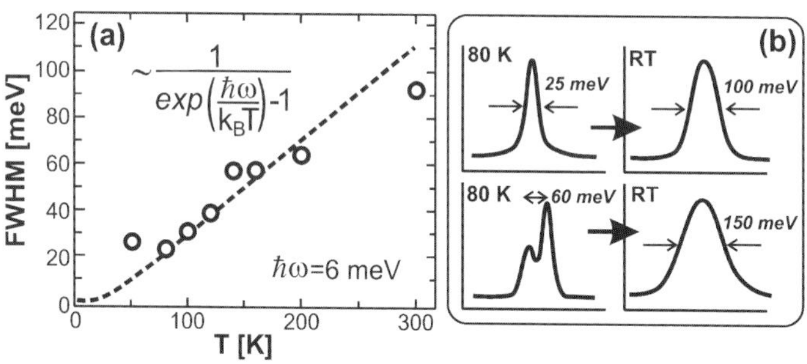

Fig. 13. (a) Temperature dependence of the PL linewidth for a dot without a TO-phonon replica.[17,19] The dashed line is a fit based on Bose statistics for a low-frequency phonon mode ($\hbar\omega$ = 6 meV). (b) Schematic representation of the observation that Si-ncs revealing TO-phonon replica have broader PL band at RT than those without TO-replica.

Finally, we have to note that the shape of PL spectra of single Si-nc may be smoothed out and broadened by the spectral diffusion revealed in measurements illustrated in Fig. 8. The detection time of a spectrum must be at least 10 min (typically ~30 min), therefore we cannot reduce the possible influence of spectral diffusion. The shift of the emission line due to variations of the local field was observed in II-VI semiconductor QDs where the line-width was found to be strongly dependent on the detection time even at low temperatures.[3]

4.3. *Photoluminescence intermittency – ON-OFF blinking*

4.3.1. *Blinking of NPSi nanocrystals*

Fluctuation of the PL signal from single Si-ncs in NPSi samples was studied by repeated detection of PL images with exposure duration of the order of tens of seconds. The overall signal from a single emitting spot was then extracted and plotted versus time. The selection of appropriate detection time for a studied single Si-nc depends on its PL intensity and blinking frequency. Too short time slot will produce low signal-to-noise ratio while too long slot will smooth out fluctuations and generate low number of statistical data. Unfortunately, there are many single Si-ncs whose PL fluctuation cannot be statistically treated, for example those emitting only in very short and rare flashes (see Fig. 14a). Therefore

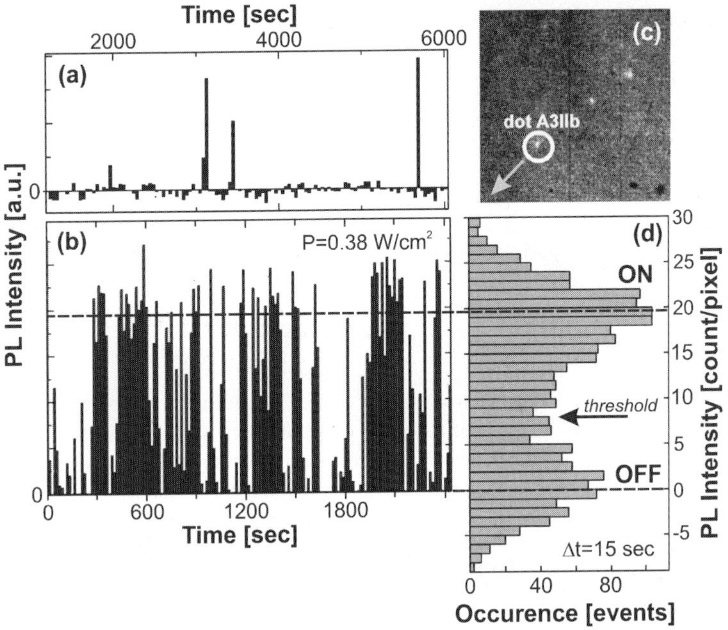

Fig. 14. PL fluctuation for a Si-nc with very rare flashes (a) and with roughly equal dark and bright intervals (b).[18,19] The detection time interval used was 40 and 15 sec, respectively. The panel (c) shows one of PL images of the Si-nc treated in (b) and (d). The bottom right panel (d) is a histogram of signals from (b) showing a double-peak structure with indicated threshold between *ON* and *OFF* states.

most of results come from single Si-ncs spending roughly equal time in bright and dark state (see Fig. 14b and Fig. 17a). If the chosen time-slot length is appropriate, the histogram of detected signals (Fig. 14d) must clearly reveal double-peak structure and a signal threshold between *ON* and *OFF* states can be established. Then the duration of *ON* and *OFF* intervals is calculated and statistically represented.

The distributions of *ON* and *OFF* interval length for a single Si-nc shown in Fig. 14b are plotted in Fig. 15. Both distributions may be well fitted by an exponential function, i.e. random switching model. It means that the probability of a Si-nc to stay in *ON* or *OFF* state is described by the equation

$$p_{ON,OFF}(t) \sim \exp\left(-\frac{t}{\tau_{ON,OFF}}\right) \quad .$$

The calculated values of characteristic *ON* and *OFF* time $\tau_{ON,OFF}$ are very long, 53 and 28 sec, respectively.[18]

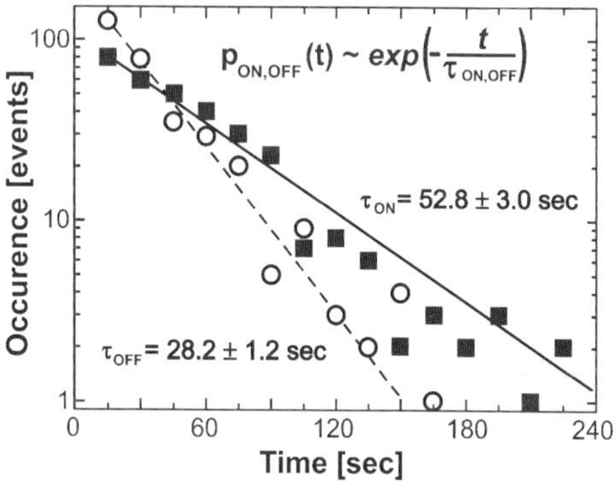

Fig. 15. The analysis of blinking data from Fig. 14b (excitation intensity 0.38 W/cm^2). *ON* (dark squares) and *OFF* (open circles) interval distributions are fitted with an exponential function (solid and dashed line).[18] Values of characteristic time are given.

In order to track changes of the blinking process with increasing excitation it is convenient to use switching rate for *ON*→*OFF* and *OFF*→*ON* events rather than the average time in a certain state. Using single-exponential distributions one can define switching rates:

$$f_{10,01} \equiv \frac{\sum_{n=1}^{\infty} \exp(-n/\tau_{ON,OFF})}{\sum_{n=1}^{\infty} \exp(-n/\tau_{ON,OFF}) \cdot n} \cdot \frac{1}{\Delta t} \approx \left[1 - \exp\left(-\frac{1}{\tau_{ON,OFF}}\right)\right] \cdot \frac{1}{\Delta t}$$

where $f_{10,01}$ are the switching rates for *ON*→*OFF* and *OFF*→*ON* processes correspondingly, [Hz]; $\tau_{ON,OFF}$ are the characteristic times in *ON* and *OFF* states, [interval]; Δt interval length, [sec]. The denominator in this formula stands for the total number of switching events, while the numerator accounts for the overall number of time intervals spent in the corresponding state. In the limit of long average times in *ON* and *OFF* states the switching rate becomes simply inverse of the corresponding average time: $f_{10,01} \approx 1/(\tau_{ON,OFF} \cdot \Delta t)$. The latter value was used in Ref. 4 and referred to as *characteristic switching rate*. The switching rates for blinking of the Si-nc from Fig. 14b are shown in Fig. 16 as a function of excitation power density. Corresponding average times in *ON* and *OFF* states for different excitation regimes were found from the threshold approach analysis of the experimental data.[18] The switching rate for *ON*→*OFF* process is strongly dependent on the excitation intensity, while the inverse process has rather weak dependence (Fig. 16). Note that 0.6 W/cm^2 roughly corresponds to occupancy of one exciton per nanocrystal suggesting that a Si-nc may switch to a dark state when more than one exciton is present, most likely as a result of an Auger recombination event.

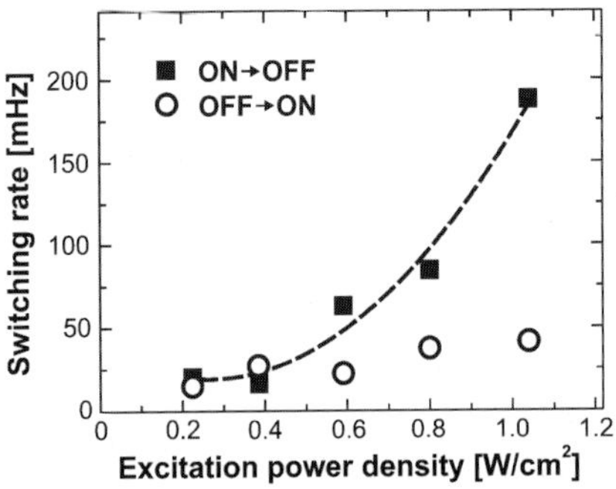

Fig. 16. Switching rates for the Si-nc under different excitation regimes. *ON*→*OFF* switching rate increase is fitted with a square dependence on excitation power (dashed line). Adapted from Ref. 18.

4.3.2. *Blinking of PSiG nanocrystals*

Blinking of Si-nc in PSi grains was studied by different groups using a confocal microscope.[9,10,13] In Fig. 17, we present PL blinking of a single PSi grain excited in the epifluorescence arrangement by a cw diode-laser (444 nm). Here arrival times of detected photons (events) at two APD detectors[c] were recorded and later binned per selected time slots. We found for PSi grains detected under these experimental conditions that the best time binning is over intervals around 300 ms. When proper binning is applied, the histogram of PL intensity per bin (see Fig. 17b) reveals a clear multilevel structure. The most abundant intensity levels are found by fitting a histogram with several Gaussians (see Fig. 17b). Thresholds between two emission levels were calculated using the equation

$$\left(I_{th} - I_i\right)/\sqrt{I_i} = \left(I_j - I_{th}\right)/\sqrt{I_j}$$

that defines the threshold I_{th} between lower I_i and higher I_j intensity levels. Such a threshold is unbiased, i.e. it has equal distances – measured in standard deviations – from the low and high levels.[26] For the case given in Fig. 17, the intensity peaks of the *OFF* (background signal), single and double *ON* states are 1334, 2779, and 4046 cps and the corresponding unbiased thresholds are 1925 and 3353 cps.

When the proper bin lengths and intensity thresholds are defined the statistics of dwell times of blinking particle in each state is calculated. In

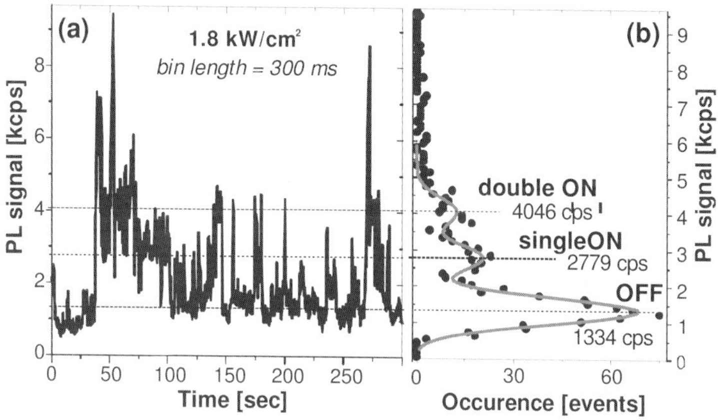

Fig. 17. (a) The time trace obtained from a single PSi grain by binning of detection events per 300 ms slots. (b) The occurrence histogram of intensities taken from the left panel. The gray line is a fit with three Gaussian peaks. Adapted from Ref. 13.

---

[c] Two APD detectors were used only because they are built in the detection system. In fact, one (low-noise) APD detector is able to detect photon flux from a single Si-nc as the emission rate is low and the detector dead-time will have negligible effect.

Fig. 18 we plot distributions of dwell times measured under 0.6 kW/cm^2 excitation and using 100 ms bin time. Experimental points for *OFF* time distributions are well fitted by a power-dependence with the exponent around −1.38. For the single and double *ON* states the data follows a power-dependence of $t^{-1.5}$, but for shorter times the decrease is faster following an exponential decay. Therefore the distribution of *ON* states on the full experimental time range is fitted by the combined function

$$P(t) = const \cdot t^{-1.5} \cdot \exp\left(-\frac{t}{\tau}\right) .$$

The characteristic exponential decay time τ for the single *ON* state is about 2.3 sec and becomes shorter (~0.96 sec) for the double *ON* state.

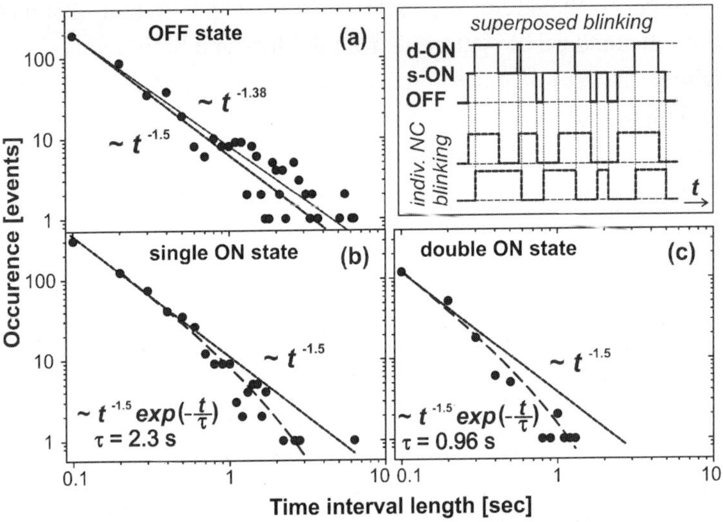

Fig. 18. Histogram of distribution of dwell times for *OFF* state and two ON states (0.6 kW/cm^2 excitation power and 100 ms bin length). Experimental points are fitted with the power-dependence lines (black straight lines) whose exponents are indicated. In addition, single and double ON state distributions are also fitted with a combination of $t^{-1.5}$ power-dependence and an exponential bending tail (dashed curves). The upper right panel illustrates the phenomenon of dwell times shortening due to combination of two (independent) blinking traces. Adapted from Ref. 13.

We have to note that blinking studies of NPSi samples[d] showed exclusively blinking between two states (see Fig. 14) in contrast to the

---

[d] In electroluminescence,[20] however, it was found that at higher bias more than one nc could be addressed yielding multiple peaks as in Fig. 17b.

multilevel blinking of PSiG samples (Fig. 17). Also the work of Buratto's group indicated presence of several *chromophores* (as they called it) in a PSi grain by calculating a histogram of PL intensities of many single emitting spots (ensemble averaging) and they observed also multilevel blinking of single spots.[9] Such multilevel blinking can be either due to several luminescence centers within one Si-nc or due to overlapping contributions of several Si-nc emitting simultaneously (possibly independently) within an optically resolved spot. Even a simple superposition of signals from a few Si-nc (without interaction) should affect the distribution of *ON* and *OFF* states, causing mainly shortening of all dwell times, as schematically illustrated by the inset in Fig. 18.

Another difference between NPSi and PSiG samples is the shape of the dwell time distributions. It was found to be exponential for NPSi [18] (indicating random telegraph switching model)[34] while the inverse power-dependence $\sim t^{-\alpha}$ (with α between 1.3 and 2.2) is typical for PSi samples.[10] This difference may be, however, not due to different material properties but simply due to different detection time scales, eventually excitation conditions. Indeed, with respect to the NPSi blinking experiments excitation was about 1000 stronger in the PSiG case and thus multiple exciton occupancy was most likely the case in these experiments. In Fig. 18 we show that PSi grains may have a combined statistics: The power-law dependence with α close to 1.5 is observed for *OFF* interval distribution and for *ON* distribution at short time scale (below 400 ms). For longer *ON* times the distribution becomes exponential (especially for multiplied *ON* states). In the case of NPSi samples we were able to detect blinking statistics only with poor time resolution, intervals from about 10 sec and longer.[18] Therefore we can observe only exponential part of the distribution.

The combined power- and exponential-dependence (observed also in nanocrystals of II-VI semiconductors)[35] may be a key indication for understanding of the blinking mechanism. The theoretical model predicting exactly this kind of blinking distribution in semiconductor QDs was developed by Marcus' group.[36,37,38,39] It is based on a diffusion-controlled electron-transfer (DCET) reaction model in which a diffusive process occurs along free-energy potentials. The DCET model predicts the −1.5 power-law decay for both *ON* and *OFF* statistics as well as its breakdown with an exponential tail for *ON* time distribution and eventually also for *OFF* time distribution at much longer times. Anomalous diffusion could cause the exponent to deviate from −1.5. A schematic diagram of this model is plotted in Fig. 19. Other models frequently used to explain QD intermittency are based on an assumption of a distributed reaction rate due to e.g. an exponential distribution of trap depths or a distribution of tunneling distances between the QD core and interface trap states.[40] However, these models are less successful in explaining our experimental results.

We have to note that the non-stationary blinking phenomena are closely related to an apparent bleaching of the PL in ensemble measurements of Si-nc.[10,36] This bleaching is reversible (after switching off the excitation) on a long time scale.

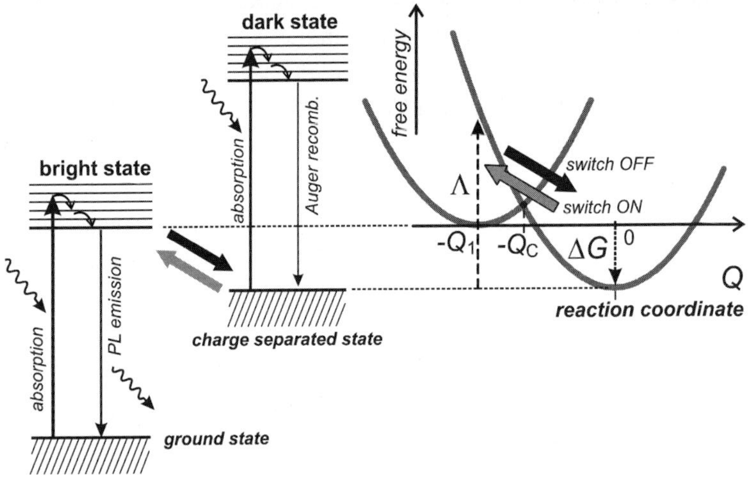

Fig. 19. Schematic illustration of the blinking processes. The left-hand side shows the energetic four-level system used in the DCET model of Tang *et al.*[36] The right-hand side of the image shows the diffusion process on the parabolic potential surfaces across a sink at the energy level crossing (intermittency phenomenon).

5. Discussion

Luminescence spectroscopy of single Si-nc in NPSi and PSiG samples indicates that the PL spectrum of a single Si-nc at RT is a single Gaussian peak with a FWHM of about 120 meV (100-150 meV). This relatively broad width is due to participation of momentum-conserving phonons in transitions, mainly the low-frequency phonons (~6 meV) and (for a fraction of Si-ncs) the TO-phonon (~60 meV). In the case of PSiG the PL spectrum at RT is sometimes more complicated because we observe within an optically resolved spot a cluster of several Si-ncs not a proper single Si-nc.[e] In spite of this fact, it is possible to observe optical signatures of single emitting Si-nc within PSi clusters, especially at low excitation power. The number of simultaneously emitting Si-ncs is limited by long *OFF* periods and possible presence of efficient non-radiative centers that prevents radiative recombination in some Si-ncs. In conculsion, PL spectra of individual Si-ncs indicate a common mechanism of the S-band emission for all studied Si nanostructures.

Significant support for the possible common mechanism of PL in Si nanostructures comes from the following comparison of the *ON* state emission rates of Si-ncs reported in literature. Apparently contradictory

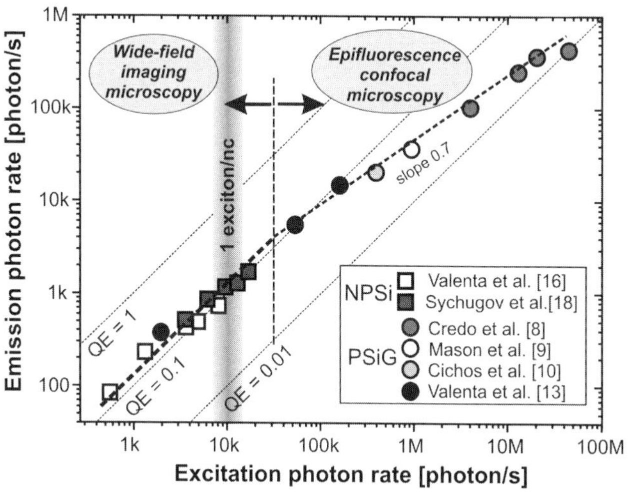

Fig. 20. Collected literature data on the single *ON* state signal represented as a dependence of emission photon rate on the excitation photon rate. The dashed line shows a linear and sublinear dependence R_{em}= const · $R_{ex}^{0.7}$. The excitation rate of about 10 k photons/sec is expected to create one exciton per nanocrystal.

---

[e] Probability to break PSi layer into individual Si-ncs using sonication or any other common separation techniques is very low. In addition the clustering may take place also during the colloid deposition on a substrate.

results become uniform when the strong dependence on excitation intensity and wavelength is taken into account. This fact is markedly illustrated by Fig. 20, where data from all available (to the authors knowledge) published works on SNS of Si-ncs are collected. The excitation photon rate is calculated as a product of excitation intensity [photon s^{-1} cm^{-2}] and excitation cross section [cm^{-2}] taken from Kovalev *et al.*[41] The emission photon rate is the detection rate [counts per second – cps] of the single *ON* state signal (it means that the effect of *OFF* periods is excluded) divided by the overall detection efficiency (detected count per emitted photon). We have to note that part of necessary data is not known precisely and therefore the calculated rates and quantum efficiency may be subject of significant inaccuracy (probably as high as 50%). Especially, the calculated maximal quantum efficiency of about 20% is several times lower than the value reported in literature[8,42,43] and also for our NPSi and PSiG samples.[13,15]

All data in Fig. 20 fit in a linear dependence that becomes slightly sublinear (in the log-log scale, $R_{\mathrm{em}} \sim R_{\mathrm{ex}}^{0.7}$) when the pumping limit creating average population of one exciton per nanocrystal is exceeded (assuming a lifetime of about 100 μs this would correspond to the excitation rate of ~10^4 photons/s). This indicates that some non-radiative recombination channel reduces the quantum efficiency (QE) from about 0.2 at very low excitation to below 0.02 for the highest excitation. If we take the definition of the QE of PL

$$\eta = \frac{1/\tau_r}{1/\tau_r + 1/\tau_{nr}} \quad ,$$

we can find the relation between radiative and non-radiative lifetime τ_r and τ_{nr}. For η=0.2 and 0.02 we obtain $\tau_{nr} = \tau_r/4$ and $\tau_{nr} = \tau_r/49$, respectively. For data plotted in Fig. 20, this means that while the emission rate increases almost four orders of magnitude the ratio of radiative and non-radiative lifetime increases only about one order of magnitude. In addition there is no clear saturation at least up to an excitation rate of 3×10^7 photons/s. Hence the performance of Si-nc would be excellent if no transition to dark *OFF* state exists. In reality the increase of dark *OFF* periods with growing excitation power is faster than the increase of emission rate in the *ON* state, therefore the overall QE of PL decreases significantly with increasing pumping.

In Fig. 20, we indicate also what experimental technique was used for measurements. It is evident that epifluorescence confocal microscopes using APDs for detection cannot work at very low excitation conditions due to a high dark count rate of several tens cps. On the other hand wide-field imaging with off-axis excitation and low-noise detection with a CCD is well suited to measure single Si-nc spectra at low excitation, but has poor time resolution to detect fast blinking. Also the focusing of an excitation beam is weaker, so the achievable intensity is much lower than for a confocal point-like excitation.

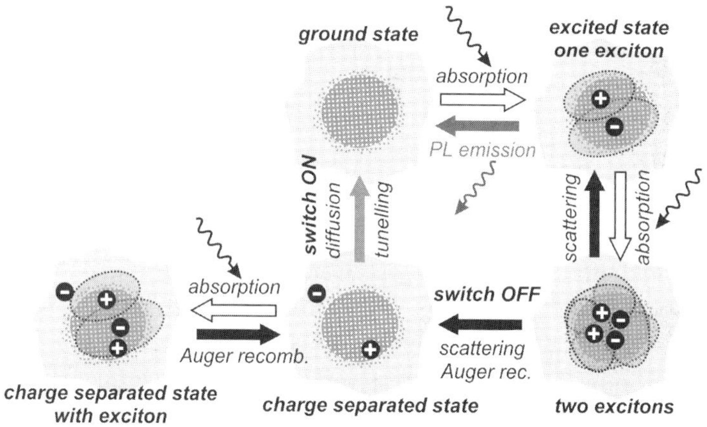

Fig. 21. Schematic illustration of processes involved in the radiative and non-radiative recombinations of a single Si-nc.

The final point of this discussion is the suggestion of a microscopic model of the radiative and non-radiative processes in an individual Si-nc (Fig. 21). After absorption of a photon an electron-hole pair – exciton – is formed in a Si-nc. In small Si-nc (diameter below 5 nm), the exciton is strongly confined (Bohr radius of exciton in bulk Si is 4.9 nm). The electron and hole "feel" strongly the amorphous-like interface between the Si core and the SiO_2 matrix and the related specific vibrations – phonons that take part in recombination processes (influencing the spectral shape). The exciton is diffusing in the complex energy landscape. Considering the very slow PL decay typical for measurements in ensemble of Si-ncs and the relatively high quantum efficiency of radiative recombination of single Si-nc in the *ON* state, there is a high probability of generation of a second exciton (or even multiple excitons for strong excitation conditions) during the lifetime of the first one. We propose that exciton-exciton scattering as well as Auger recombination are crucial phenomena in radiative and non-radiative recombination processes of Si-ncs. It can induce shortening of the radiative lifetime, energy transfer from one to another exciton as well as non-radiative recombination of the exciton. The scattering can eventually also lead to a charge separation. In the charge-separated state the Auger recombination effectively quenches created excitons, hence causing persistence of the Si-nc in the dark *OFF* state until the charges recombine back.

6. Conclusions

We have given a review on single Si-nc spectroscopy experiments performed in NPSi and PSiG samples using both wide-field imaging micro-spectroscopy and confocal microscopy. The main findings are summarized in a few points:

- The PL spectrum of a single Si-nc at RT is formed by a single Gaussian band with FWHM between 100-150 meV – the broad shape is caused by participation of momentum-conserving phonons in transitions. The cryogenic experiments performed down to 30 K, reveal "atomic-like" narrow line (most probably due to zero-phonon transitions) and a low-frequency (6 meV) phonon replica that is mainly responsible for temperature broadening. For a fraction of dots also TO-phonon replicas contribute to the additional broadening.
- The PL emission of a single Si-nc shows intermittence - *ON/OFF* blinking (in case of PSiG samples it may be multilevel blinking) - typically on the time scale of a fraction of second or longer. The distribution of *ON* and *OFF* intervals is described by combined power- & exponential-dependence.
- A possible origin of the PL blinking is the diffusion-controlled electron-transfer reaction model developed by Marcus et al. [36,37]. The dark state of a Si-nc is probably a charge separated state in which radiative recombination is quenched by Auger recombination.

- All available data on PL spectroscopy of single Si-ncs at RT follow a relation between emission (in the single *ON* state) and excitation photon rate $R_{em} \sim R_{ex}^{0.7}$, at least for excitations stronger than about 1 exciton/nanocrystal average occupancy. Below this a linear dependence is expected. The quantum efficiency decreases from about 0.2 to 0.02 when excitation increases from 10^3 to 10^7 photon/s.

From the point of view of applications of Si-ncs in light-emitting devices, there are both good and bad news. The PL performance of Si-ncs in *ON* state is excellent, but the transition to dark *OFF* states and their long duration strongly limits the overall emission rate. Therefore the blinking phenomenon must be studied and understood in details, in order to eventually find a way how to inhibit or reduce the *OFF* periods.

Acknowledgements

The authors would like to acknowledge contributions of many current and former colleagues throughout many years of work on single Si nanocrystal spectroscopy: I. Sychugov, A. Galeckas, R. Juhasz, and N. Elfström at KTH; A. Fucikova, J. Hala, and M. Vacha at Charles University; I. Pelant, K. Dohnalova, K. Herynkova, and K. Kusova at Institute of Physics ASCR Prague; F. Vacha and F. Adamec at University of South Bohemia Budweis; J. Humpolickova and M. Hof at Institute of Physical Chemistry ASCR Prague; F. Cichos, J. Martin, and Ch. von Borczyskowski at Technical University Chemnitz and Jun Lu at Uppsala University (TEM, Fig. 2). Partial funding was received from the KTH Faculty, National Swedish Research Counsel (VR), Royal Swedish Academy of Sciences. JV acknowledges support from the GACR project 202/07/0818, research centre LC510, KAN400100701 FUNS project, and the research plan MSM 0021620835 granted by the Czech Ministry of Education, Youth, and Sports.

References

1. A. Gustafsson, M-E. Pistol, L. Montellius and L. Samuelson, *J. Appl. Phys.* **84**, 1715 (1998).
2. S.A. Empedocles, R. Neuhauser, K. Shimizu and M.G. Bawendi, *Adv. Mater.* **11**, 1243 (1999).

3. S.A. Empedocles, D.J. Norris and M.G. Bawendi, *Phys. Rev. Lett.* **77**, 3873 (1996).
4. M-E. Pistol, P. Castrillo, D. Hessman, J.A. Prieto and L. Samuelson, *Phys. Rev.* **B59**, 10725 (1999).
5. D. Kovalev, K. Heckler, G. Polisski and F. Koch, *Phys. Stat. Solidi* **B215**, 871 (1999).
6. J. Linnros, N. Lalic, A. Galeckas and V. Grivickas, *J. Appl. Phys.* **86**, 6128 (1999).
7. M.D. Mason, G.M. Credo, K.D. Weston and S.K. Buratto, *Phys. Rev. Lett.* **80**, 5408 (1998).
8. G.M. Credo, M.D. Mason and S.K. Buratto, *Appl. Phys. Lett.* **74**, 1978 (1999).
9. M.D. Mason, D.J. Sirbuly, P.J. Carson and S.K. Buratto, *J. Chem. Phys.* **114**, 8119 (2001).
10. F. Cichos, J. Martin and Ch. von Borczyskowski, *Phys. Rev.* **B70**, 115314 (2004).
11. F. Cichos, J. Martin and Ch. von Borczyskowski, *J. Luminescence* **107**, 160 (2004).
12. J. Martin, F. Cichos and Ch. von Borczyskowski, *J. Luminescence* **108**, 347 (2004).
13. J. Valenta, A. Fucikova, F. Vacha, F. Adamec, J. Humpolickova, M. Hof, I. Pelant, K. Dohnalova, K. Kusova and J. Linnros, *Adv. Funct. Mater.*, in press.
14. D.S. English, L.E. Pell, Z. Yu, P.F. Barbara and B.A. Korgel, *Nano Lett.* **2**, 681 (2002).
15. J. Valenta, R. Juhasz and J. Linnros, *Appl. Phys. Lett.* **80**, 1070 (2002).
16. J. Valenta, R. Juhasz and J. Linnros, *J. Luminescence* **98**, 15 (2002).
17. I. Sychugov, R. Juhasz, J. Valenta and J. Linnros, *Phys. Rev. Lett.* **94**, 087405 (2005).
18. I. Sychugov, R. Juhasz, J. Linnros and J. Valenta, *Phys. Rev.* **B71**, 115331 (2005).
19. I. Sychugov, R. Juhasz, A. Galeckas, J. Valenta and J. Linnros, *Opt. Mater.* **27**, 973 (2005).
20. J. Valenta, N. Lalic and J. Linnros, *Appl. Phys. Lett.* **84**, 1459 (2004).
21. J. Valenta, J. Linnros, R. Juhasz, F. Cichos, J. Martin, in *Towards the first silicon laser*, Ed. L. Pavesi et al., NATO Sci Series (Kluver Acad. Publishers, 2003) p. 89.
22. I. Sychugov, J. Lu, N. Elfström and J. Linnros, *J. Luminescence* **121**, 353 (2006).
23. H.I. Liu, D.K. Biegelsen, F.A. Ponce, N.M. Johnson and R.F.W. Pease, *Appl. Phys. Lett.* **64**, 1383 (1994).
24. J. Valenta, P. Janda, K. Dohnalová, D. Nižňanský, F. Vácha and J. Linnros, *Opt. Mater.* **27**, 1046 (2005).
25. I. Sychugov, R. Juhasz, J. Valenta, M. Zhang, P. Pirouz and J. Linnros, *Appl. Surf. Sci.* **252**, 5249 (2006).
26. M. Lippitz, F. Kulzer and M. Orrit, *Chem. Phys. Chem.* **6**, 770 (2005).
27. J. Linnros, N. Lalic, A. Galeckas and V. Grivickas, *J. Appl. Phys.* **86**, 6128 (1999).
28. D. Kovalev, H. Heckler, M. Ben-Chorin, G. Polisski, M. Schwartzkopff and F. Koch, *Phys. Rev. Lett.* **81**, 2803 (1998).
29. M.S. Hybertsen, *Phys. Rev. Lett.* **72**, 1514 (1994).

30. C. Delerue, G. Allan and M. Lannoo, *Phys. Rev.* **B64** 193402 (2001).
31. T. Takagahara, *Phys. Rev. Lett.* **71**, 3577 (1993).
32. T. Takagahara, *J. Luminescence* **70**, 129 (1996).
33. M. Fujii, Y. Kanzawa, S. Hayashi and K. Yamamoto, *Phys. Rev.* **B54**, R8373 (1996).
34. A.L. Efros and M. Rosen, *Phys. Rev. Lett.* **78**, 1110 (1997).
35. K.T. Shimizu, R.G. Neuhauser, C.A. Leatherdale, S.A. Empedocles, W.K. Woo and M.G. Bawendi, *Phys. Rev.* **B63**, 205316 (2001).
36. J. Tang and R.A. Marcus, *J. Chem. Phys.* **123**, 054704 (2005).
37. J. Tang and R.A. Marcus, *Phys. Rev. Lett.* **95**, 107401 (2005).
38. P.A. Frantsuzov and R.A. Marcus, *Phys. Rev.* **B72**, 155321 (2005).
39. J. Tang and R.A. Marcus, *J. Chem. Phys.* **125**, 044703 (2006).
40. M. Kuno, D.P. Fromm, H.F. Hamann, A. Gallagher and D.J. Nesbit, *J. Chem. Phys.* **112**, 3117 (2000).
41. D. Kovalev, J. Diener, H. Heckler, G. Polisski, N. Kunzner and F. Koch, *Phys. Rev.* **B61**, 4485 (2000).
42. G. Ledoux, J. Gong, F. Huisken, O. Guillois and C. Reynaud, *Appl. Phys. Lett.* **80**, 4834 (2002).
43. J.S. Biteen, D. Pacifici, N.S. Lewis and H.A. Atwater, *Nano Lett.* **5**, 1768 (2005).

CHAPTER 8

SILICON NANOCRYSTAL MEMORIES

Panagiotis Dimitrakis, Pascal Normand and Dimitris Tsoukalas*

Institute of Microelectronics, NCSR "Demokritos"
P.O. Box 60228, 153 10 Ag. Paraskevi, Greece
E-mail: pdimit@imel.demokritos.gr

School of Applied Sciences, National Technical University of Athens
157 80 Zografou, Greece

This chapter reviews current research and development in the field of nanocrystal memories with emphasis on Si-nc non-volatile memories. The operation principles and advantages of nc memories together with the issues associated with nc fabrication, device architecture and implementation in manufacture are presented comprehensively. Original experimental results and data analysis are reported in order to give a detailed picture of the physical mechanisms that determine the program/erase operations as well as the reliability characteristics of nc memories. The current state-of-the-art and prospects for future development of nc memory devices are also summarized.

1. Introduction

Nanoparticles are some of the most important vehicles of the blossom of applied nanotechnology. These are mainly clusters of metal or semiconductor atoms with dimensions in the range of a few nanometers. For electronic applications they should have preferentially a crystalline structure and their size should be below 10nm (hence the name "nanocrystals"). The nanocrystals (ncs) are treated by researchers as quantum-dots (i.e., systems with zero dimensions). The increase of band gap as the nc diameter decreases, the quantization of energies and of density of states outside of the energy band gap and Coulomb blockade are the most important quantum mechanical phenomena that can be exploited for nc-based-electronics. More details on the electronic structure and the properties of nanocrystals can be found in Refs. 1-3.

In the present chapter, the application of nanocrystals for the fabrication of nonvolatile memory devices is examined. More emphasis will be paid to the use of silicon nanocrystals (Si-nc). Si-nc memory is based on discrete charge storage nodes and is fully compatible with CMOS technology. According to the International Technology Roadmap of Semiconductors,[4] nc memories belong to emerging memory structures for future integrated circuit generations because they have good potential for pushing the scaling limits of nonvolatile memories (NVM). The unique properties of the nanocrystal memory devices compared to the conventional (i.e., present technology) polysilicon floating gate (FG) NVM devices will be described comprehensively. Next, the operation principles of Si-nc memories and the electrical characterization procedures of single memory cells will be reviewed. The performance and reliability features of Si-nc single memory cells are then analyzed highlighting the most important alternatives proposed for the optimization of the memory characteristics. Finally, the current status of Si-nc memories and the open issues of this technology will be summarized.

2. Silicon Nanocrystals in Memory Technology

2.1. *The limitations of current memory technology*

In the present technology of NVM a single Floating Gate (FG) metal oxide semiconductor field effect transistor (MOSFET) is used as the primary cell for information storage. Nonvolatility requires memory cells that retain their data for at least 10 years. In the simplest device architecture, the lowest amount of stored data is one bit per transistor. However, sophisticated device architectures and operation methods have been applied for multi-bit storage per cell (i.e., 2-bit or 4-bit/cell) which are called multilevel cells (MLC).[5-7]

The basic structure of the standard FG NVM cell is presented in Fig. 1a.[5-7] The injection or tunnel oxide is a thin (≥ 8 nm) thermal SiO_2 layer above which a conductive polysilicon layer is deposited which will act as floating gate. Between the polysilicon control gate and the floating gate a dielectric is formed; the so-called interpoly dielectric (IPD). In state-of-the-art FG NVM cell, IPD comprises of a tri-layer dielectric structure oxide/nitride/oxide (ONO) having an equivalent oxide thickness in the range of 13-20 nm. The operation principle of FG NVM is essentially based on the switching of the MOSFET's threshold voltage V_{th} between a low and a high value. This is achieved by storing an amount of charge into the FG. Electron storage shifts the threshold voltage to a more positive value according to:

$$V_{thW} = V_{th0} + Q_W / C_{IPD} \tag{1}$$

where V_{thW} and V_{th0} are the threshold voltage values after and before electron storage respectively, Q_W is the amount of stored electrons (in Cb) and C_{IPD} is the capacitance (in F) of the interpoly dielectric. This high V_{th} state is commonly named "Write" or "Program" state. If the stored amount of charge is decreased to Q_E ($<Q_W$) then the threshold voltage of the new state V_{thE} is proportionally shifted to a more negative value:

$$V_{thE} = V_{th0} + Q_E / C_{IPD} \tag{2}$$

This new state is usually named "Erase" state and the difference between V_{thW} and V_{thE} is called memory window

$$\Delta V_{th} = V_{thW} - V_{thE} = Q / C_{IPD} \tag{3}$$

Another very important design parameter of a FG NVM cell is the coupling factor α between the gate and the FG:

$$\alpha = \frac{C_{IPD}}{C_{TO} + C_{IPD}} \tag{4}$$

which is reduced to $\alpha = t_{TO} / (t_{IPD} + t_{TO})$ in case the TO and IPD are made of the same material and C_{IPD} (C_{TO}) and t_{IPD} (t_{TO}) are the capacitance and the thickness of the IPD (TO) layer respectively.

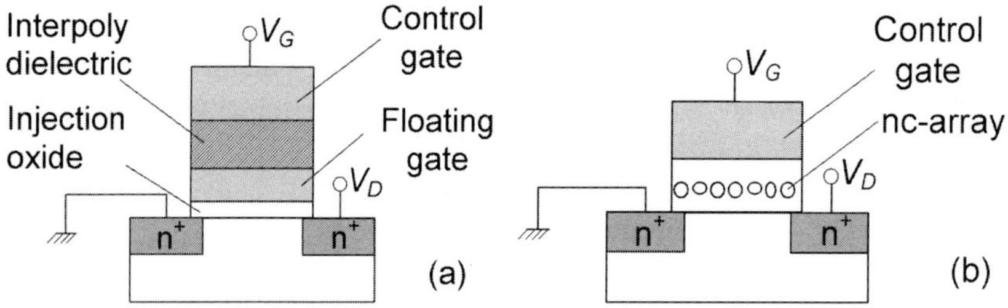

Fig. 1. Schematic representation of (a) a standard FG NVM cell and (b) a NC NVM cell. Both are demonstrated using NMOS transistor.

The individual state of a cell, corresponding to the logic “0” or “1”, is detected by applying a gate voltage in between the V_{thW} and V_{thE} values. Specific circuits sense the current flowing from the source to drain, I_{DS}, and indicate whether the transistor is “ON” or “OFF”, i.e., whether the cell is in the erase or the program state, respectively.[8] It should be noticed here that any applied voltage during the program and erase operation is a square voltage pulse and the operation speed of the memory cell is determined by the pulse duration.

The charges are injected to and ejected from the floating gate by Fowler-Nordheim (F-N) tunneling and/or hot-carrier injection mechanism.[5-7] Due to the high conductivity – metallic behavior – of the FG material, the stored charge is spread all over the layer. In both procedures, highly energetic charges are injected/ejected via quantum mechanical tunneling through the tunnel oxide (see Fig. 1a) to the FG. Obviously, the repetition of the charge movement forced by the external applied high electric field through the tunnel oxide (TO) generates oxide defects and consequently degrades its insulating properties, i.e., the electrical isolation between the FG and the channel of the transistor.

The most common memory array organization schemes are the NAND and the NOR architectures[5-7] and both are integrated on a Si wafer following the Flash memory system.[9] For data storage, NAND architecture is more preferable while NOR is more suitable for code storage. Furthermore, NAND memory arrays are programmed and erased by F-N tunneling mechanisms, while NOR arrays are programmed by hot-carriers and erased by F-N tunneling mechanism.

According to the ITRS, there is a gap between the technology nodes of logic devices (45 nm has been achieved) and Flash NVM memories (moving towards 90 nm). Obviously, a design of an integrated circuit with embedded logic and memory devices cannot be realized following the design rules of the smallest node. This technology node mismatch is due to several scaling limitations of the conventional memory device structure. Nc NVM is a potential candidate to replace the FG NVM at 45 nm technology node and below. In the following paragraph we present, a brief discussion of the scaling limitations of conventional Flash FG NVM.

The shrinkage of MOSFET dimensions improves not only the density of the fabricated devices but primarily enhances their performance. Moving from a technology node to a smaller one, several physical parameters of the MOSFET should be downscaled.[10] Among them the most important are the thickness of the SiO_2 gate insulator and the geometrical characteristics of the source and drain junctions. Both parameters, affect the performance of conventional FG NVM.[6,11] As previously mentioned, the gate oxide thickness in these devices seems almost impossible to be reduced below 7-8 nm, because the stored charge into the FG can easily leak to the channel, via oxide defects created after many program/erase cycles or by quantum mechanical tunneling under the influence of the internal electric field. In Figure 2a, the equivalent oxide thickness (EOT) of the gate insulator for logic and high-performance devices at different technology nodes is compared with that of FG NVM devices.[4] Obviously, the characteristics of the source and drain junctions affect the channel hot-carrier injection mechanism. Following the market demand for portable applications, the operation voltages of conventional FG NVM should be reduced and their speed should be increased. In view of that, the IPD thickness should be reduced[12] according to the downscaling design rules (see Fig. 2b).[4] In this case, the memory window and the coupling factor α become smaller. At wafer level, the reduction of the distance between two adjacent cells is reduced causing the increase of the capacitive coupling between the FGs of the cells.[6,12] The coupling between the FGs of adjacent cells induces significant cross-talk effects, i.e., the program or erase operation on a cell affects the memory state of the neighboring cells.

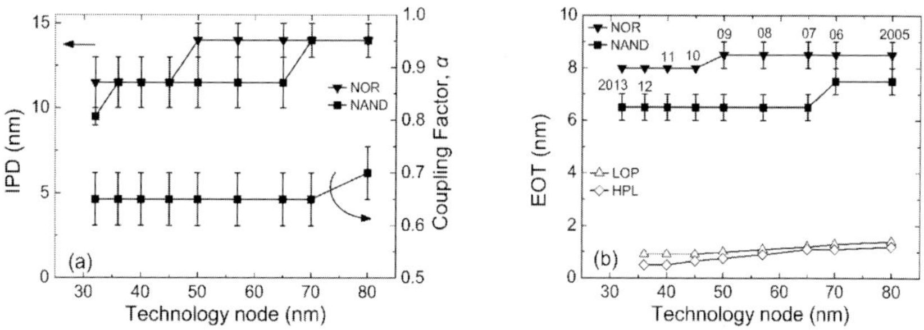

Fig. 2. ITRS target values for (a) the EOT of the gate dielectric and (b) the thickness of the IPD layer and coupling factor at different technology nodes of FG NVMs are presented. For the sake of comparison, the same values for logic and high-performance devices are shown in (a).

To sustain continuous downscaling of the Flash FG NVMs novel and revolutionary changes in the device architecture and operation principle should be adopted.

2.2. *Nanocrystal floating gate vs polysilicon floating gate memories*

The nc memory concept has been introduced by IBM researchers about ten years ago[13] to overcome some of the above limitations. The major innovations in nc memories are (a) the nano-granular FG and (b) the discrete charge storage in quantum dots. As described in more details in section 2.3 spherical ncs can be fabricated in a 2-D array configuration over a thin tunnel oxide layer (see Fig. 1b). The density of ncs in the array should be in the 10^{11}–10^{12} cm^{-2} range and their diameter in the 3 – 10 nm range. A control oxide (CO) or another dielectric can be used to prevent charge injection from the gate to the ncs (or the inverse) that acts as IPD in conventional FG NVM. It must be also emphasized that the ncs should be isolated from each other to prevent charge exchange between them. Experimental studies reveal that nonvolatility can be achieved using a control oxide with thickness around 7 nm and an inter-nc distance of about 5 nm.[14-16] The tunnel oxide might be thinner (≥ 4 nm) than in the case of conventional FG memories. This is due to the Coulomb blockade effect and the delta function like density of states around each confined energy level.[11,16-18] Once an electron is stored, the potential energy of the nc increases by $\Delta E=[(v+1)^2-v^2]e^2/2C_\Sigma$, where v is the number of stored electrons in the nc, C_Σ is the nc self-capacitance and e the electronic charge. This energy increase prohibits the next electron to be stored in the nc. Thereby, the charge injection process is suppressed and the nc memory exhibits a self-convergent injection mechanism.[18a] In this way, the detrimental stress induced leakage current (SILC) effect shown in conventional FG memories is delayed. It is also worthwhile to notice that the presence of a simple localized defect in the tunnel oxide of a conventional FG memory results in total loss of the charge in the FG as shown in Fig. 3a, while in the case of nc memory only the nc in the defect vicinity will be discharged .

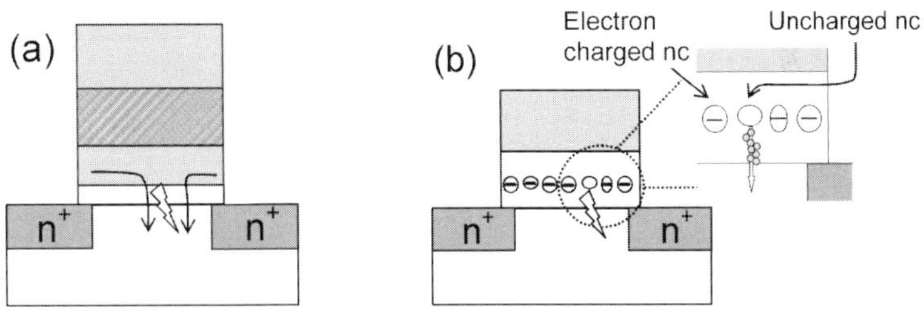

Fig. 3. (a) Conventional FG NVM and (b) nc NVM devices with tunnel oxide suffering from a defect. The stored charge in (a) is leak entirely to the channel and the stored information is lost. Contrary, in (b) only the nc above the defect is discharged and hence the stored information remains.

Consequently, nc memory cells with ~4 nm thick tunnel oxide are expected to operate at a higher speed and lower power compared to conventional FG memory cells.[19] Tunnel oxides thinner than 4 nm improve further the operation speed, but limit the charge retention in the ncs to a few hours up to a few days.[14-16] While such charge retention times are prohibitive for non-volatile memory applications, this opens an opportunity window for investigation of nc memories for low voltage, fast DRAM applications with slow refresh times.[20] Retention is extremely sensitive to the tunnel oxide thickness, due to the exponential

[a] A floating gate with similar tunnel oxide thickness would continue to charge exponentially in time, similar to a capacitor in a RC circuit with a DC voltage source.

dependence of the tunneling probability on the thickness. There is also a strong dependence on the potential barrier that electrons (holes) in the nc have to overcome for tunneling back to the Si substrate. This barrier primarily depends on the conduction (valence) band offset between the tunnel oxide and the nc, Φ_{Bnc} (see Fig. 4a). It is worthwhile to remind here that the smaller the nc, the higher the energy band gap because of the conduction (valence) band shifts to higher (lower) values respectively;[2,20] consequently Φ_{Bnc} decreases for both electrons and holes. The lower the barrier height, the higher the escape probability of a carrier. An ideal condition for retention is that nc material exhibits a barrier $\Phi_{Bnc} \geq \Phi_{BS}$, where Φ_{BS} is the potential barrier between the semiconductor substrate and the tunnel oxide (e.g., for Si Φ_{BS} is ~3.1eV for conduction band and~4.8eV for valence band) (see Fig. 4b).

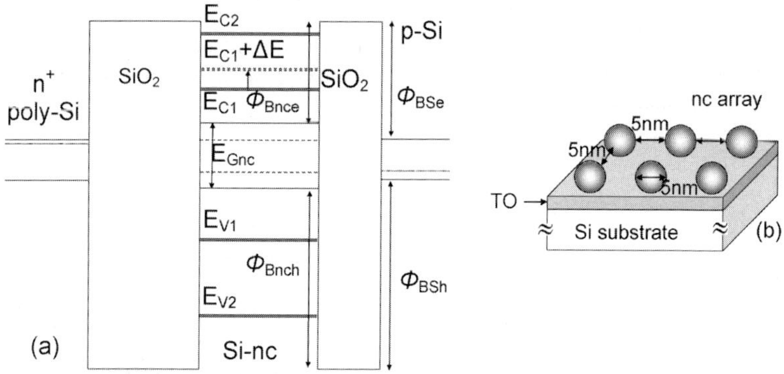

Fig. 4. (a) Energy band diagram of a Si-nc between the tunnel and control oxides, with a poly-Si gate. The discrete allowed energy levels in conduction and valence bands of the Si-nc are shown. For clarity, the energy modification (ΔE) of the E_{C1} level due to Coulomb blockade effect is presented. The energy levels inside the band gap of the Si-nc denote electron and hole trapping levels. (b) Considering, a distribution of spherical ncs with 5nm diameter where each one has a distance 5nm from its first neighbors, the surface density is 10^{12} ncs/cm^2.

Finally, Coulomb blockade for nanocrystals larger than ~10 nm is reduced allowing storage of many electrons/holes and leading to a significant increase of the ncs potential energy. This increase of the potential energy favors the probability of the stored carriers to tunnel back to the substrate.[20,21] The ideal size of the Si-ncs for memory applications is ~5 nm because the Φ_{Bnc} is high enough and Coulomb blockade effect is still pronounced.[21]

From an integration point of view, the floating gate coupling among neighboring cells in nc memories is eliminated, as explained in Fig. 5, and so device density can be increased with the minimum crosstalk effects. In addition, less processing steps are required to fabricate nc memory cells compared to conventional FG NVM.[22]

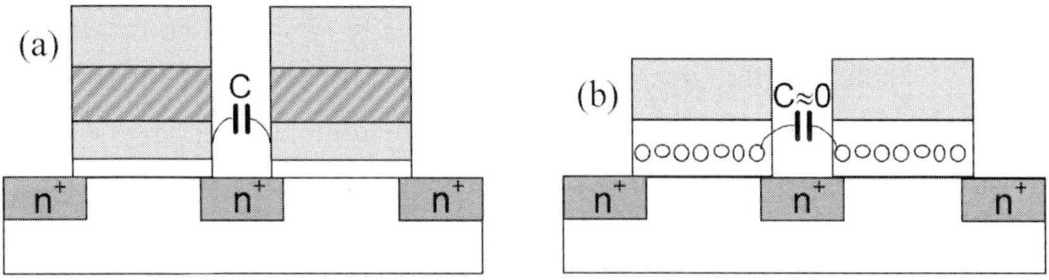

Fig. 5. Schematic representation of capacitive coupling between adjacent NVM cells realized by (a) continuous FG and (b) nanocrystal concepts. The coupling is minimized in case (b) due to nanometer size of the coupling elements.

Finally, metal-ncs may be more favorable compared to semiconductor-ncs, primarily to Si-ncs, since they offer a great variety of different materials with higher Φ_{Bnc} as compared to Si and a larger density of states around the Fermi level which result in more carriers per nc.[22,23] Since this review is focused on silicon nanocrystal memories we shall not extend our discussion to metal nanoparticle memories further.

2.3. *Fabrication of silicon nanocrystals embedded in gate dielectrics*

Progress in nanofabrication techniques based on self-organization of materials is a very important parameter for the large variety of nc applications. In fabrication of integrated circuits, the introduction of a new process, like the formation of ncs, should fulfill several requirements in order to prevent contamination and deviations from the CMOS process thermal budget.

Several methods for growth, deposition and synthesis of ncs have been successfully applied to fabricate memory structures.[3,11] Growth techniques like molecular beam epitaxy (MBE)[24] and chemical vapor deposition (CVD) have been used to fabricate semiconductor and metal ncs on tunnel oxides or other dielectrics.[21] Similarly, direct deposition techniques like aerosol,[11] sputtering[25] techniques have been also utilized to fabricate semiconductor and metal ncs. Finally, ion beam synthesis techniques, like ion implantation, plasma immersion ion implantation[26] have been also used to synthesize semiconductor and metal ncs. All these nc formation techniques are combined with thermal annealing processes before or after the deposition of control dielectric, where furnace or laser radiation sources can be used for heating. Finally, it should be mentioned that the shape of the ncs (e.g., hemispherical, spherical, pyramidal) depends on the formation mechanism and the fabrication method. Theoretical studies and simulation of the charging/discharging properties revealed that the shape of the ncs affects significantly the charge storage efficiency and the operation characteristics[27] of the nc NVM cells.

3. Operation, Memory Characteristics and Reliability Aspects of Si-nc Nonvolatile Memories

3.1. *Operation principles of Si-nc memory devices*

Nc memory devices have been demonstrated for both NAND and NOR architectures.[27] As mentioned before, NOR is less favorable than NAND mainly due to (a) the hot-carrier injection to program the cells which enhances the tunnel oxide degradation and is limited by the source/drain formation technology and (b) causes severe read disturb to adjacent cells. Nevertheless, it has been demonstrated experimentally that another major problem of scaled NOR FG NVM cells, the drain-turn phenomenon, is sufficiently suppressed.[29] Statistics and projections of the market sales and demands show that NAND memories will be the major application.[7,30,31]

Primarily, nc memory cells are designed and tested for NAND applications. Hence, the basic program/erase operations are obtained by F-N tunneling (see Fig. 6): the source, drain and body are grounded and the gate is pulsed. Positive voltage pulses $V_G >> V_{th0}$ are used to program the cell, while negative pulses are used to erase the cell. During erase, electrons are ejected from the ncs and tunnel back to the Si substrate.

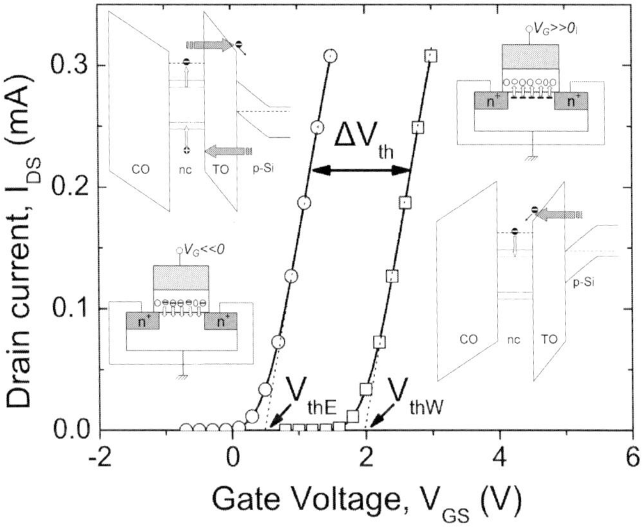

Fig. 6. Schematic representation of the program/erase operation principles of a nc memory cell. Insets describe the physical mechanisms of the charge injection for each operation.

When an erasing pulse offers enough energy to the accumulated holes to tunnel through the injection oxide, hole-tunneling current might be flow from the hole accumulation layer towards the ncs. Under this situation, positive charge might be stored into the ncs, leading to an increase of the memory window ΔV_{th}. Following this operation scheme, (F-N for programming and erasing), two-terminal devices like MOS capacitors (MOSC) can be used for a quick test procedure to estimate the charge storage characteristics. In this case, an appropriate voltage pulse is applied to the gate electrode, while the substrate electrode is grounded. After, each pulse the capacitance of the device is recorded by sweeping the gate voltage in short range from inversion to accumulation. Then, the difference between the flat-band voltage, V_{FB}, under positive and negative voltage pulses is calculated, which is directly proportional to the ΔV_{th}. An alternative quick characterization method of the charge trapping efficiency of nc floating gate devices is to measure the hysteresis in C-V characteristics of nc MOSCs after sweeping the voltage from inversion to accumulation and return back to inversion.[32] This procedure should be repeated for various voltage sweep limits. Obviously, following this later method the response of the memory structure on gate pulse excitation can be estimated, while the extracted memory window is over estimated.

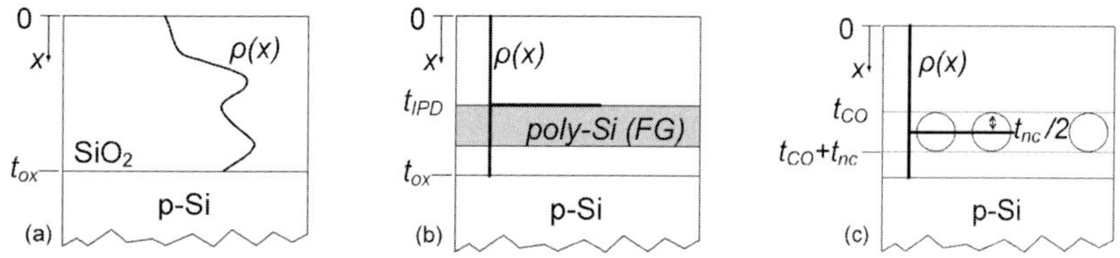

Fig. 7. Schematic representation of three different gate dielectric charge distribution ρ(x): (a) arbitrary oxide charge distribution, (b) charge stored in a poly-Si FG and (c) charge stored in nc.

The relation that determines the memory window ΔV_{th} in the case of semiconductor-ncs memory can be derived, in a first approximation, by the generic equation[33] of the threshold shift due to the presence of an arbitrary distribution $\rho(x)$ oxide trapped charges (see Fig. 7a)

$$\Delta V_{th} = \pm\gamma \frac{Q_{ot}}{C_{ox}} \tag{5}$$

where $\gamma = \dfrac{\int_0^{t_{ox}} \frac{x}{t_{ox}} \rho(x)dx}{\int_0^{t_{ox}} \rho(x)dx}$,

Q_{ot} is the volume concentration of oxide trapped charges, C_{ox}; t_{ox} is the capacitance and the thickness of the oxide layer. In case charge is stored in a FG (highly conductive poly-Si) the $\rho(x)$ distribution can be approximated as $\rho(x)$=const. when $x=t_{CO}$ and 0 elsewhere (see Fig. 7b) and hence ΔV_{th} is described by Eq. (5). If the charge is stored in the center of ncs then $\rho(x)$=const. when $x = t_{CO} + \varepsilon_{ox} t_{nc} / 2\varepsilon_{Si}$ and 0 elsewhere (see Fig. 7c) and the upper limit of the γ integration is $t_{ox}^{eq} = t_{CO} + \varepsilon_{ox} t_{nc} / 2\varepsilon_{Si} + t_{TO}$ where t_{nc} is the diameter of the ncs. Subsequently, the threshold voltage shift relation is given by:

$$\Delta V_{th} = \frac{e\nu n_{nc}}{\varepsilon_{CO}}\left(t_{CO} + \frac{1}{2}\frac{\varepsilon_{CO}}{\varepsilon_{nc}}t_{nc}\right) \tag{6}$$

where n_{nc}, t_{nc} and ε_{nc} are the surface density, the diameter and the dielectric constant of the ncs, t_{CO} and ε_{CO} are the thickness and the dielectric constant of the control oxide respectively. The integer ν denotes the number of stored electrons per nc. The first term on the right-hand-side of Eq. (6) is the same with that described previously by Eq. (3) (i.e., in case of standard FG NVM cell) considering that $Q = e\nu n_{nc}$ and the IPD is a single oxide layer. So, nc memory cells provide always a higher memory window than the equivalent FG cells. In Fig. 8, the ratio of Eq. (6) over Eq. (3) plotted as a function of the dielectric constant ratio $\varepsilon_{CO}/\varepsilon_{nc}$ for different values of the t_{nc}/t_{CO} ratio. Obviously, for $\varepsilon_{CO} \leq \varepsilon_{nc}$ no significant memory window increase can be achieved for any value of the t_{nc}/t_{CO} ratio.

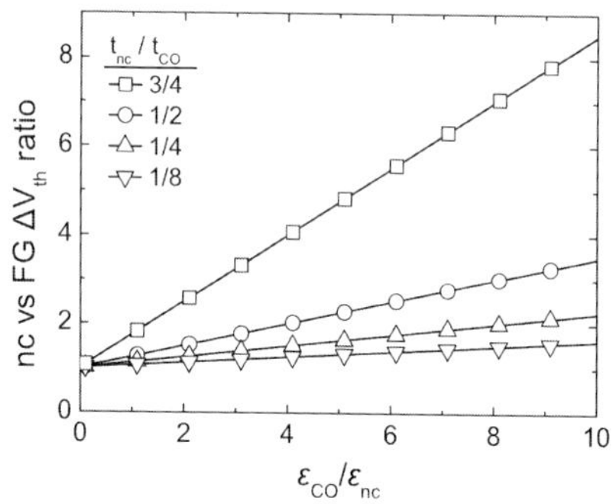

Fig. 8. Plot of the ratio between the memory window of a nc cell and the memory window of an equivalent FG cell.

Eq. (6) has been extensively used in the literature to calculate the surface density of ncs, n_{nc}, assuming that $\nu=1$ (i.e., one electron is stored in each nc) or the number of stored electrons per nc, ν, assuming that n_{nc} is known (e.g., by AFM or TEM studies). However, the calculated values for both quantities may be different because of the following two reasons. First, it is very difficult to know if all the ncs are occupied with electrons and second because the assumption about a constant $\rho(x)$ distribution at position $x = t_{CO}+\varepsilon_{CO}t_{nc}/2\varepsilon_{nc}$ is an approximation only. In order to face the discrete nature of the electron storage at the above x position and correct Eq. (6), the surface coverage R (i.e., the percentage of the gate area covered by the ncs) should be taken into account.

In this direction, Ishii *et al.*[34] proposed a model where the memory window is proportional not only to the density of the ncs n_{nc} but also to the ratio R which is proportional to the density n_{nc} and the diameter t_{nc} of the ncs, $R = n_{nc} A_{nc} = n_{nc} \pi t_{nc}^2 / 4$, where A_{nc} is the area occupied by a nc in the plane. According to this model the memory window is given by:

$$\Delta V_{th} = R \frac{Q_{nc}}{C'_{ox}} \tag{7}$$

where $C'_{ox} = \varepsilon_{ox} S(1-R)/(t_{CO}/t_{TO})$, S is the gate area.

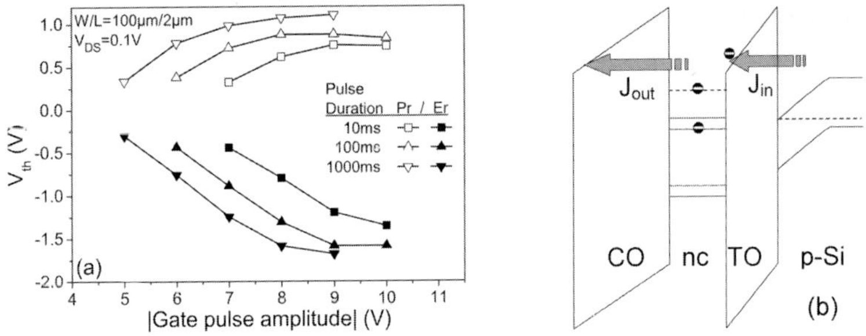

Fig. 9. (a) Experimental programming and erasing memory windows obtained from a Si-nc (~3nm, ~10^{12}cm^{-2})[35] memory transistor for different pulse durations and amplitudes. (b) Energy band diagram of a Si-nc in between TO and CO layers under programming conditions. The current densities and the possible electron states are shown.

Fig. 9a depicts the programming and erasing memory windows obtained from a prototype Si-nc memory transistor for different pulse durations and amplitudes.[35] All V_{th} curves exhibit saturation at different gate voltage depending on the pulse width. It is clear that the lower the pulse duration, the higher the pulse amplitude necessary to start injection. Consequently, V_{th} saturates at higher pulse voltages. The saturation observed during programming (erasing) may be attributed to either the fill of all available storage nodes or to holes (electrons) injected from the gate electrode that compensate any new stored electron in the Si-ncs. It might also be possible that $J_{in} = J_{out}$ where J_{in} is the current density of the injected carriers from the channel to Si-nc and J_{out} is the current density of the ejected carriers from Si-ncs to the gate. It should be also emphasized that above the saturation voltage regime memory window narrowing is observed. For example, in Fig. 9a when 100 ms programming pulses are used, the saturation voltage regime is 8-9 V. When the pulse amplitude becomes 10V then the memory window decreases. This is because the number of stored electrons decreases, which might be due either to leakage of the stored electron to the gate because of the high applied electric field or to significant hole injection from the gate electrode and their subsequent trapping into the ncs. Both these cases lead to $J_{out} > J_{in}$.

The equilibrium conditions (when $J_{in} = J_{out}$ occurs) can be used to estimate and optimize the largest memory window of Si-nc memories[36] in terms of TO and CO material properties (i. e., thickness and barrier height Φ_B).

3.1.1. *Possible source of errors in estimation of charge stored in nanocrystals*

An important parameter affecting the measured memory window of nc memory cells is the possible charge storage into oxide and interface traps.[37,38] The later, include the traps at the oxide/Si channel interface and those existing at the interface between the ncs and the surrounding oxide matrix. It should be emphasized that the curved nc surface results to a highly strained nc lattice at the surface as compared to a plane lattice and therefore the defect density at the interface between the nc lattice and the surrounding matrix is increased.[39] The excess memory window is very difficult to be extracted because this requires knowledge of the trap density and their position in the energy band-gap. However, their contribution could be roughly estimated by capacitance transient measurements. The effect of charge storage in oxide and interface traps becomes more significant when the memory window is extracted by the C-V hysteresis method. The reason is that during the forward and backward voltage sweeps, the nc MOSC structure is subjected to an electric field, which blocks the fast ejection of the excess charges back to the substrate. At this point we notice that the use of nc MOSC structures for memory window extraction, either using pulses or the hysteresis method, should be performed under conditions avoiding deep-depletion effect (DDE).[40] If DDE occurs, charge injection does not take place due to the absence of minority carriers. White light illumination of the sample during the measurements or the formation of a heavily doped region surrounding the effective area of the MOSC should be followed to preclude DDE.

In Fig. 10, the influence of the memory window extracted from the C-V hysteresis of Al gate nc MOSC on p-Si on deep-depletion effect is presented. The effect of light illumination in electron injection during C-V measurements (that determines the "programming branch", see Fig. 10a, and causes positive ΔV_{FB} values, see Fig. 10b) is clear. As expected, hole injection (majority carriers, "erasing branch") is not affected by the presence or the absence of light.

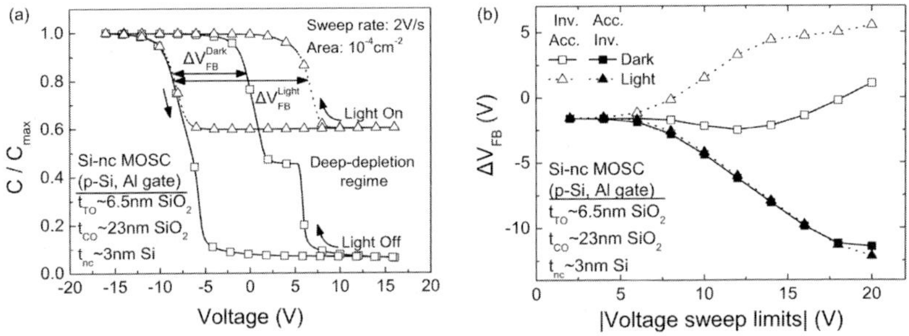

Fig. 10. (a) C-V characteristics obtained after voltage sweep +16V→-16V→+16V at a rate 0.2V/s. When light is off strong deep-depletion occurs affecting the observed hysteresis. (b) Programming and erasing memory windows for nc MOSC structure extracted from C-V hysteresis after forward and backward voltage sweep in dark and under illumination.

Finally, the extraction of ΔV_{th} using a nc memory cell may be sensitive to the method used. In memory applications, the threshold voltage of the cell is determined following the constant current method.[41,42] According to this, the threshold voltage of a MOSFET is obtained from the transfer characteristic at low drain-source potential when a specific value of the drain current is equal to the threshold current which is defined by the following relation[41,42]

$$I_{TH} = (W / L) I_0 \tag{8}$$

where $I_0 \approx$ 50-100 nA, W and L is the width and the length of the transistor's gate. Shallow trench isolation (STI) has been adopted as the standard isolation technique in Flash memory technology.[5] It has been previously investigated[43] that the FET structure in combination with the presence of STI region lead to the formation of two other "parasitic transistors" (FETp) at the channel edges (edge transistors) of the FET structure, which operate in parallel with the intrinsic transistor (FETi) at the central region of the channel (see Fig. 11a and Fig. 11b inset). Therefore, the measured transfer characteristic is the sum of the transfer characteristics of both transistors (see Fig. 11b). Because of the different V_{th} values between FETi and FETp, a "hump" appears in the measured transfer characteristic. Obviously, the extracted memory window ΔV_{th} depends on the position of the hump, i.e., on the adopted value of the threshold current.

Recently, it has been shown in the case of Si-nc memory cells that FETp could be programmed and erased like any other memory cell;[44,45] a parasitic memory transistor action occurs in parallel with the intrinsic memory cell. Furthermore, it was found that the memory characteristics (in terms of program/erase speed, electron/hole injection onset electric fields) of the parasitic transistor are different from those of FETi. The memory behavior of FETp is attributed to the presence of ncs at the channel edges in the vicinity of the STI. Lastly, the validity of this hypothesis has been reinforced by TEM studies.[46] Under this condition, the hump characteristics are partially controlled from the characteristics of the parasitic memory transistor, leading to erroneous ΔV_{th} extraction using the constant current method.

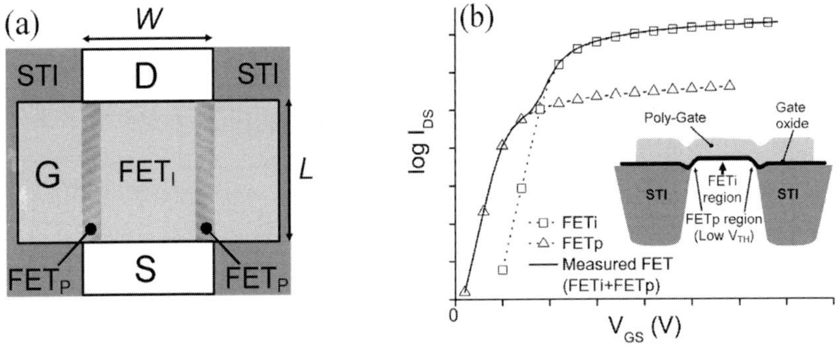

Fig. 11. Schematic representation of (a) plane-view layout of typical STI MOSFET where the major layers and regions are indicated and (b) transfer characteristics of FETi, FETp and actually measured MOSFETs. The inset represents the area of FETi action (central region) and the area of FETp action (channel edges) with lower V_{TH} value compared to that of FETi.

Additionally, studies on deep submicron devices revealed[45] that as the channel dimensions are getting smaller, the action of the parasitic memory transistors becomes stronger and thus the characteristics of the nc memory cells are determined mainly by the nc distribution at the channel edges rather than the ncs located above the central region of the channel. Thereby, significant deviations from the targeted memory windows are expected.

3.2. *Reliability considerations*

Basic aspects on reliability of nc NVM devices include: (a) endurance to repetitive program/erase (P/E) operations; a figure of merit that refers specifically to the ability of the device to withstand a large number (typically on million or more) of P/E cycles (Fig. 12a) and, (b) charge (or data) retention time (referred herein as t_R) which represents a measure of the lifetime of the stored information when the device is programmed or erased; a figure of merit that denotes the capability of the device to retain the stored charge for a time as long as 10 years for the purpose of non-volatility. Typical results on reliability aspects of nc NVM cells are presented in article.[47]

3.2.1. *Endurance of nc memory cells*

Endurance failure of nc NVM cells to repeated program and erase operations is mainly caused by degradation of the gate dielectric stack. As mentioned previously, nc NVM cells can withstand a large number of P/E cycles until malfunction operation appears, i.e., until significant degradation (e.g. closure) or drift of the memory window is detected. The latter effect commonly refers to a parallel shift of the programming (V_{thP}) and erasing (V_{thE}) threshold voltages of the cell with P/E cycling while the memory window remains the same ($\Delta V_{th} = V_{thP} - V_{thE} = $ const.).

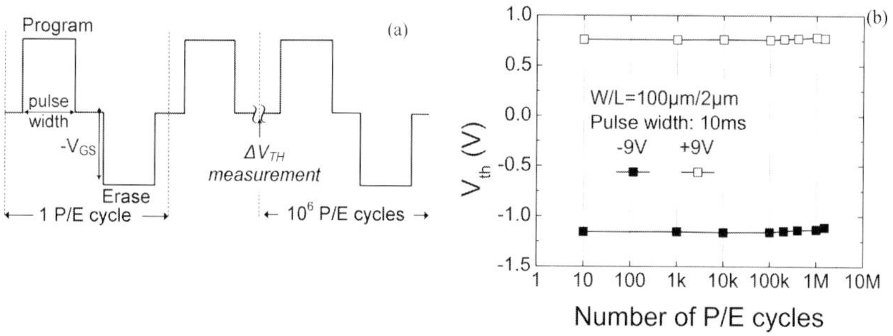

Fig. 12. (a) Schematic representation of the endurance test procedure and applied P/E cycle sequence. (b) Measured memory window after various P/E cycles on Si-nc NVM cell (t_{nc}~3nm, t_{CO}~5-6nm and t_{TO}~6-7nm). The cell was pre-stressed by 1million P/E cycles. V_{th} remains constant for program and erase states revealing neither degradation nor drift in the memory window.

In Fig. 12b, results of endurance tests performed on single IBS-nc NVM cells[35] are presented. The cell was programmed and erased using F-N mechanisms and has a 6-7 nm thick tunnel oxide. No significant V_{th} change was detected after 2×10^6 P/E cycles for both memory states; a figure that conventional Flash memory cells cannot achieved. Endurance to 10^5 P/E cycles without tail bits has been reached under NAND operating conditions for 256 Kb[29] and 1 Mb[28] array of nc NVM cells (0.15 µm technology design rules) using a 4 nm-thick tunnel oxide.

As expected, endurance failure in nc NVM cells was mainly reported under NOR operating conditions. This relates to the channel hot electron injection (CHEI) conditions applied for programming. CHEI produces a large amount of defects in the gate dielectric stack that prevent the cell to endure a large number of P/E cycles. Significant drifts in memory window after 10^5 and 5×10^3 P/E cycles have been reported for 4 Mb[22] (90nm technology node, t_{TO} = 4 nm) and 1 Mb[28] (0.15μm technology node, t_{TO}= 5.5 nm) array of nc NVM cells, respectively. In the latter case (1 Mb array), further increase in P/E cycles lead to a substantial shrinkage of the memory window, ΔV_{th}; only ~20% of the initial window remains after 10^5 P/E cycles.

In general, the observed shifts in V_{thP} and V_{thE} are attributed to the creation of interface and bulk oxide traps. Traps at the channel/tunnel oxide interface contribute to the detected increase in MOSFET subthreshold current and thus, to the increase in subthreshold swing of the transfer characteristics. Obviously, care should be paid to avoid current values in the subthreshold region for memory window calculation when the constant current method is applied for V_{th} extraction. The traps originating from stress experiments are primarily electron traps[48] and therefore, their filling introduces a repulsive electric field that degrades the effectiveness of the erase pulses (F-N operation in NAND and NOR architectures). This is a reasonable scenario to explain the memory window closure after a high number of P/E cycles.

3.2.2. *Charge retention of nc memory cells*

Charge or data retention is one of the most important characteristics of memory devices, independently if these are allocated to NVM applications or not. Retention constitutes a forth mode of operation of a memory device in addition to the other three modes of program, erase and read. The energy band-diagram of a programmed nc NVM single nMOSFET cell is shown in Fig. 13a. For an erased cell, the conduction and valence bands should have opposite slopes. According to Fig. 13a, an internal electric field develops across the tunnel and control oxide layers. The control oxide is thick enough to prevent leakage of stored electrons.

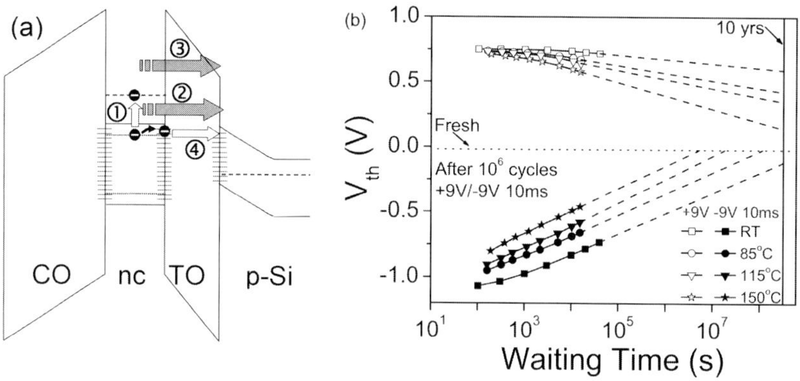

Fig. 13. (a) Schematic representation of electron storage in a nc memory cell during retention mode, i.e., no external bias is applied. The presence of an electric field across the tunnel and the control oxides is due to the internal electric field produced by the stored charges. Described mechanisms: (1) thermal de-trapping of electrons from a deep level inside the nc, (2) direct-tunneling or trap-assisted-tunneling, (3) F-N tunneling and (4) trap-to-trap tunneling or trap-to-band tunneling. (b) Experimental charge retention characteristics at elevated temperatures for the device used in Fig. 12b.

The measurement procedure of the retention time comprises three basic steps. First, the memory device under test is selected and programmed or erased. Next, the device is stored in a controlled temperature place (e.g., a furnace) with no bias supplied to the MOSFET electrodes. After a pre-determined time has elapsed, the device is get-out of the storage place and the threshold voltage is measured at room temperature. This procedure is repeated until the total storage (or waiting, or bake) time exceeds 10^4 s. Semi-logarithmic plots of V_{th} vs waiting time (t) at different temperatures are presented in Fig. 13b. V_{th} exhibits linear dependence with time, suggesting that $V_{th}(t) = K_1 - K_2\log(t)$, where K_1 and K_2 are constants. Of course, any change in V_{th} is proportional to the charge loss from the trapping sites to the substrate such as:

$$\Delta V_{th} = K \cdot Q_{nc}(t) \tag{9}$$

Although a theory concerning the program, erase and read operations of nc NVM has been already described herein, a realistic scenario regarding the storage sites of electrons and holes is still missing. There are many potential storage sites (see Fig. 13a): (i) the confined energy levels in the ncs, (ii) the interface states between the ncs and the surrounding oxides, (iii) deep traps into the ncs, and (iv) oxide bulk traps. In order to design memory cells with specific charge retention characteristics it should be clear where the charge storage sites are located and which loss mechanisms take place. This can be done only by means of comprehensive experimental studies taking into account various parameters such as baking temperature, pre-stress conditions, pre-conditioning, etc. Two common characteristics of charge retention in nc NVM devices may be distinguished from published experimental data:[20] (i) the retention time is very long (~10 years) even for highly stressed samples and (ii) the retention time strongly depends on the ambient temperature. Thus, the possibility of charge storage in the confined levels of the nc conduction band should be excluded; such confined levels are usually lying at the same or higher position than the Si conduction band causing leakage of the stored charges back to the channel via F-N or direct or trap assisted tunneling (TAT) (see Fig. 13a). Moreover, in opposition to the experimental results all these tunneling mechanisms are slightly dependent on temperature. The effect on charge loss of nc/SiO_2 and SiO_2/channel interface states and bulk oxide traps can be eliminated after forming gas annealing (FGA, e.g., 90%N_2+10%H_2) due to hydrogen passivation.[37,38,49,50] After FGA treatment, the improvement is twofold: First, the "background" memory window due to oxide and interface traps is eliminated and second, trap contribution to charge loss via trap-to-trap and/or trap-to-band tunneling mechanisms (Fig. 13a) during retention is minimized. Consequently, the experimentally observed long retention times should be attributed to the presence of deep-traps inside the ncs.[20,37-39] A more detailed description of the nature and characteristics of traps inside Si-ncs and Ge-ncs can be found elsewhere.[38,50] Very few attempts to model the retention characteristics of nc NVM devices have been presented.[20,51,39,52]

The probability for stored carriers to tunnel-back to the channel by trap-to-trap or trap-to-band in case of a high quality tunnel oxide, without suffering from high density of interface states, is extremely low. Such a scenario where mechanism 4 (see Fig. 13a) is dominating could explain the observed discharging process only at room temperature. On the other hand, electrons stored at discrete energy levels in the conduction band of the ncs can easily tunnel-back due to the strong internal electric field. At elevated temperatures the trapped carriers gain enough thermal energy to be de-trapped depending on the activation energy E_A of the trap, i.e., on the relative position of the trapping level with respect to the conduction or valence band edges. The emission rate of carriers from a deep-trap at temperature T is generally described by the following equation:[38,53]

$$e_{n,p}(T) = \gamma_{n,p} T^2 \exp(-qE_A / kT) \quad (10)$$

where the subscript n (p) is used when the emitted carriers are electrons (holes), $E_A = E_C - E_T$ ($= E_T - E_V$) is the electron (hole) trap activation energy and

$$\gamma_{n,p} = 4\sqrt{6}\sigma_{n,p} g k^2 m^*_{e,h} / (\pi\hbar^2)^{3/2} \quad (11)$$

where σ is the capture cross section of the trap, g is the degeneracy of the trap level, m^* is the carrier effective mass in the oxide. Thus, the charge loss rate is directly proportional to the probability of an electron to escape from the nc per unit time which depends on the emission rate.[20,38] Baik *et al.*[39] proposed a method to calculate the activation energy of deep traps inside Si-ncs from charge retention data. The method is based on the temperature-dependent-transient of the I_{DS} current in a Si-nc NVM single cell after programming or erasing until $I_{DS}(t) = 0.9 I_{DS}(\infty)$; $I_{DS}(\infty)$ denotes the current value at steady state. The inverse of the transient time constant τ_E is directly proportional to the discharging current, assuming that the injected charge is the same at each temperature. Therefore, the activation energy of the traps, where the carriers were stored and emitted from, can be calculated by the Arrhenius plot $1/\tau_E T^2$ versus $1/T$. The application of this method to the retention characteristics of the program state for Si-nc[39] and Ge-ncs[38] NVM single cells revealed the presence of two different slopes in the Arrhenius plots, indicating the presence of two different mechanisms. The first mechanism appears at low temperatures and has an insignificant dependence on the temperature while the second one occurs at highest temperatures with activation energy of 0.2 eV and 0.13 eV for Si-ncs and Ge-ncs NVM devices, respectively. It is claimed that the first mechanism relates to the tunneling of carriers directly from their trapping sites inside the ncs to the TO/channel interface states (mechanism 4, in Fig. 13a) and the second one is linked to the thermal de-trapping of carriers and subsequent, internal-electric-field-assisted-tunneling to the channel (mechanism 1+2 in Fig. 13a). It should be here emphasized that it is necessary for transient recording of I_{DS} to bias the gate and drain of the transistor; a condition that may affect the charge de-trapping process.

A more sophisticated approach to model the temperature dependence of the charge loss has been proposed by Schuler *et al.*[54] for conventional FG NVM cells. This approach is based on the multi-phonon-assisted tunneling mechanism that considers an electron-phonon coupling model. According to this model, the charge loss occurs via tunneling to different trapping levels present into the TO (e.g., after cycling) and the capture and emission rates of these traps depend on the phonon transition probability.

In Fig. 14a, the charge retention behavior of Si-nc NVM devices at elevated temperatures is presented. The retention characteristics for the program state at both 200°C and 250°C exhibit two charge loss regimes and reveal a slowing down of the discharging process after long waiting times. This is probably attributed to the contributions of both the internal electric field forced by the trapped charges and the defects in the TO. It has been shown for FG NVM[54] that the slow-down behavior of the charge loss process cannot be explained assuming trap-assisted-tunneling with a single trapping level in the TO. However, this could happen assuming an electron-phonon coupling model with the participation of more than one trapping level in the TO. A possible scenario to explain the slow-down behavior of the retention characteristics is the following: At early stages of charge loss the internal electric field is strong enough for the trapping level Φ_{t1} (Fig. 14b) to be aligned with the trap site inside the ncs. So, discharging might occur through a trap-assisted tunnelling. Simultaneously, the thermally de-trapped electrons (mechanism 1, in Fig. 13a) can tunnel-back to the Si substrate due the strong electric field via a trap-assisted conduction mechanism. After long waiting times, the internal electric field is significantly reduced causing a misalignment between the trapping level into the ncs and the traps in the TO. In this case, only the thermally excited electrons have the possibility to gain enough energy to tunnel into the TO defects. As a consequence the resulting charge loss rate is reduced.

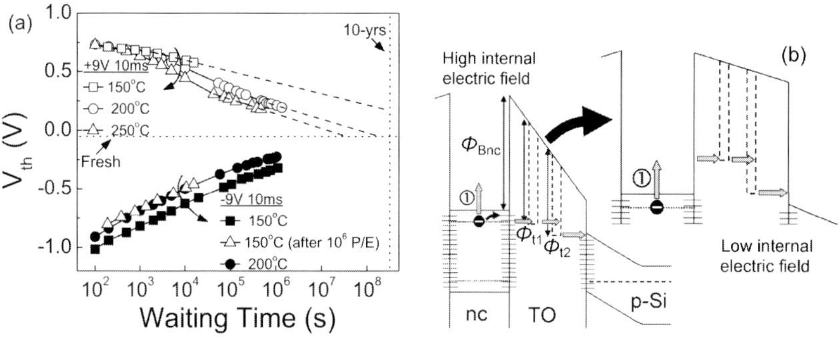

Fig. 14. (a) Charge retention characteristics obtained at elevated temperatures for both program and erase states. The dotted lines represent the extrapolation curves to complete loss of the stored electrons following two different loss rates indicating the slow-down of the charge loss process. (b) Energy band diagrams according to the model adopted to explain the reduction of the charge loss rate after 5×10^4 s.

Contrary to electron de-trapping, the mechanism responsible for hole de-trapping, that controls the retention time of the erase state, presents a poor dependence on temperature (Fig. 13a).[35,44] This observation suggests that holes are stored in deep traps into the ncs and leak to the substrate via trap-to-trap tunneling (mechanism 4, in Fig. 14a). This also suggests a high concentration of hole traps in the TO. Such a hole trap concentration appears to be independent on P/E cycling as shown in Fig. 14a where the same charge loss rate at 150°C is detected for cycled and un-cycled devices. In addition, the absence of charge loss rate reduction for high baking times indicates that the hole traps are located deeper inside the energy band gap of the ncs than the electron traps, i.e., higher thermal energy should be transferred to the trapped holes in order to observe thermal de-trapping.

3.3. *Optimization of memory characteristics*

Research on nc NVM technology revealed a number of limitations coming either from the nc fabrication techniques or from the gate stack architectures. As the feature sizes (< ~10 nm) of functional ncs lie well below the current photolithography resolution, research in fabrication focused on the development of self-assembly techniques. Many synthesis techniques (see section 2.3) have been explored and tested to fabricate nanocrystals in dielectrics. Despite undeniable advancements, these techniques fail to fulfill basic requirements, especially regarding the fabrication and control of high-density arrays of ncs uniform in size and interspacing. As a result, these techniques cannot avoid fluctuations in device performance and fail to exploit size-dependence effects. An attractive alternative to meet the above nc requirements is through the use of self-assembly templates as it has been recently demonstrated in the case of diblock copolymer-assisted-patterning.[55,56] However, fabrication of high-density arrays of ncs with size smaller than 10nm has not been demonstrated yet and remains a tough challenge for this technique.

Another alternative in fabrication is the introduction, after nc synthesis of additional processing for surface coverage and nc size optimization. In this direction, various thermal oxidation processes have been demonstrated to mutually isolate neighboring Si-ncs and to separate elongated Si-ncs. Application of thermal treatments in oxidizing ambient requires a tight control of the oxidation rate of the ncs. Such a condition can be fulfilled by oxynitridation processes under NO or N_2O as it has been shown in the case of CVD Si-ncs.[57] Compared to pure oxygen atmosphere, oxidation of Si-ncs is significantly retarded under NO or N_2O ambient, probably due to the formation of high-density Si-N bonds that block the diffusion of oxygen species. Another alternative has been proposed by Normand *et al.*[58] in the case of ion-beam-synthesized Si-ncs in SiO_2.This approach makes use of thermal treatments in N_2 diluted O_2. TEM studies[59] in conjunction with electrical investigations[60] have further revealed that such a process (a) reduces the Si-nc size, (b) separates the elongated Si-ncs, (c) enhances the Si-nc size uniformity and (d) induces passivation of the Si-nc/SiO_2 interface. The latter decreases the density of states at the Si-nc/SiO_2 interface and thereby, improves the charge retention time. Further oxidation experiments[61] in N_2 diluted O_2 or pure O_2 for various temperature and time regimes have shown that under appropriate conditions, the oxidation of Si-ncs is a self-limiting process. Charge retention improvement has been also demonstrated after Si-nc/SiO_2 interface passivation by plasma nitridation[62] or suitable annealing in N_2.[29] Such processing allows formation of a thin silicon nitride layer with high-density of intrinsic defects (i.e. a high-density of trapping sites) at the nc outer interface.

According to Eq. (6) and Fig. 8, replacement of the control silicon oxide with an insulating material (or a stack of materials) of higher dielectric constant would increase significantly the memory window, ΔV_{th}. In addition,[20] as the physical thickness of a control dielectric layer with higher dielectric constant is larger compared to that of the SiO_2, keeping C_{CO} constant, the new control dielectric prevents further the leakage current from the ncs to the gate electrode. This has been demonstrated in the case of control dielectrics made of Oxide-Nitride-Oxide (ONO) stacks[46] and various high-k dielectrics.[36,63] The main drawbacks in utilizing high-k dielectrics as control dielectrics are: (a) the lower barrier height Φ_B and (b) the relative high density of bulk traps compared to SiO_2. These two parameters increase, at high gate voltage pulses, the probability to inject charge from the gate electrode which subsequently may be trapped into the high-k dielectric layers.

On the other hand, high-k dielectrics may have significant advantages when used as tunneling dielectrics. As an example, the lower potential barrier for a larger physical thickness of high-k dielectric materials compared to SiO_2 can be successfully exploited for charge retention improvements and faster programming at lower voltages in the case where programming refers to electron injection from the channel. A comprehensive study on that point is presented in Ref. 64. This study examines the major high-k dielectrics as single layers or multilayer stacks in order to find the most suitable tunnel dielectric barrier for optimum F-N program/erase operations and charge retention. It is concluded that tri-layer stacks made of $Si_3N_4/Al_2O_3/Si_3N_4$ and $ZrSiO_x/Al_2O_3/ZrSiO_x$ could improve both the operation speed and retention time and therefore are potential candidates as tunneling dielectrics for nc NVM devices. Nitride/Oxide/Nitride (NON) tri-layer tunneling stack has been already demonstrated[39] as a highly effective tunnel barrier for nc NVM cells.

Many memory structures using ncs embedded in high-k gate dielectrics have been reported in the literature. Among them, the formation of Ge-ncs in HfAlO[65] and their integration in an industrial

Flash CMOS process should retain particular attention. HfAlO material exhibits a high crystallization temperature and therefore, may sustain the high-temperature annealing regimes required for nc formation. Promising results in terms of operating voltage, speed, endurance and retention have been obtained for single Ge-ncs-HfAlO memory cells under the F-N program/erase (NAND) scheme. It should be noticed that Hf-based dielectrics are more favorable materials for the formation of Ge ncs compared to Si-ncs. This relates to the change in Gibbs formation energy of GeO_2 that is smaller than SiO_2 and HfO_2.[65] Formation of functional high-k Si-ncs structures has been demonstrated mainly for silicon nitride materials using high-thermal budget techniques like LPCVD[66] or IBS[67] and, in the case of HfO_2 by remote plasma enhanced CVD (~300°C).[68] Finally, it should be emphasized that due to their low-temperature production process, metallic materials (e.g. Ni by sputtering[69] or physical vacuum deposition[70]) would be preferred to Si for nc formation in high-k dielectrics. More recently, (PVD) of magnetron sputtering-generated-ncs (room-temperature).[70]

Another approach aiming at nc NVM optimization has been proposed by Ohba *et al.*[71] It relates to the fabrication by CVD of two Si-nc layers where the Si-ncs of one layer are vertically aligned to those of the second layer. The first layer, the closest one to the substrate, should have smaller ncs than the second one. The tunnel oxide thickness and the oxide thickness between the two nc-layers might be as low as 3nm to preserve the nc NVM advantage of fast programming speeds at low voltages. Due to energy confinement effects, the conduction band of a Si-nc in the second layer is lower than the one of a Si-nc located in the first layer; a feature that allows rapid transfer of charge stored in the first layer to the second one. This also insures long retention times since a possible tunnel-back mechanism of the electrons stored in the second layer (at retention mode) requires a transfer to a higher energy level (first nc layer) through a tunneling barrier that is a process with negligible probability. The double nc-layer approach has been applied also in the case of metal ncs embedded either in SiO_2[72] or Si_3N_4.[73] More recently, promising nc NVM devices using aligned double layers of nickel-silicide ncs with 2.7×10^{12} cm^{-2} surface density have been demonstrated.[74]

4. State of the Art, Novel Devices and Open Issues

Manufacturability is a critical step for ruling on the maturity and viability of the nc NVM technology. The most convenient and easiest approach for ncs integration is to keep the same device architecture and process flow as used in conventional Flash and replace the steps of poly-Si by those of ncs fabrication. Several manufacturers have followed this approach that required few modifications in already established technologies. Freescale[22] demonstrated 4Mb Si-ncs NVM cells at 90nm technology node and claimed that the company is ready to commercialize 24-Mb nc NMV products.[75] According to reference,[22] fabrication of a floating-gate made of ncs instead of poly-Si decreases significantly the number of processing steps and therefore, reduces manufacturing complexity and fabrication cost. STMicroelectronics presented[28] 1Mb test arrays following 0.15-0.18 μm design rules and examined the performance of both NAND and NOR architectures. For NOR devices, the storage of two bits per cell has been demonstrated using CHE injection into ncs located in a narrow region close to the drain. Such a local charge trapping, a feature that cannot be achieved in the case of conventional poly-Si FG, affects the channel conductivity only underneath the nc charged area. Therefore, different I_{DS} values can be read if source and drain electrodes are mutually interchanged. Although, promising device results have been presented, this work emphasized the critical issue of nc-size and surface coverage non-uniformity which affects the V_{th} shift either for single or double bit storage per cell. Following the same cell fabrication process but using a different nc synthesis technique, more uniform memory windows have been achieved across a 200-mm wafer.[44] More recently STMicroelectronics[46] demonstrated a 16-Mb NOR Si-nc NVM by modifying substantially conventional poly-Si FG processing in order to optimize the density of Si-ncs, program/erase operations and retention time. Lately, ATMEL[76] presented 32 Mb Si-nc NOR Flash memory products processed in 0.13 μm technology platform, using 5 nm and 10 nm thick SiO_2 as tunnel and control oxide respectively. Comprehensive statistical evaluation and benchmarking among samples with different nc sizes revealed the robustness of nc memory devices for NVM Flash applications.

In spite of many research and development efforts, it is still unclear which type of memory will be used below the 45 nm node. In this direction, the promising results obtained in nc-NVM technology have prompted many research institutions and NVM manufacturers to integrate this technology in new device architectures. Interesting SOI laboratory prototypes have been demonstrated by Hiramoto *et al.*[77] Using Ultra-thin body Silicon-On-Insulator (SOI) substrates, Si-nc NVM MOSFETs having a second layer of Si-ncs embedded at the buried interface of the buried oxide were fabricated. These double-gate SOI transistors are used as single memory cells where more than two bits can be stored depending on the charge state of the nc layer in the buried oxide.

Recently, nc-NVM devices using FinFET technology, an emerging MOSFET architecture for the 45nm node and beyond, have been reported. Choe *et al.*[78] demonstrated FinFET nc NVM devices (smallest fin's width: 20 nm) in a manufacturing environment using state-of-the-art semiconductor processing and technology like TaN metal gates, atomic layer deposition (ALD) for Al_2O_3 blocking dielectrics, post-deposition anneal (PDA) and plasma doping (PLAD) to provide product-wise solutions for the next generation NAND Flash memory. They succeeded in realizing memory structures having a memory window of more than 8V that is applicable for MLC NAND operation. FinFET Si-ncs NVM devices with similar performance have been achieved by Gerardi *et al.*[79] using ONO stack as control oxide.

Finally, three other innovative nc NVM device architectures and concepts should deserve particular attention. First, the split-gate architecture that was successfully demonstrated in manufacturing environment (90 nm node) for Si-nc NVM devices.[80-82] Such devices exhibit reliable performance with promising scalability features. Second, the optical Si-nc NVM concept for which programming and erasing are performed using optical signals.[83] This original operation scheme in nc NVMs is particularly interesting for applications in optical communication systems. At last, it should be mentioned the recent nc-NVM device concept based on a nano-electro-mechanical system (NEMS) reported in Ref. 84. The resulting memory switching devices utilize a MOSFET structure where the gate dielectric is a free to bend SiO_2 membrane with embedded Si-ncs.

5. Summary

The unclear future of conventional poly-Si FG NVM devices below the 45 nm technology node forces many research groups and companies to develop alternative memory device technologies. The nc NVM approach that makes use of ncs as discrete and mutually isolated charge storage nodes is appealing in this regard. Improved endurance to repetitive cycling and charge retention under harsh operating conditions of nc-NVM devices compared to conventional FG devices have been demonstrated. After more than ten years since the date of the first publication, a fairly comprehensive picture has emerged of the physical mechanisms related to charging, discharging and charge storage. At integration level it has been found that formation of ncs can be achieved in a great variety of dielectric materials and MOSFET architectures, like high-k and metal gate MOSFETs, FinFETs, split-gate MOSFETs. The number of memory manufacturers involved in nc NVM fabrication increases progressively and functional memory arrays as large as 32 Mb are now demonstrated. Recent statistical data indicate significant improvement in reproducibility and reliability of Si-nc NVM devices for the 45 nm technology node and beyond. In regard to its flexibility and advantages, nc NVM technology is also attractive for new device concepts (e.g. optical memory) and applications like optical communication systems and NEMS.

References

1. T. Bukowski and J. H. Simmons, *Critical Reviews in Solid State and Materials Sciences* **27**, 119 (2002).
2. P. Harrison, *Quantum Wells, Wires and Dots (2nd Ed.)*, Wiley (2005).
3. E. Kapetanakis, P. Normand, K. Beltsios and D. Tsoukalas, Encyclopedia of Nanoscience and Nanotechonology, Ed. H. S. Nalwa, Vol. 6, 321, ASP (2003).
4. http://public.itrs.net (2006).
5. R. Bez, E. Camerlenghi, A. Modelli and A. Visconti, *Proc. IEEE* **91**, 489 (2003).
6. A. Fazio, *MRS Bulletin* **29**, 814 (2004).

7. C. Clementi and R. Bez, *Materials and Processes for Nonvolatile Memories*, Ed. A. Claverie, D. Tsoukalas, T-J. King and J. M. Slaughter (Mater. Res. Soc. Symp. Proc. **830**, Warrendale, PA , 2005), D1.2
8. T. P. Haraszti, *CMOS Memory Circuits*, Kluwer, New York (2002).
9. R. S. Shen, *et al. Flash Memories*, in *The VLSI Handbook*, ed. W.-K.Chen, Boca Raton, CRC Press LLC, (2000).
10. P. A. Gargini, *J.Nanoparticle Res.* **6**, 11 (2004).
11. J. De Blauwe, *IEEE Trans. Nanotechnology* **1**, 72 (2002).
12. B. Govoreanu, D. P. Brunco and J. Van Houdt, *Solid-State Electronics* **49**, 1841 (2005).
13. S. Tiwari, F. Rana, H. Hanafi, A. Hartstein, E. F. Crabbe and K. Chan, *Appl. Phys. Lett.* **68**, 1377 (1996).
14. J. J. Welser, S. Tiwari, S. Rishton, K. Y. Lee and Y. Lee, *IEEE Electr. Dev. Lett.* **18**, 278 (1997).
15. H.-S. P. Wong *et al.*, *Proc. IEEE* **87**, 537 (1997).
16. S. Tiwari, J. Welser and F. Rana, *Proc. Symp. VLSI Technology*, 133 (1997).
17. S. Tiwari, *Appl. Phys. A* **71**, 403 (2000).
18. S. Tiwari, F. Rana, K. Chan, L. Shi and H. Hanafi, *Appl. Phys. Lett.* **69**, 1232 (1996).
19. H. Silva, M. K. Kim, U. Avci, A. Kumar and S. Tiwari, *MRS Bulletin* **29**, 845 (2004).
20. M. She and T.-J. King, *IEEE Trans. Elec. Dev.* **50**, 1934 (2003).
21. R. Rao *et al.*, *Solid-State Electr.* **48**, 1463 (2004).
22. R. Muralidhar *et al.*, *IEDM Tech. Digest* (IEEE), 601 (2003).
23. Z. Liu, C. Lee, V. Narayanan, G. Pei and E. C. Kan, *IEEE Trans. Electr. Dev.* **49**, 1606 (2002).
24. A. Kanjilal *et al.*, *Appl. Phys. Lett.* **82,** 1212 (2003).
25. L. W. Teo, W. K. Choi, W. K. Chim, V. Ho, C. M. Moey, M. S. Tay, C. L. Heng, Y. Lei, D. A. Antoniadis and E. A. Fitzgerald, *Appl. Phys. Lett.* **81**, 3639 (2002).
26. C. Bonafos *et al.*, *J. Appl. Phys.* **95**, 5696 (2004).
27. J. S. de Sousa, J.-P. Leburton, A. V. Thean, V. N. Freire and E. F. da Silva, Jr. *Appl. Phys. Lett.* **82**, 2685 (2003).
28. B. De Salvo *et al.*, *IEDM Tech. Digest* (IEEE), 597 (2003).
29. S. Lombardo, B. De Salvo, C. Gerardi and T. Baron, *Microelec.Eng.* **72**, 388 (2004).
30. K. Kim, *IEEE Asian Solid-State Circuit Conference 2005*, 7-11 (2005).
31. C.-Y. Lu, T.-C. Lu and R. Liu, *13th International Symposium on the Physical and Failure Analysis of Integrated Circuits 2006*, (IEEE), 18-23.
32. P. Dimitrakis *et al.*, *Mat. Scienc. Eng. B* **101**, 14 (2003).
33. D. Schroder, *Semiconductor material and device characterization*, 2nd Ed., Wiley (1998).
34. T. Ishii, T. Osabe, T. Mine, F. Murai and K. Yano, *IEDM Tech. Digest 2000*, 305.
35. P. Dimitrakis *et al.*, *Solid-State Electronics* **48**, 1511 (2004).

36. C. M. Compagnoni, D. Ielmini, A. S. Spinelli and A. L. Lacaita, *IEEE Trans. Electr. Dev.* **52**, 2473 (2005).
37. Y. Shi, K. Saito, H. Ishikuro and T. Hiramoto, *J. Appl. Phys.* **84**, 2358 (1998).
38. B. H. Koh, E. W. H. Kan, W. K. Chim, W. K. Choi, D. A. Antoniadis and E. A. Fitzgerald, *J. Appl. Phys.* **97**, 124305 (2005).
39. S. J. Baik, S. Choi, U-In Chung and J. T. Moon, *Solid-State Electron.* **48**, 1475 (2004).
40. S. M. Sze, *Physics of semiconductor devices*, 2nd Ed., Wiley (1981).
41. T. Hori, *Gate dielectrics and MOS ULSI: principles, technologies and applications*, Springer (1997).
42. A. Ortiz-Conde, F. J. Garcia Sanchez, J. J. Liou, A. Cerdeira, M. Estrada and Y. Yue, *Microelectron. Reliab.* **42**, 583 (2002).
43. S. C. Lin, J. B. Kuo, K. T. Huang and S. W. Sun, *Solid-State Electron.* **42**, 1871 (1998).
44. P. Dimitrakis and P. Normand, *Mater. Res. Soc. Symp. Proc.* **830**, D5.1 (2005).
45. P. Dimitrakis and P. Normand, *Solid-State Electron.* **51**, 125 (2007).
46. C. Gerardi *et al.*, *IEEE Trans. Elect. Dev.* **54**, 1376 (2007).
47. B. Salvo *et al.*, *IEEE Trans. Dev. Mater. Realiab.* **4**, 377 (2004).
48. H. P. Belgal, N. Righos, I. Kalastirsky, J. J. Peterson, R. Shiner and N. Mielke, *Inter. Reliab. Phys. Sympos.* (IEEE), 7 (2002).
49. A. R. Wilkinson and R. G. Elliman, *Appl. Phys. Lett.* **83**, 5512 (2003).
50. Y. Shi, K. Saito, H. Ishikuro and T. Hiramoto, *Jpn. J. Appl. Phys.* **38**, 425 (1999).
51. D. Ielmini, C. M. Compagnoni, A. S. Spinelli, A. L. Lacaita and C. Gerardi, *Int. Reliability Physics Symp.* (IEEE), 515 (2004).
52. T.-H. Hou, C. Lee, V. Narayanan, U. Ganguly and E. C. Kan, *IEEE Trans. Elect. Dev.* **53**, 3095 (2006).
53. M. Lanoo and J. Bourgoin, *Point Defects in Semiconductors II: Experimental Aspects*, Springer-Verlag, NY (1981).
54. F. Schuler, R. Degraeve, P. Hendrickx and D. Wellekens, *Inter. Reliab. Phys. Sympos.* (IEEE), 26 (2002).
55. K. W. Guarini, C. T. Black, Y. Zhang, I. V. Babich, E. M. Sikorski and L. M. Gignac, *IEDM Tech. Digest* (IEEE), 541 (2003).
56. D. Shahrjerdi, D. I. Garcia-Gutierrez and S. K. Banerjee, *IEEE Elect. Dev. Lett.* **28**, 793 (2007).
57. K. C. Scheer, R. A. Rao, R. Muralidhar, S. Bagchi, J. Conner, L. Lozano, C. Perez, M. Sadd and B. E. White, Jr., *J. Appl. Phys.* **93**, 5637 (2003).
58. P. Normand *et al.*, *App. Phys. Lett.* **82**, 168 (2003).
59. C. Bonafos *et al.*, *Solid-State Electron.* **49**, 1734 (2005).
60. D. Tsoukalas, P. Dimitrakis, S. Koliopoulou and P. Normand, *Mater. Sci. Engin. B* **124**, 93 (2005).
61. H. Coffin, C. Bonafos, S. Schamm, N. Cherkashin, G. Ben Assayag, A. Claverie, M. Respaud, P. Dimitrakis and P. Normand, *J. Appl. Phys.* **99**, 044302 (2006).

62. S. Huang, K. Arai, K. Usami and S. Oda, *IEEE Trans. Nanotech.* **3**, 210 (2004).
63. E. Spitale, D. Corso, I. Crupi, S. Lombardo and C. Gerardi, *Microelectr. Reliab.* **45**, 895 (2005).
64. J. D. Casperson, L. D. Bell and H. A. Atwater, *J. Appl. Phys.* **92**, 261 (2002).
65. J. H. Chen, Y. Q. Wang, W. J. Yoo, Y.-C. Yeo, G. Samudra, D. S. H. Chan, A. Y. Du and D.-L. Kwong, *IEEE Trans. Elect Dev.* **51**, 1840 (2004).
66. R. F. Steimle, M. Sadd, R. Muralindhar, R. Rao, B. Hradsky, S. Straub and B. E. White, Jr., *IEEE Trans. Nanotech.* **2**, 335 (2003).
67. V. Ioannou, P. Dimitrakis, V. E. Vamvakas, P. Normand, C. Bonafos, S. Schamm, A. Mouti, G. Ben Assayag and V. Paillard, *Nanotechnology* **18**, 215204 (2007).
68. P. Punchaipetch, Y. Uraoka, T. Fuyuki, A. Tomyo, E. Takahashi, T. Hayashi, A. Sano and S. Horii, *Appl. Phys. Lett.* **89**, 093502 (2006).
69. J. J. Lee and D.-L. Kwong, *IEEE Trans. Elect. Dev.* **53**, 507 (2005).
70. E. Verrelli, D. Tsoukalas, K. Giannakopoulos, D. Kouvatsos, P. Normand and D. E. Ioannou, *Microelectron. Eng.* **84**, 1994 (2007).
71. R. Ohba, N. Sugiyama, K. Uchida, J. Koga and A. Toriumi, *IEEE Trans. Elect Dev.* **49**, 1392 (2002).
72. C. Lee, A. Gorur-Seetharam and E. C. Kan, *IEDM Tech. Dig.*, 557 (2003).
73. C. Lee, T.-H. Hou and E. C. Kan, *IEEE Trans. Elect. Dev.* **52**, 2697 (2005).
74. W.-R. Chen, T.-C. Chang, P.-T. Liu, P.-S. Lin, C.-H. Tu and C.-Y.Chang, *Appl. Phys. Lett.* **90**, 112108 (2007).
75. "World's first 24-Mbit silicon nanocrystal memory", *European Semiconductor*, 8th December 2005.
76. S. Jacob *et al.*, *Proc. of ESSDERC*, 410 (2007).
77. T. Hiramoto, I. Kim, M. Saitoh and K. Yanagidaira, *Mater. Res. Soc. Symp. Proc.*, Vol. **830**, D1.5.1 (2005).
78. J.-D. Choe, Y. J. Ahn, S.-H. Lee, D. Jang, Y.-B. Yoon, J. J. Lee, I. Chung, K. Park and D. Park, *Electron. Lett.* **43**, 545 (2007).
79. C. Gerardi *et al.*, *IEEE Non-Volatile Semicon. Mem. Work.*, 44 (2007).
80. J. A. Yater, T. Kirichenko, E. J. Prinz, M. Sadd, R. Steimle, C. T. Swift and K-M. Chang, *IEEE Non-Volatile Semicon. Mem. Work.*, 60 (2006).
81. J. A. Yater, S. T. Kang, R. Steimle, C. M. Hong, B. Winstead, M. Herrick and G. Chindalore, *IEEE Non-Volatile Semicon. Mem. Work.*, 77 (2007).
82. C. M. Hong, J. Yater, S.-T. Kang, H. Gasquet and G. Chindalore, *IEEE Non-Volatile Semicon. Mem. Work.*, 75 (2007).
83. R. J. Walters, P. G. Kik, J. D. Casperson, H. A. Atwater and R. Lindstedt, *Appl. Phys. Lett.* **85**, 2622 (2004).
84. S. Oda S. Y. Huang, M. A. Salem, D. Hippo, A. Tanaka, Y. Tsuchiya and H. Mizuta, *Int. Conf. Solid-State Integ. Circ. Technol.* (IEEE), 1045 (2006).

CHAPTER 9

ENGINEERING THE OPTICAL RESPONSE OF NANOSTRUCTURED SILICON

Joachim Diener

Technische Universität München, Physik-Department E16, Garching, Germany
E-mail: jdiener@ph.tum.de

Minoru Fujii

Department of Electrical and Electronic Engineering,
Graduate School of Engineering, Kobe University,
Rokkodai, Nada, Kobe 657-8501, Japan
E-mail: fujii@eedept.kobe-u.ac.jp

Dmitri Kovalev

Department of Physics, University of Bath, Bath BA2 7AY, United Kingdom
E-mail: D.Kovalev@bath.ac.uk

The preceding miniaturization in microelectronics connected with the application of optical signals for information transfer demands silicon-based optical devices. Almost all the various photonic devices integrated on silicon wafer: e.g. silicon based optical waveguides, splitters, couplers, modulators, photonic crystals, detectors, *etc.* have already been demonstrated. These advances are based on well-developed silicon technology such as lithography, selective etching and silicon oxidation. This chapter describes another approach of engineering different silicon-based optical devices: it is based on in-depth and in-plane nanostructuring of silicon wafers. In-depth variation of the refractive index is performed via porosification of silicon wafers aiming to achieve the desired spectral response of these devices while in-plane nanostructuring results in their polarization-dependent optical properties.

1. Introduction

Despite of the well-known advantages of established silicon (Si)-based manufacturing processes, in recent years, some concerns about the evolution of Si industry have been raised which are related to fundamental materials and processing aspects. Some examples are: the limitations of the operating speed of microelectronic devices due to interconnect bottleneck, the length of the interconnects on a single chip and power dissipation relevant to the length of interconnect. Finally the complexity of the architecture has been recognized to be also a limiting factor. Si micro/nano-photonics is the optical equivalent of integrated microelectronic circuits. Integration of various optical devices with and within integrated circuit electronics is the overall goal of this technology. However, a key limitation to apply Si for active and passive optical devices stems from its indirect band gap and highly isotropic cubic lattice structure.

Silicon photonics mainly aims to use silicon-based optical devices fabricated compatible with standard methods for integrated circuits. There was large optimism in the early nineties that silicon based optoelectronics will become a reality within approximately 10 years.[1,2] Indeed, all the main components have already been demonstrated e.g., silicon based optical waveguides, splitters, couplers, modulators, resonators, photonic crystals, detectors[3] etc. All of them are based on well-defined geometrical profiling of Si or SiO_2-based structures. The first essential component in Si photonics is the medium through which light propagates: the waveguide. To realize low losses optical waveguides, various approaches have been followed;[3] low dielectric mismatch structures (e.g. doped silica or silicon nitride on oxide) or high dielectric mismatch structures (silicon on oxide).

Low loss silica waveguides are characterized by large dimensions, typically 50 μm of thickness, due to the low refractive index mismatch and cannot be integrated on microscale. On the other extreme, silicon on insulator or polysilicon-based waveguides exhibits a large refractive index mismatch and, therefore, small size waveguides in the sub-micrometer range. This allows a large number of optical components to be integrated within a small area or volume. A large number of other photonic components in silicon-on-insulator systems have been demonstrated and commercialised (for detail see Ref. 3). On the other hand it has been demonstrated that transmitted optical signals can be modulated using a variation of the refractive index due to the presence of free carriers[4] and the optical signals can be converted into an electrical signals by using silicon based photodetectors. Most of mentioned Si-based optical devices are based on their well-defined shape and spatial modulation of the refractive index, for example by using alternative layers of high and low refractive index materials.

Nanostructuring is a fast growing field in modern solid state physics. Among other techniques it enables to overcome restrictions entailed with the natural bulk materials via tuning their physical properties as desired for applications. As an example we refer to photonic crystals, artificially created structures consisting of spatially ordered well-defined fragments having sizes comparable to the wavelength of light.[5,6] A strong wavelength-dependent scattering of light leads to the appearance of photonic bandgaps. But even when the wavelength of light is much longer than the constituent's size of a photonic system the propagation of light can be noticeably influenced.

Over the past ten years, there were rising numbers of works devoted to different Si-based nanostructures. Some examples are: SiGe and Si/SiO_2 superlattices[7,8] or Si nanocrystal assemblies.[9] Specifically, most research efforts were devoted to light emission properties of the porous modification of bulk Si produced by electrochemical etching (porous Si, PSi, for details see reviews in Refs. 10, 11, and 12). PSi is prepared via anodization of bulk Si wafers in HF-based solutions. The anodic etching of bulk Si in hydrofluoric acid results in a sponge-like network of interconnected Si nanowires with sizes (from micrometers to nanometers) that depend on the doping level of the Si substrate, the etching-current density, and the concentration of the etching solution used.[10,12] This preparation procedure has attracted much interest due to its simplicity. Specific dielectric properties of PSi differ from those of bulk Si. The absence of a large volume fraction of Si results in a reduced value of the dielectric constant ε and the refractive index n.

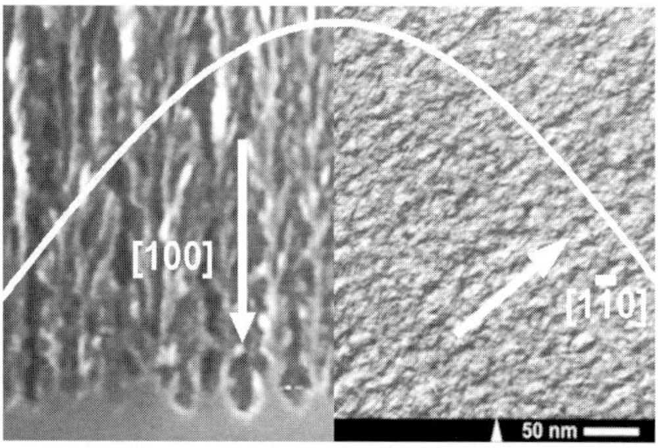

Fig. 1. Left part: HRTEM image of the cleaved edge of a porosified (100) p^{++} Si wafer. Pore and Si nanowire alignment corresponds to the [100] crystallographic direction. Right part: HRTEM image of the surface of porosified (110) P^{++} Si wafer (top view). Note the alignment of Si nanowires and pores in the indicated $[1\bar{1}0]$ crystallographic direction. The white curve demonstrates a characteristic photon wavelength in the visible spectral range with respect to structural inhomogeneities of PSi structures.

For an etched (100) Si surface the preferential alignment of nanocrystals is in the [100] growth direction while for low symmetry [110] Si wafer the pores, which are oriented along the [100] and [010] crystallographic directions are tilted to the (110) surface[13,14] and their projection on the (110) surface is aligned along the $[1\,\bar{1}\,0]$ direction (see Fig. 1). In microscopic scale the spatial modulation of the dielectric constant is very large but since the wavelength of light significantly exceeds the typical dimensions of the nanowires and the pores (2–10 nm) an electrostatic approach can be used to analyze the macroscopic dielectric response of this system.[15] Probably the most adequate effective medium model applicable to PSi is the Bruggeman approximation.[15] In this model the effective dielectric constant ε_{eff} of a mixture of small spherical inclusions in a host medium (*i* components with the *i*th fraction having the dielectric constant ε_i and the relative volume fraction f_i) can be calculated according to:

$$\varepsilon_{eff} = \sum_i f_i \frac{\varepsilon_i - \varepsilon_{eff}}{\varepsilon_i + 2\varepsilon_{eff}}$$

This model successfully describes the reduction of the refractive index of PSi compared to that of bulk Si with increasing porosity.

2. Optical Devices Based on PSi Layers

The first reliable optical devices made from PSi were based on the controlled variation of the refractive index of PSi layers in depth. PSi can easily be produced with a wide variation of porosities via changing the etching current density and the doping level of bulk Si substrates. The increase of the porosity results in a reduction of the refractive index of the PSi layers. This allows to modify their dielectric properties in a controlled manner. Due to the extremely small size of Si nanocrystals, in the range of nanometers, the sponge-like network of remaining Si and pores is a quasi-continuous medium. This concept is the basis of the variety of dielectric PSi structures like antireflection coatings for Si solar cells,[16] waveguides,[17] microcavities,[18] distributed Bragg reflectors (DBRs),[19] rugate filters,[10] Fibonacci quasicrystals,[21] omnidirectional mirrors,[22] lateral superlattices,[23] Thue-Morse structures,[24] etc.

In all devices mentioned the refractive index is varied in depth via changing the etching current in time. One important advantage of this method is that the already etched part of the layer is unaffected by subsequent current changes; thus porosity can be modulated in depth. This procedure results in a stack of alternating layers with different refractive indices and thicknesses. It has been demonstrated that these devices can be used as optical sensors for various types of chemical compounds since a filling of the pores with dielectric substances modifies the average refractive index of each layer and thus the optical response of the stack.[25-28] The simplest, one dimensional photonic structure is the Bragg mirror which is made of alternating layers of low (n_1) and high (n_2) refractive index, whose thicknesses satisfy the Bragg condition: $n_1d_1 + n_2d_2 = m\lambda/2$ where m denotes the order of the optical response. Another planar photonic structure is known as a microcavity. Usually it consists of a $\lambda/2$ layer placed between two distributed Bragg reflectors made from pairs of alternating $\lambda/4$ layers of high and low refractive indices. It acts as a multipass cell confining radiation and causing field enhancement inside. The interference of the reflected waves

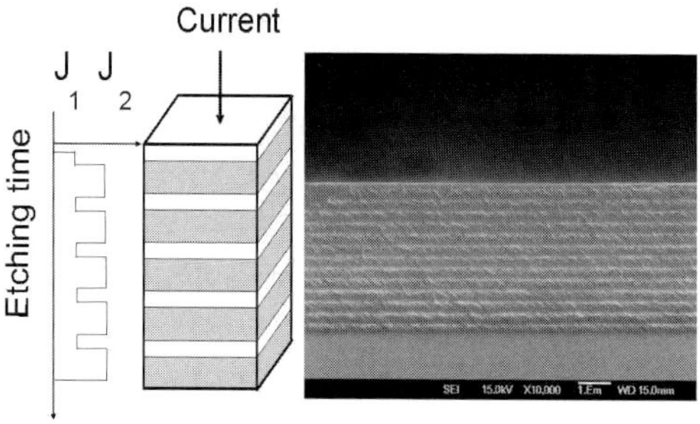

Fig. 2. Left side: sketch of the electrochemical etching procedure used for growing stacks of PSi consisting of layers having different refractive indices. Two different current densities (J_1 and J_2) and etching duration times have been used. Longer etching time result in a thicker layer and higher current density implies higher porosity and lower value of the refractive index. Right side: typical scanning electron microscope (SEM) image of cleaved edge of a dielectric stack of PSi layers. Roughness of PSi layers is due to cleaving procedure.

of the two DBRs leads to a transmission maximum within a spectral region of high reflection (stop band). Microcavities are of special interest for nonlinear or laser applications since the electric field of light inside the central layer can be significantly enhanced.

Depending on the desired parameters of PSi-based optical structures (achieved by the dielectric in-depth profile) different current modulation regimes have been used. In most of experiments periodic step-like variation of the etching current densities have been performed (see sketch in the left side of Fig. 2). However it results in a step-like variation of the refractive index and causes complex behavior of optical structures, for instance, in the presence of reflection sidebands. However, if the current is modulated gradually, a smooth index profile can be achieved, giving rise to so-called rugate filters. Berger *et al.* reported the first porous silicon-based rugate filter in 1997.[29] The main difference of these structures from those where abrupt variation of the refractive index has been used, is that higher harmonics can be completely eliminated. For instance, to suppress dielectric contrast the authors of Ref. 20 performed PSi-based interference filter apodization, i.e. the refractive index contrast was modulated with a smooth envelope function to reduce the index contrast to zero at the two boundaries (surface and substrate) of the filter, which successfully attenuated the sidelobes at both sides of the photonic stop band. They achieved successful reduction of interference ripples through the insertion of index-matching layers on the first and last interfaces.

The right side of Fig. 2 shows a SEM picture of a PSi multilayer structure which has been prepared according to the etching procedure sketched in the left side. The contrast in the image is due to different porosities of the layers. Good quality of the interface between layers and their thickness uniformity are evident. The theoretically obtainable refractive index modulation makes PSi microcavities an ideal structure for generating strong optical confinement. However, homogeneous growth of PSi layers in depth and flatness of interface between dielectric layers are crucial for the performance of the optical device. The authors of Ref. 30 have demonstrated that microcavities having subnanometer linewidths can be fabricated from PSi. They employed the recipe which has been known before: the quality of the PSi thin film interfaces is improved by decreasing the temperature of the electrolytic solution.[31] At low temperatures, a decrease in the ion mobility causes a higher rate of electropolishing, which tends to smooth out any inhomogeneities at the interfaces.

In our opinion at this point, it is very important to mention two major obstacles in engineering of PSi based optical devices. Indeed, the refractive index of PSi layers, their thickness and optical quality of their interfaces can be controlled very precisely. However, PSi, similarly to bulk Si, absorbs light through entire near-infrared and visible range and reflection/transmission coefficients of any type of optical devices will be significantly affected. Another omnipresent complication is the strong spatial variation of the refractive index (silicon/air) through PSi layers assembling optical devices. Despite it appears on microscopical scale and the wavelength of the propagating light is much longer than characteristic scale of the spatial variation of the refractive index, Rayleigh scattering still cannot be ignored. For PSi the typical scattering length is in the range of 100 μm and it reduces the quality of PSi-based devices and limits their length, for instance for waveguide applications, or thickness of planar structures.

3. Polarization-Dependent Optical Properties of PSi

If a unit cell of crystalline lattice has uniaxial symmetry, for example, hexagonal lattice structure two distinct components of the refractive index value along specific crystallographic axes appear. A crystal having such anisotropic dielectric properties is said to be intrinsically birefringent.

However, Si having a highly symmetric diamond-type unit cell is optically isotropic. Lorentz has demonstrated that for cubic lattices the refractive index value only very weakly depends on the light wave vector direction.[32] For bulk Si the maximum value of birefringence for light propagating along the [110] direction was found to be extremely small: $\Delta n = n_{(110)} - n_{(100)} = 5\times10^{-6}$, where subscripts denote the direction of the electric field vector.[33] Optical anisotropy requires the reduction of a cubic lattice symmetry and can be achieved by applied or growth induced uniaxial stress as well as by the reduction of crystal dimensionality (e.g., growing quantum well slabs). This results in a difference between the two principal components of the refractive index $\Delta n = n_{par.} - n_{per.}$ where $n_{par.}$ and $n_{per.}$ are values of the refractive index for light polarized parallel and perpendicular to the perturbation direction of the crystal symmetry.[34] When the wavelength of light λ is much longer than the constituents (size a) of a photonic system the propagation of light can be noticeably influenced. Two scenarios can be distinguished: either the constituents exhibit an intrinsic optical anisotropy or optically isotropic constituents are arranged spatially anisotropically. Examples are bulk materials assembled by elongated molecules[34] or atoms in a noncubic lattice,[35] respectively. When $\lambda >> a$ it is not possible to distinguish between the scenarios mentioned above for composite materials.

Already in 1904, Braun suggested that birefringence can be achieved in a composite system by an anisotropic spatial arrangement of its optically isotropic constituents.[36] This effect is referred as form birefringence and anisotropic optical properties of the system arise from polarization-dependent screening induced by the electric field of the propagating light wave.

3.1. *In-plane birefringence of porosified (110) Si wafers*

TEM studies show that the propagation of pores during the etching process always takes place in equivalent [100] crystallographic directions independent of the wafer orientation.[10] Si nanowires that remain after the etching process are aligned along the same crystallographic directions. For (100)-oriented wafers this alignment results in a preferential direction of the pores and nanowires perpendicular to the wafer surface. The application of an electric field to a spatially anisotropic dielectric structure (e.g., to a dielectric nanowire) results in a different dielectric constant values depending on the polarization direction of the applied electric field as a result of the difference in the screening of the applied electric field by induced surface polarization charges.[37] Therefore, for (100)-oriented Si wafers one expects an anisotropy of the refractive index of light that is incident upon a cleaved edge of the porous layers, as has been verified experimentally.[38,39] However, for the realization of polarization-sensitive optical devices, in-plane anisotropy of the refractive index is required. This anisotropy can be realized by etching of Si wafers with surface symmetries lower than (100), e.g., (110). The pores, which are oriented along the [100] and [010] crystallographic directions, are tilted to the (110) surface,[40] and their projection on the (110) surface is aligned along the $[1\bar{1}0]$ direction (see sketch of crystallographic directions on the left side of Fig. 3). Etching of (110) Si wafers results in a reduction of the symmetry of the optical properties from an isotropic cubic to an in-plane uniaxial symmetry and gives different refractive indices, $n[1\bar{1}0]$ and $n[001]$ for light polarized along the $[1\bar{1}0]$ and [001] crystallographic directions, respectively. Therefore artificial in-plane uniaxial symmetry can be achieved in a cubic silicon crystal via its anisotropic dielectric nanostructuring.

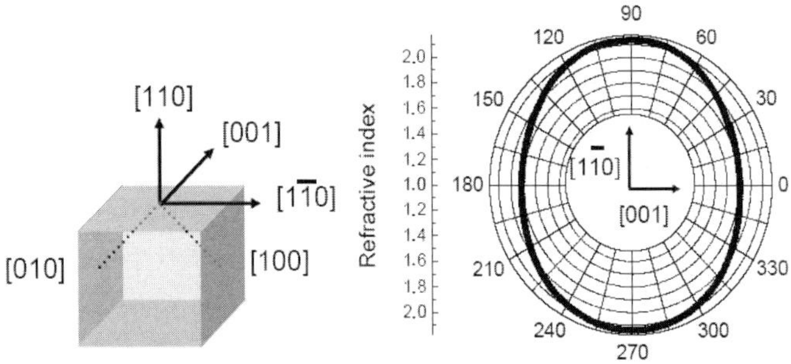

Fig. 3. Left side: sketch of crystallographic directions for a (110) bulk Si wafer. Right side: polar plot of the refractive index dependence on the polarization direction of the incident light for a (110) surface of a PSi layer obtained from reflectivity measurements of linearly polarized light ($\lambda = 514.5$ nm).

The right side of Fig. 3 shows the dependence of the refractive index value of a (110) p^{++} PSi layer on the angle between the crystallographic axes and the polarization direction of incident light. The uniaxial symmetry of the refractive index is evident; the largest value is for the light polarized in the [1$\bar{1}$0] crystallographic direction. Thus, the layer is a negative uniaxial crystal. We would like to note here that the optical anisotropy parameters of bulk systems are fixed. The advantage of the PSi layers is that these parameters can be adjusted to a desirable level via modification of their morphology. In Ref. 41 it has been demonstrated that depending on morphology of PSi layers, mainly on the level of spatial alignment of remnant Si nanocrystals (defined by doping level of bulk Si wafers and current densities used during anodization) the value of Δn can be continuously varied from 0.01 up to 0.3.

While light travels through the optically anisotropic PSi layer, the components of the electric field parallel to the [1$\bar{1}$0] and [001] crystallographic directions experience a phase shift dependent on the thickness of layers. To demonstrate the capability of birefringent PSi layers to control the polarization state of light, the thickness of layers was intentionally chosen to achieve a desired retardation value (e.g., $\lambda/4$ or $\lambda/2$) in the transparency range of optical communication lines. Figure 4 (left side) shows how linearly polarized incident light having a wavelength of 1.53 μm transforms into circularly polarized light. The

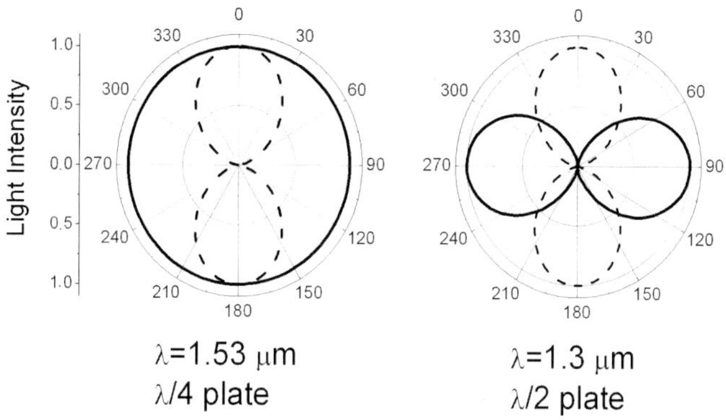

Fig. 4. Polar plots of the intensity of incident (dashed lines) and transmitted (solid lines) light as a function of the angle of the analyzer. Measurements are performed at λ/4 (left side) and λ/2 (right side) retardation values.

right side shows how a $\lambda/2$ retardation condition can be achieved at $\lambda =$ 1.3 μm. The analyses of the angular polarization pattern of the transmitted light shows high retardation accuracy: the deviation from the ideal circularly or linearly polarized light is of the order of 0.1% for $\lambda/4$ and $\lambda/2$ plates conditions. This quality is comparable or even better than that of standard retardation plates produced from crystalline quartz. The transmittance of layers is even better than that of bulk wafers due to the reduced refractive index. The refractive index can be easily adjusted to desired values by varying the layer porosity. Retardation values are well controlled and can be tuned to cover most of the visible, infrared and far infrared spectral ranges due to non-polar crystalline structure of Si. All commercially available retarders have much narrower spectral range of operation.

3.2. *Polarization-sensitive Bragg reflectors based on (110) PSi layers*

As it has been mentioned previously, most of optical devices made from PSi were based on the controlled variation of the refractive index of PSi layers in depth. In Ref. 42 an alternative approach for dielectric patterning combining the variation of the etching rate in depth with an in-plane crystallographic direction-sensitive etching has been proposed.

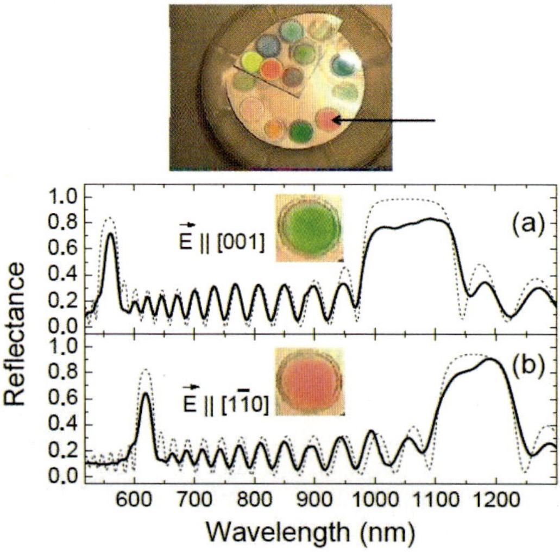

Fig. 5. Upper part: the image of different PSi-based Bragg reflectors at certain polarization direction of the incident light. These Bragg reflectors have anisotropic polarization response. Lower part: spectrally and polarization-resolved reflection of a (110) distributed Bragg reflector. Direction of the electric field of the incident light is indicated. Color insets: real images of a (110) distributed Bragg reflector (indicated by the arrow) for two orthogonal light polarizations (detected through linear analyzer). Dashed lines represent results of the simulated optical response.

This allows to produce a stack of PSi layers with modulated refractive index in depth with each layer being dielectrically anisotropic in the surface plane, i.e., to vary the refractive index in three dimensions. These photonic structures yield optical effects which are difficult to achieve with conventional stacks of isotropic dielectrics. Upper part of Fig. 5 shows the image of different PSi-based Bragg reflectors having anisotropic polarization response at certain polarization direction of the incident light. The Bragg condition: $n_1d_1 + n_2d_2 = m\lambda/2$, where m denotes the order of the optical response, is still valid for this device but since refractive index depends on the direction of the electric field of the incident light it is necessary to treat the dielectric properties of a (110) PSi-based Bragg reflector separately for two orientations of the electric field.

Figures 5a and 5b demonstrate the reflectivity spectra of a Bragg reflector at almost normal incidence. Contrary to standard PSi (100) Bragg reflectors the spectral position of reflection bands depends on the polarization direction of the incident light. For the electric field direction parallel to the [001] crystallographic direction first order band appears around 1060 nm, but it is significantly shifted towards longer wavelength (1170 nm) when electric field is parallel to the $[1\bar{1}0]$ direction. Based on the etching times and the measured etching rates for the different current densities the thickness of the layers is estimated to be approximately 90 nm and 200 nm, respectively.

For a high quality dichroic device it is essential that the two reflection bands do not overlap what is not the case for the first order Bragg bands in Fig. 5. However, the second order maxima in the visible spectral range are fully spectrally separated. The slightly reduced value of the reflectivity with respect to the first order is caused by self-absorption of visible light in the porous layers. This dichroic behavior can be easily seen as a change of color of the reflected light from green (~560 nm) to red (~620 nm) depending on the orientation of the polarization direction of the incident light. Compared to the first order Bragg bands in the near infrared region the second order bands are twice narrower as expected. The dispersion of the dielectric constant of bulk Si leads to their spectral positions which are not exactly the half of the first order. Their spectral positions are almost coinciding with those predicted by the effective medium approach (~567 and 622 nm, respectively).

3.3. *Polarization-sensitive microcavities based on (110) PSi layers*

Another specific type of PSi multilayer structures are microcavities grown, again, via electrochemical etching of Si wafers. Microcavity is realized by inserting a layer with *nd* equal to $\lambda_{\text{microcavity}}/2$ between two symmetric Bragg mirrors each one assembled from alternative layers having high and low refractive indices with $d_1n_1 = d_2n_2 = m\lambda_{\text{microcavity}}/4$, where m denotes the order of the optical response. The interference of the reflected waves of the two Bragg reflectors leads to a transmission maximum at $\lambda_{\text{microcavity}}$ within the spectral region of the high reflection stop band. Microcavities are of special interest for nonlinear or laser

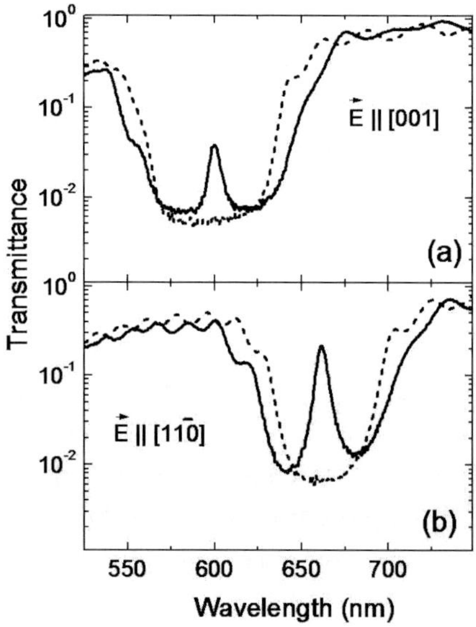

Fig. 6. Spectrally and polarization-resolved transmission through a (110) distributed Bragg reflector (dashed lines) and the corresponding (110) microcavity (thick solid lines). Polarization directions of the incident light are indicated.

applications since the electric field of the light inside the central layer can be significantly enhanced. The main requirement for high performance of microcavity is high quality of reflectors. Figure 6 shows the transmission spectra of a free standing (110) Bragg reflectors for the two principal polarization directions of the transmitted light (dashed lines). Since the suppression of the transmitted light for the first order Bragg peak is on the order of 99% and the edges of the stop bands are well pronounced the optical performance of the Bragg reflector is high. A $\lambda/2$ layer has been introduced in an identical dielectric stack of layers by doubling of the etching time of the central layer. The thick solid lines in Fig. 6 show the polarization-resolved transmission spectra of a (110) microcavity. For both polarization directions a sharp cavity mode approximately in the center of the stop band can be seen.

The finesse F of a microcavity or Fabry–Perot filter is defined as the wavelength position of $\lambda_{\text{microcavity}}$ divided by the full width at half maximum of the cavity mode $\Delta\lambda$. For the microcavity in Fig. 6, F for light polarized in the $[1\bar{1}0]$ direction is $\lambda_{microcavity}/\Delta\lambda$ = 661 nm / 7 nm = 94.4, which satisfies the requirements for optical grade quality. Due to the difference in Δn for different current densities the best performance of the Bragg reflectors can be achieved for one polarization direction only. To get the best optical quality Bragg reflectors usually d_1n_1 is chosen to be equal to d_2n_2 with a value of $m\ \lambda_{microcavity}$ /4. However, for (110) Δn is not the same for different current densities. Therefore this condition can only be fulfilled for only one polarization direction. Since the microcavity has been optimized for light polarized in the $[1\bar{1}0]$ direction, F for light polarized in the [001] crystallographic direction is smaller (600 nm / 7 nm = 85.7). The intensities of the transmitted light differ as well. For the $[1\bar{1}0]$ polarization direction 22% of the initial light intensity is transmitted while it is only 4% of the light polarized along the [001] crystallographic direction. The intensity of the transmitted light is limited by self-absorption in the PSi layers in this spectral range. For (100) mesoporous Si having 50% porosity the absorption coefficient at 660 nm is about 950 cm^{-1} (~1500 cm^{-1} at 600 nm).[43] For a given total thickness of the microcavity structure (~3–3.5 μm) with an average porosity of ~50% up to 30% of the light is absorbed at 660 nm and 40% at 600 nm. The field enhancement at 660 nm in the central layer for light polarized along the $[1\bar{1}0]$ crystallographic direction is estimated to be around 3. This value can be significantly enlarged by an increase of the dielectric contrast between the layers. Therefore thin (110) PSi layers having a larger variation of their average porosity and smaller interface roughness are desired and the preparation procedure still has to be improved. We would like to note that all degrees of freedom for tuning the spectral position and the polarization state of the cavity mode are identical to those demonstrated for the Bragg reflectors.[44]

3.4. *Plane polarizers based on (110) PSi layers*

Later a more complex dielectric nanostructuring of Si wafers has been performed to realize Si-based planar polarizers. Two approaches towards realizing Si-based planar devices that can control the wavelength and polarization state of the transmitted light have been suggested.

Combining in one structure a dichroic Bragg reflector with a second reflector or with a polarization-sensitive microcavity with different spectral responses allow planar polarizers to be produced. Again, multilayer structures were prepared by etching of (110) bulk Si wafers at alternating low and high current density. In Fig. 7b, we illustrate the operation principle of the device. The thicknesses of the upper 20 pairs of PSi layers were adjusted to act as a dichroic Bragg reflector. Dichroic Bragg reflectors produced from (110)-oriented bulk Si substrate exhibit two reflection bands being dependent on the polarization direction of the incident light. Because of the smaller refractive index for light polarized along the [001] crystallographic direction, the first-order reflection band is significantly shifted toward shorter wavelengths with respect to that measured for light polarized along $[1\bar{1}0]$ (dotted curves). The same dichroic behavior holds for the second reflector (10 pairs) whose reflection bands were shifted to shorter wavelengths by a proper decrease in the layer thickness (solid curves). Therefore the following situation can be realized: the transmittance range of both reflectors for a certain polarization direction of the incident light coincides with the reflection band of one of them for the orthogonal polarization. Consequently, the transmitted light is linearly polarized and the whole multilayer structure acts as a narrow spectral band plane polarizer. In Fig. 7a, the polarization-resolved transmission spectra of a dielectric stack consisting of 60 PSi layers (20 pairs with thicknesses of 92 and 96 nm, 10 pairs with thicknesses of 68 and 81 nm) are shown.

As it is evident from Fig. 7a, the transmitted light intensity has a maximum in the wavelength range around 685 nm when the polarization direction of the detected light is parallel to the $[1\bar{1}0]$ crystallographic direction, and it is almost zero for light polarized in the [001] direction. Because the $[1\bar{1}0]$ and [001] directions are orthogonal in the (110) Si surface plane, the transmitted light in this spectral range is linearly polarized. To illustrate the optical performance, Fig. 7c depicts the polarization-resolved transmission spectra in the wavelength range from 680 to 690 nm. The transmission value is 0.7 (solid line) with discrimination of the orthogonally polarized light component of better than 99% (dotted curve). The transmission of the structure is limited since 30% of the light is absorbed/reflected at this wavelength. We

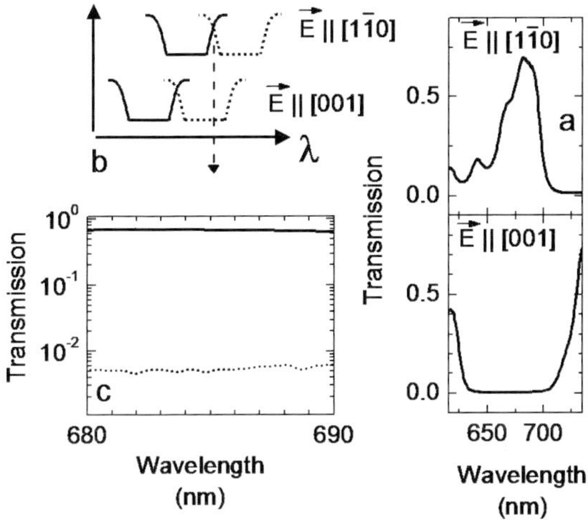

Fig. 7. (a) Polarization-resolved transmission spectra of the device described in the text. (b) Operation principle of a Si-based polarizer assembled from two dichroic reflectors with different spectral responses. (c) Polarization-resolved transmission spectra of the same structure in the spectral range from 680 to 690 nm (note the logarithmic scale). The solid line (dotted curve) corresponds to incident light polarized along the [1$\bar{1}$0] ([001]) direction.

would like to mention that the reflectors exhibit additional reflection sidebands arising from multiple interference of light in the multilayer structure with finite thickness. Therefore the intensity of the transmitted light depends significantly on its wavelength. This seems to be an inherent limitation for the operation bandwidth of Si-based planar polarizers.

The achieved full width at half maximum (FWHM) of the transmission band in Fig. 7a is 35 nm within a 100-nm-wide stop band for light polarized along the [001] direction. For a variety of optical applications it is desired to split a given spectral range in a maximum number of information channels. Specifically for laser applications, e.g., polarization-sensitive mirrors, a spectrally sharp transmission band is required, and consequently *F* must be as large as possible. This can be achieved with another specific type of PSi multilayer structure: a dichroic Bragg reflector combined with a dichroic microcavity. The interference of the reflected waves of the relectors results in a sharp transmission maximum (cavity mode) with large F within a spectral range of a high reflection band.

It has been previously demonstrated that microcavities etched from (110)-oriented Si substrate are dichroic. They show spectrally separated cavity modes that are dependent on the direction of the polarization vector of the incident linearly polarized light (schematically shown by solid curves in Fig. 8a). To block one of these two transmission bands of the microcavity (here polarized along the [001] direction) a concept similar to the one described before was applied. An additional dichroic Bragg reflector with reflection bands shown by the dotted curves in Fig. 8a was added to a dichroic microcavity in a stack of dielectric layers. Its reflection band for light polarized along the [001] direction was adjusted to cover the spectral range of the transmittance band of the microcavity for light polarized along the same direction. For this specific spectral

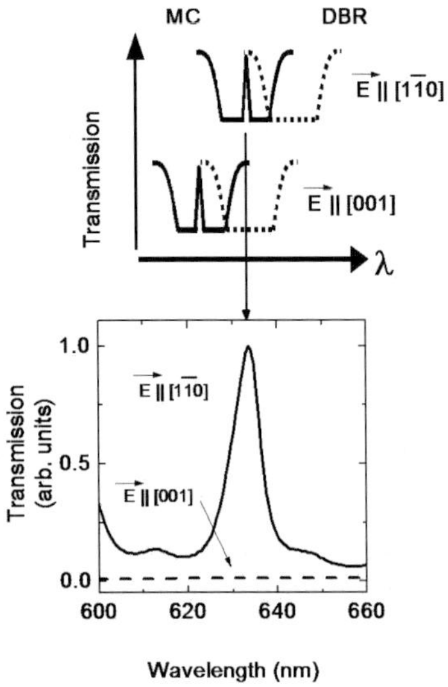

Fig. 8. (a) Sketch of the operation principle of a Si-based polarizer (combination of dichroic Bragg reflector and microcavity). (b) Polarization-resolved transmission spectra of the device in arbitrary units. MC denotes microcavity and DBR Bragg reflector, respectively.

range, light polarized along the [001] direction is reflected by the Bragg reflector.

Light polarized along the $[1\bar{1}0]$ direction passes the whole structure at the spectral position of the cavity mode. Therefore this dielectric stack also acts as a narrow spectral band polarizer. In Fig. 8b polarization-resolved transmission spectra of a multilayer structure assembled from 81 layers of PSi are shown. The first 20 pairs (73 nm / 109 nm) acted as a dichroic Bragg reflector. Thereafter a dichroic microcavity was formed by 10 pairs (66 nm / 99 nm), a 110-nm-thick central layer, and 10 more pairs (66 nm / 99 nm) of PSi slabs. The transmission band at 633 nm (Fig. 8b, solid curve) was linearly polarized with a FWHM of 8 nm (*F* ~80).

Therefore, multilayer structures based on anisotropically nanostructured Si can act as efficient planar polarizers for visible and infrared light at normal incidence. The presented devices offer additional design freedom for integrated optoelectronics, because they allow the wavelength, the spectral width, polarization, and intensity of the transmitted light to be controlled. Furthermore, contrary to all other existing polarizing elements, the presented structures do not absorb light in the entire infrared or far-infrared range, because they are based on a non-polar material: silicon.

4. Conclusions

We have demonstrated that optical devices based on three-dimensional patterning of the refractive index of Si can exhibit functionality which is difficult or impossible to achieve with conventional stacks of dielectric layers. Beside obvious device applications they can be employed as optical sensors.[46] It is also found that the optical anisotropy of PSi layers are crucial for nonlinear wave interactions. The anisotropy level is high enough to compensate the normal dispersion of bulk Si and to achieve phase matching conditions for nonlinear optical wave interactions.[47] However, an ideal structure for nonlinear optical applications has to exhibit additionally light localization properties for the fundamental and higher harmonics of the propagating wave. For instance, a microcavity acts as a multipass cell confining radiation and causing field enhancement inside. Therefore optical devices based on a stack of PSi layers acting as a microcavity with each layer exhibiting an in-plane birefringence level which is high enough to achieve phase-matching conditions i.e., having variation of the refractive index in three dimensions seem to be almost ideal candidates for nonlinear optical interactions. Finally birefringent PSi layers can play a role of phase-matching matrix for substances having large high order dielectric susceptibilities incorporated in the pores.

Acknowledgements

We thank many colleagues and co-workers who have contributed ideas, theoretical and experimental work and are not formally authors of this review. In particular, Nicolai Künzner, Egon Gross, and Viktor Timoshenko deserve to be mention here.

References

1. R. A. Soref, *IEEE* **81**, 1687 (1993).
2. O. Bisi, S. U. Campisano, L. Pavesi and F. Priolo, *Silicon Based Microphotonics: from Basics to Applications* (IOP Press, Amsterdam, 1999).
3. Silicon Photonics, in *Topics of Applied Physics*, Springer-Verlag (Berlin, Heidelberg, New York, 2004), Ed. By L. Pavesi and D. J. Lokwood, and references therein.
4. A. Liu, R. Jones, L. Liao, D. Samara-Rubio, D. Rubin, O. Cohen, R. Nicolaescu and M. Panicci, *Nature* **427**, 615 (2004).
5. S. W. Leonard, H. M. van Driel, K. Busch, S. John, A. Birner, A.-P. Li, F. Müller, U. Gösele and V. Lehmann, *Appl. Phys.Lett.* **75**, 3063 (1999).
6. J. D. Joannopolous, R. D. Maede and J. N. Winn, *Photonic Crystals: Molding the Flow of Light* (Princeton University Press, Princeton, NJ, 1999).
7. A. Zrenner, B. Fröhlich, J. Brunner and G. Abstreiter, *Phys. Rev.* **B52**, 16608 (1995).
8. L. Tsybeskov, K. D. Hirschman, S. P. Duttagupta, M. Zacharias, P. M. Fauchet, J. P. McCaffrey and D. J. Lockwood, *Appl. Phys. Lett.* **72**, 43 (1998).
9. L. Brus, in *Light Emission in Solids, Semiconductors and Semimetals*, edited by D. J. Lockwood (Academic, New York, 1996).
10. A. G. Cullis, L. T. Canham and P. D. J. Calcott, *J. Appl. Phys.* **82**, 909 (1997).
11. D. Kovalev, H. Heckler, G. Polisski and F. Koch, *Phys. Stat. Sol. (a)* **215**, 871 (1999).

12. O. Bisi, S. Ossicini and L. Pavesi, *Surf. Sci. Rep.* **38**, 1 (2000).
13. N. Künzner, D. Kovalev, J. Diener, E. Gross, V. Yu. Timoshenko, G. Polisski, F. Koch and M. Fujii, *Opt. Lett.* **26**, 1265 (2001).
14. V. Kochergin, M. Christophersen and H. Föll, *Appl. Phys. B* **79**, 731 (2004).
15. D. A. G. Bruggeman, *Ann. Phys. (Paris)* **24**, 636 (1935).
16. P. Menna, G. Di Francia and V. La Ferrara, *Solar Energy Materials and Solar Cells* **37**, 13 (1995).
17. V. P. Bondarenko, A. M. Dorofeeva and N. M. Kazuchits, *Microelectronic Engineering* **28**, 447 (1995). H. F. Arrand, T. M. Benson, A. Loni, M. G. Krueger, M. Thönissen and H. Lüth, *Electron. Lett.* **33**, 1724 (1997). P. Ferrand, R. Romestain and J. C. Vial, *Phys. Rev.* **B63**, 115106 (2001).
18. G. Vincent, *Appl. Phys. Lett.* **64**, 2367 (1994). L. Pavesi, *Riv. Nuovo Cim.* **20**, 1 (1997), and references therein. M. Ghulinyan, C. J. Oton, G. Bonetti, Z. Gaburro and L. Pavesi, *J. Appl. Phys.* **93**, 9724 (2003).
19. L. Pavesi, C. Mazzoleni, A. Tredicucci and V. Pellegrini, *Appl. Phys. Lett.* **67**, 3280 (1995). M. G. Berger, R. Arens-Fischer, M. Thönissen, M. Krüger, S. Billat, H. Lüth, S. Hilbrich, W. Theiss and P. Grosse, *Thin Solid Films* **297**, 237 (1997).
20. E. Lorenzo, C. J. Oton, N. E. Capuj, M. Ghulinyan, D. Navarro-Urrios, Z. Gaburro, and L. Pavesi, *Appl. Opt.* **44**, 5415 (2005). S. Ilyas, T. Böcking, K. Kilian, P. J. Reece, J. Gooding, K. Gaus and M. Gal, *Opt. Materials* **29**, 619 (2007).
21. L. Dal Negro, C. J. Oton, Z. Gaburro, L. Pavesi, P. Johnson, A. Lagendijk, R. Righini, M. Colocci and D. S. Wiersma, *Phys. Rev. Lett.* **90**, 055501 (2003).
22. E. Xifré-Pérez, L. F. Marsal, J. Ferré-Borrull and J. Pallarès, *J. Appl. Phys.* **102**, 063111 (2007).
23. G. Lérondel, R. Romestain, J. C. Vial and M. Thönissen *Appl. Phys. Lett.* **71**, 196 (1997). N. Nagy, J. Volk, A. Hámori and I. Bársony, *Phys. Stat. Sol. (a)* **202**, 1639 (2005).
24. L. Moretti, I. Rea, L. De Stefano and I. Rendina, *Appl. Phys. Lett.* **90**, 191112 (2007).
25. V. S.-Y. Lin, K. Motesharei, K. S. Dancil, M. J. Sailor and M. R. Ghadiri, *Science* **278**, 840 (1997).
26. P. A. Snow, E. K. Squire, P. St. J. Russel and L. T. Canham, *J. Appl. Phys.* **86**, 1781 (1999).
27. T. A. Schmedake, F. Cunin, J. R. Link and M. J. Sailor, *Adv. Mater.* **14**, 1270 (2002).
28. J. Gao, T. Gao, T and M. J. Sailor, *Appl. Phys. Lett.* **77**, 901 (2000).
29. M. G. Berger, R. Arens-Fischer, M. Thönissen, M. Krüger, S. Billat, H. Lüth, S. Hilbrich, W. Theiss and P. Grosse, *Thin Solid Films* **297**, 237 (1997).
30. P. J. Reece, G. Lerondel, W. H. Zheng and M. Gal, *Appl. Phys. Lett.* **81**, 4895 (2002).
31. S. Setzu, G. Lerondel and R. Romestain, *J. Appl. Phys.* **84**, 3129 (1998).

32. H. A. Lorentz, *Collected Papers* (Martinus Nijhoff, The Hague, 1936), **Vol. II**, p. 79.
33. J. Pastrnak and K. Vedam, *Phys. Rev.* **B3**, 2567 (1971).
34. W. L. Bragg and A. B. Pippard, *Acta Crystallogr.* **6**, 865 (1953).
35. Consider for example atoms in a tetragonal or hexagonal crystal lattice.
36. F. Braun, *Phys. Z.* **8**, 199 (1904).
37. E. M. Lifshitz, L. P. Pitaevskii and L. D. Landau, *Electrodynamics of Continuous Media* (Elsevier, New York, 1985).
38. D. Kovalev, M. Ben-Chorin, J. Diener, F. Koch, Al. Efros, M. Rosen, A. Gippius and S. G. Tikhodeev, *Appl. Phys. Lett.* **67**, 1585 (1995).
39. I. Mihalcescu, G. Lerondel and R. Romestain, *Thin Solid Films* **297**, 245 (1997).
40. S. F. Chuang, S. D. Collins and R. L. Smith, *Appl. Phys. Lett.* **55**, 675 (1989).
41. N. Künzner, J. Diener, E. Gross, D. Kovalev, V. Yu. Timoshenko and M. Fujii, *Phys. Rev.* **B71**, 195304 (2005).
42. J. Diener, N. Künzner, D. Kovalev, E. Gross, V. Yu. Timoshenko, G. Polisski and F. Koch, *Appl. Phys. Lett.* **78**, 3889 (2001).
43. D. Kovalev, G. Polisski, M. Ben-Chorin, J. Diener and F. Koch, *J. Appl. Phys.* **80**, 5978 (1996).
44. J. Diener J, N. Kunzner, D. Kovalev, E. Gross, F. Koch and M. Fujii, *J. Appl. Phys.* **91**, 6704 (2002).
45. J. Diener, N. Künzner, E. Gross and D. Kovalev, *Opt. Lett.* **29**, 195 (2004).
46. E. Gross, D. Kovalev, N. Künzner, V. Yu. Timoshenko, J. Diener, F. Koch, *J. Appl. Phys.* **90**, 3529 (2001).
47. L. A. Golovan, V. Yu. Timoshenko, A. B. Fedotov, L. P. Kuznetsova, D. A. Sidorov-Biryukov, P. K. Kashkarov, A. M. Zheltikov, D. Kovalev, N. Künzner, E. Gross, J. Diener, G. Polisski and F. Koch, *Appl. Phys.* **B73**, 31 (2001).

CHAPTER 10

GUIDING AND AMPLIFICATION OF LIGHT DUE TO SILICON NANOCRYSTALS EMBEDDED IN WAVEGUIDES

Tomáš Ostatnický, Martin Rejman, and Jan Valenta

Department of Chemical Physics and Optics, Faculty of Mathematics and Physics, Charles University, Ke Karlovu 3, CZ-121 16 Prague 2, Czech Republic

Kateřina Herynková, Ivan Pelant

Institute of Physics, Academy of Sciences of the Czech Republic, v.v.i., Cukrovarnická 10, CZ-162 53 Prague 6, Czech Republic

In this chapter, we summarize our progress in both experimental and theoretical investigations of spontaneous emission of silicon nanocrystals embedded in planar waveguides. We give an overview of our photoluminescence measurements which reveal peculiar phenomena like spectral filtering of the signal or polarization splitting of the spectra. We discuss possible interpretations of the observed events and we present a model which is capable of explanation of all the peculiarities. Finally we apply the model in order to investigate the consequences of the waveguiding effects for the light amplification, propagation and spectroscopy measurements.

1. Introduction

Nanocrystalline silicon attracts attention in many research groups because of its prospective applicability in optoelectronic devices as an amplifier or a source of light. High concentrations of Si nanocrystals and high optical quality of the active media are desired to get a positive net optical gain. To accomplish the requirements, samples prepared by using various methods have been checked in experiments addressing optical

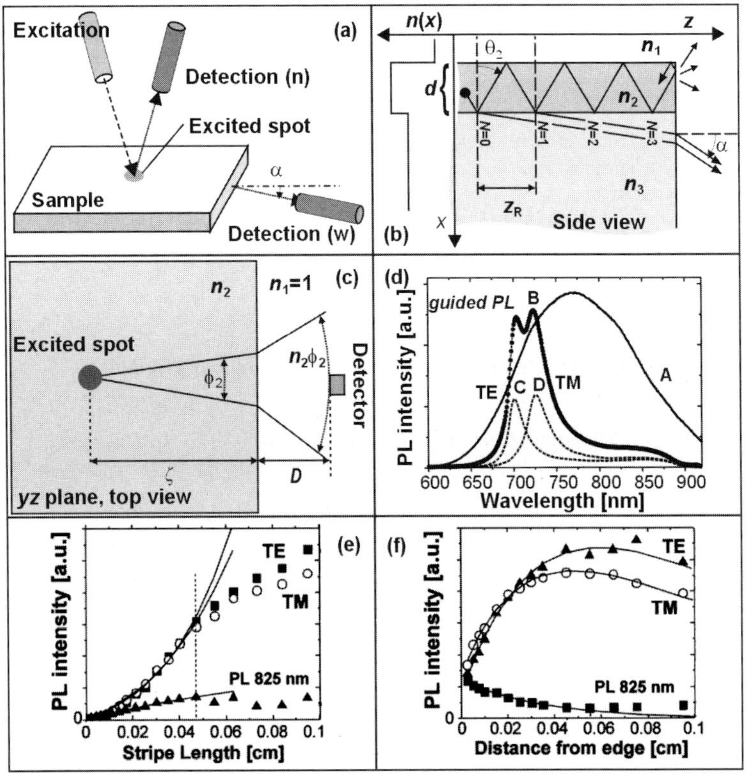

Fig. 1. (a) Experimental setup with two possible detection positions: standard (n) and waveguiding (w). (b) Propagation and decoupling of substrate modes; (c) sketch for derivation of coupling coefficients; (d) measured PL in standard (A) and waveguiding geometry (B) and in the waveguiding geometry with a linear polarizer (C,D); (e) measured ASE intensity at various spectral positions: at maxima of TE and TM modes and at 825 nm; (f) measured SES signal at the same spectral positions.[2]

gain: Si^+ ion implantation,[1,2] plasma–enhanced chemical vapour deposition,[3] reactive Si deposition,[4] magnetron sputtering[5] or others.[6] The samples prepared by these methods have a form of a thin SiO_2 layer (few micrometers thick) doped with Si nanocrystals, located on an optically thick Si or silica substrate. The thin "sheet" which contains nanocrystals then may reveal waveguiding properties because of the refractive index mismatch between the sheet and the SiO_2 matrix.

The waveguiding property and the "planar" method of fabrication are advantageous for integration of components in optoelectronics, however, it was found that the characteristics of a photoluminescence (PL) emission of the samples are strongly affected by the presence of the waveguide[2,7-12] and depend upon many parameters of the sample preparation.[9,13,14] We have investigated the PL emission of the Si nanocrystals embedded in waveguides both theoretically and experimentally and we summarize our findings in this chapter.

Effects connected with waveguiding in the implanted samples were firstly discussed in detail by our group[2] and confirmed by an independent observation of the same phenomenon on samples prepared by reactive deposition[7] — in the experiments, photoluminescence from the active layer was collected in the "waveguiding" configuration along the layer with Si nanocrystals (see Fig. 1a for the experimental setup). Compared to a PL spectrum collected in the standard configuration (perpendicular to the sample plane), the spectra reveal a disturbance of the wide PL band and a narrowing to one or more pairs of modes resolved in linear polarization (Fig. 1d). The unusual behaviour of the PL spectra was observed in measurements of amplified spontaneous emission (ASE) using the variable stripe length (VSL) method[2,15] where the shape of the ASE curves strongly depended on the wavelength of detected radiation (see Figures 1e-1f).

Numerical analysis of the problem shows that the spectrally narrow modes are formed very close to the cut-off frequency of the waveguide. This fact has led us to the interpretation[8,16,17] of our observations, namely, that the modes are formed by the waves leaking from the waveguide core slightly below the critical angle for the total reflection on the core/substrate boundary. Another interpretation was given by Khriachtchev *et al.*[7,9,10] in terms of particular guided modes which are spatially delocalized near the cut-off frequency and therefore their losses due to absorption in the core are lower than losses of the ordinary guided modes.

In this chapter, we develop a consistent model which predicts the observed behaviour and we discuss both hypotheses (delocalization and wave leaking) in order to find which of them plays a major role in formation of the PL spectra. We then show the consequences of the waveguiding and related effects for spectroscopic measurements, applications in optics and optoelectronics, gain measurements and construction of a lasing device. After a short introduction to waveguide theory in section 2, our model is discussed in more details in section 3. Since the original model does not take into account propagation of the modes along the waveguiding layer, we present its phenomenological extension in section 4, and in section 5 we present numerical calculations of the wave propagation in active waveguides.

2. Characterization of Waves in Waveguides

Discussion of propagation of the optical field through *passive* waveguides is usually limited to the discussion of the guided waves, *i.e.* waves (modes) which are spatially confined in the waveguide core. In the *active* devices, however, the sources of photons are located inside the guiding layer and therefore the amount of energy coupled to the guided modes may be comparable or smaller than the energy coupled to other modes. On that account, all waves must be considered when dealing with active devices.

In this section, we review the general theory of planar (slab) waveguides.[18–20] We consider a structure with three layers[a] of a dielectric described by a (generally complex) refractive indices n_1 (cladding), n_2 (core) and n_3 (substrate), see Fig. 1b; we assume for the sake of clarity for the real parts of the refractive indices $n_2 > n_3 > n_1$. The thickness of the core d is comparable with or larger than the vacuum wavelength of the optical radiation λ. The cladding and the substrate are optically thick and therefore considered to be infinite. The z axis is the propagation axis and the x axis is perpendicular to the waveguide layer.

Electromagnetic field from a source within a waveguide is coupled only to the states of a field which are allowed by boundary conditions — to the modes of the waveguide. According to the symmetry of the

[a] The structure with a step-like profile of the refractive index is taken into account for the sake of clarity. The theory may be, however, generalized to an arbitrary profile of refractive index including a continuous graded profile. Comparison of calculated PL spectra with our experimental data has been done considering the real parameters of our samples, *i.e.* graded profile of refractive index across the waveguide.

waveguide, the intensity of electric field of a TE mode may be written as $\vec{E}(x,z,t) = \hat{y}E(x,z)\mathrm{e}^{-\mathrm{i}\omega t}$, where $\hat{y}$ is a unit vector in the y direction, $\omega = 2\pi c / \lambda$ is the frequency of the mode, c is the light velocity and

$$E(x,z) = \{A_{m\beta}\exp[\alpha_{m\beta}(x - x_m)] + B_{m\beta}\exp[-\alpha_{m\beta}(x - x_m)]\}\mathrm{e}^{\mathrm{i}\beta z} \quad (1)$$

in every layer denoted by the index m (m=1,2,3). The modes are fully characterized by the propagation constant β, which may be interpreted in terms of the angle of propagation θ_2 (see Fig. 1b) of a ray: $\beta = (2\pi n_2 \sin\theta_2)/\lambda$. A variable $\alpha_{m\beta} = [\beta^2 - (2\pi n_m/\lambda)^2]^{1/2}$ determines the profile of the mode in the x direction, x_m is a coordinate of the boundary between the m-th and (m+1)-st layer and the variables A and B are constants for every layer and mode, fully determined by the boundary conditions and normalization.[20] The axial system has been adapted by rotation in order to ensure $\vec{E} \parallel \hat{y}$. In our case of a three-layer structure, the modes may be resolved into three groups depending on their spatial profile in the direction of the x axis, hence depending on the values of the coefficients $\alpha_{m\beta}$. Let us consider for instance a lossless structure, *i.e.* n_m real. The maximum value of β for propagating waves is clearly $2\pi n_2/\lambda$, therefore $\alpha_{2\beta}$ is always imaginary and the field intensity in the core is nonzero. Other α's, on the contrary, may be real or imaginary.

For both $\alpha_{1\beta}$ and $\alpha_{3\beta}$ imaginary, the mode takes the form of a propagating wave in all three layers and it is called the *radiation mode*. In terms of the ray optics, a ray emitted by a source inside the waveguide is refracted either to the substrate or to the cladding and only a small portion of its energy is reflected back to the core. The energy of the mode is therefore radiated from the waveguide core on short distances compared to the waveguide thickness d.

If the coefficients $\alpha_{1\beta}$ and $\alpha_{3\beta}$ are, on the contrary, both real, the profile of the mode in the cladding and the substrate given by Eq. (1) takes the form of evanescent waves and the mode energy is localized predominantly in the core. The mode is bound to the layer with the highest refractive index and propagates along the z axis, and it is called the *guided mode*. In terms of the ray optics, a ray is incident on both boundaries core/cladding and core/substrate at an angle larger than the critical angle for total internal reflection, therefore it is totally reflected and its energy cannot leave the core. Near the critical propagation constant $\beta_C = 2\pi n_3 / \lambda \leq \beta$, the coefficient $\alpha_{3\beta}$ is close to zero and thus the attenuation of the evanescent wave in the substrate is very weak and the wave penetrates into the substrate to depths comparable or even larger than the core thickness.[21]

Substrate radiation modes[b] *(substrate modes)* are the last kind of the waveguide modes. They are characterized by a real coefficient $\alpha_{1\beta}$ and an imaginary coefficient $\alpha_{3\beta}$. The wave may propagate in the substrate but it is totally reflected at the core/cladding boundary. The substrate modes are very similar to the radiation modes since the energy radiated by an emitting nanocrystal to that mode may leave the core on short distances in the z direction, however the energy of the modes with the propagation constant near the critical propagation constant, *i.e.* $\beta \leq \beta_C$, may reside inside the core layer at large distances from the source, by orders of magnitude larger than the core thickness. The propagation angle θ_2 of such mode is near the critical angle for total reflection on the core/substrate boundary, thus the reflectivity is slightly below 1 and only a small portion of energy is lost during one round-trip of the ray across the core layer (*cf.* Fig. 1b). The angle of propagation of the mode in the substrate is near π/2 implying that the mode propagates along the core/substrate boundary and in many experiments these modes cannot be distinguished from regular guided modes. Since the substrate modes propagate mostly outside the core, their characteristics are determined by the optical constants of the substrate. Nevertheless, the aforementioned substrate modes with $\beta \approx \beta_C$ are affected by the core properties near the source. Therefore they behave like guided modes at the distances smaller than a characteristic distance[18] of their leakage to the substrate but at the larger distances they behave more like free plane waves in the substrate.

[b] In our previous publications, substrate radiation modes have been inaccurately denoted as "leaky" waves or leaky modes. The label *leaky modes* should be rather reserved for a special sort of unphysical solutions of the wave equation with boundary conditions determined by the symmetry of a waveguide.[20]

3. Spectral Filtering of the Modes

In this section, we present a simple model which explains how the TE and TM modes are formed and which types of the waveguide modes take part in the spectral filtering process. We consider a large planar waveguide with the radiating nanocrystals uniformly distributed in the *yz* plane and therefore the system is assumed to have a full translational symmetry in both *y* and *z* directions. This model is therefore not able to simulate phenomena which arise from the wave propagation (for appropriate models, see sections 4 and 5) since the symmetry-breaking due to the sample edge is not included in the description, however the principle of the peak formation in the PL spectra is best illustrated.

The main idea of the model follows from the aforementioned fact that the substrate modes with the propagation constant slightly below the critical value undergo several round-trips in the core before their energy leaves the core. Due to the interference effects (similarly to the Fabry–Perot interferometer),[22] the nanocrystals couple effectively only to the resonant modes, *i.e.* to the modes whose phase is reconstructed after one round-trip. These modes, in addition, refract to the substrate under an angle near $\pi/2$, therefore they are hardly distinguishable from the guided modes in experiments and we observe a superposition of the guided and the substrate modes in the PL spectra. Using the detection system with a small numerical aperture,[2] substrate modes with a fixed propagation angle are selected (*cf.* Fig. 1b) and the resonances arise for certain wavelengths only. These resonances are then observed as sharp peaks in the PL spectra.

To give a more quantitative description, let us consider nanocrystals randomly oriented and randomly located in the waveguide core which radiate the optical field with intensity of electric field $E_0(\lambda)$ under an angle θ_2. The total field at the position of the nanocrystal is evaluated as a sum of the field E_0, the field after one round-trip in the core, the field after two round-trips *etc.* Mathematically, we express the field as an infinite series:

$$E_{\text{internal}}(\lambda,\theta_2)=E_0(\lambda)\{1+r_{21}r_{23}\exp[\mathrm{i}4\pi n_2 d\cos\theta_2/\lambda]+ \\ +(r_{21}r_{23}\exp[\mathrm{i}4\pi n_2 d\cos\theta_2/\lambda])^2+\ldots\}=\frac{E_0(\lambda)}{1-r_{21}r_{23}\exp[\mathrm{i}4\pi n_2 d\cos\theta_2/\lambda]}, \tag{2}$$

where r_{21} and r_{23} represent the reflection coefficients at the core/cladding and the core/substrate boundaries, respectively, which depend upon the propagation angle θ_2. Note that the summation is possible in lossless structures only if $|r_{21}r_{23}|<1$, therefore Eq. (2) is valid only for the radiation and the substrate modes, the guided modes (with $|r_{21}|=|r_{23}|=1$) will be discussed hereafter. Function $|E_{\text{internal}}(\lambda,\theta_2)|^2$ defined by Eq. (2) stands for light intensity in the core of the waveguide and has maxima at the resonances for the wavelengths λ for which the condition $\mathrm{Im}(r_{21}r_{23}\exp[\mathrm{i}4\pi n_2 d\cos\theta_2/\lambda])=0$ is fulfilled.

3.1. *Substrate and radiation modes*

Considering only the substrate modes, $|r_{21}|=1$ and $|r_{23}|<1$. The more is the value $|r_{23}|$ close to unity, the narrower are the peaks in the intensity $|E_{\text{internal}}(\lambda)|^2$. Additionally, the peak value increases with increasing $|r_{23}|$ and thus the substrate modes are most visible for angles θ_2 approaching the critical angle for total internal reflection on the core/substrate boundary. For radiation modes, the value $|r_{23}|$ is usually much less than unity and thus the spectra of the radiation modes reveal only broad peaks with a small peak-valley contrast.

In experiments, the detector has a nonzero numerical aperture and therefore radiation from a nonzero spatial angle is collected. The total intensity recorded by the detector is then

$$I_S(\lambda,\alpha)=\int_{\alpha-\Delta\alpha/2}^{\alpha+\Delta\alpha/2}|E_{\mathit{internal}}(\lambda,\theta_2)|^2\,T_{23}(\theta_2)T_S(\alpha')\frac{\mathrm{d}\theta_2}{\mathrm{d}\alpha'}\mathrm{d}\alpha', \tag{3}$$

where the detection angle is given by its mean value α (*cf.* Figs. 1a-b) and $\Delta\alpha/2$ denotes the numerical aperture of the detector. The symbols T_{23} and T_S stand for the respective transmittances at the core/substrate boundary and at the sample edge, which are given by the Fresnel formulae. The detection angle α' and the propagation angle θ_2 are connected by the Snell's law

$$n_2^2 \sin^2 \theta_2 = n_3^2 - \sin^2 \alpha' . \tag{4}$$

The expression $\mathrm{d}\theta_2 / \mathrm{d}\alpha'$ in Eq. (3) stands for the angular mode density and reflects the fact that the nanocrystals radiate isotropically in the core.

3.2. *Guided modes*

For the guided modes, the field intensity $|E_0(\lambda)|^2$ in the core is enhanced similarly to Eq. (2), however losses due to radiation at the sample edge should be included in order to get a physically correct formula. Such derivation is beyond this simple model and therefore we consider that the intensity of the electric field is given by the formula

$$E_{\text{guided}}(\lambda, \theta_2) = F(\lambda, \theta_2) E_0(\lambda), \tag{5}$$

with $F(\lambda,\theta_2)$ being the enhancement factor similar to the factor in Eq. (2). The detected intensity is

$$I_{\mathrm{G}}(\lambda, \alpha) = \int_{\alpha-\Delta\alpha/2}^{\alpha+\Delta\alpha/2} \mathrm{d}\alpha' \int_0^{\pi} \mathrm{d}\theta_2 \left| E_0(\lambda) F(\lambda, \theta_2) \right|^2 T_{\mathrm{G}}(\lambda, \theta_2) D_{\mathrm{G}}(\lambda, \theta_2, \alpha'), \tag{6}$$

where $T_G(\lambda,\theta_2)$ stands for the transmittance of the mode denoted by its wavelength λ and propagation angle θ_2 through the sample edge and $D_G(\lambda, \theta_2, \alpha')$ is the diffraction efficiency from the mode to the diffraction angle α'. The enhancement factor $F(\lambda,\theta_2)$ has sharp maxima, *i.e.* only the guided modes with well resolved propagation angle are allowed by the waveguide, therefore we may simplify[c] Eq. (6):

$$I_{\mathrm{G}}(\lambda, \alpha) = \int_{\alpha-\Delta\alpha/2}^{\alpha+\Delta\alpha/2} \left| E_0(\lambda) F(\lambda) \right|^2 T_{\mathrm{G}}(\lambda) D_{\mathrm{G}}(\lambda, \alpha') \mathrm{d}\alpha' . \tag{7}$$

Decoupling of electromagnetic waves from a dielectric waveguide to free space (evaluation of the functions $T_G(\lambda)$ and $D_{\mathrm{G}}(\lambda, \alpha')$) cannot be described simply by the Snell's law but we must take into account diffraction and full boundary conditions. Such approach is, however, beyond the scope of this paper, therefore we assess the situation only qualitatively. It is well known that diffraction on a slit with aperture

---

[c] We consider here for the sake of simplicity one guided mode for all wavelengths, however Eq. (7) may be extended to a general case of multimode guiding.

width comparable to the light wavelength produces a wide diffraction maximum in the Fraunhoffer limit, thus we may conclude that the function $D_G(\lambda,\alpha')$ has a broad peak[16] in the second variable. The guided modes, unlike the substrate modes, cannot be distinguished one from the other by selecting the angle of detection.

The magnitude of the transmittance T_G may be estimated from the Fresnel formulae; we assume that for the angle of incidence at the waveguide edge well above the critical value for the total internal reflection at the boundary the transmittance is low while for smaller angles, the modes are decoupled more effectively. Simple derivation gives that the guided modes near the cut-off wavelength are only weakly decoupled for $n_2^2 - n_3^2 > 1$. In waveguides where $n_2^2 - n_3^2 < 1$, all guided modes may be radiated from the waveguide core with high transmittance.

3.3. *All modes together, comparison with experiment*

The total intensity at the detector is

$$I(\lambda) = I_S(\lambda) + I_G(\lambda) + I_R(\lambda). \tag{8}$$

The symbol $I_R(\lambda)$ denotes the spectrum of the radiation modes which is almost flat and always less intense when compared to the spectrum of the substrate modes as seen from Eq. (2). Therefore it plays a minor role and hence it is not discussed from now on. The detected PL spectrum then consists of the two major components which compete with each other: the broad spectrum of the guided modes and the spectrum of the substrate modes with narrow peaks. Visibility of the respective components is determined by the numerical aperture of the detection system and mainly by the geometrical properties of the sample since these determine the strength of decoupling of the guided modes.

The spectra of the substrate modes reveal one more property which has not been discussed yet. The phase of the reflection coefficient r_{23} in Eq. (2) does not vary with the incident angle θ_2 when considering substrate and radiation modes, however the phase of the coefficient r_{21} does change from 0 to π in the region of substrate modes (see Fig. 3 in Ref. 17). The phase is, in addition, dependent upon the mode polarization and therefore the peaks of the TE and TM modes split in the PL spectra.

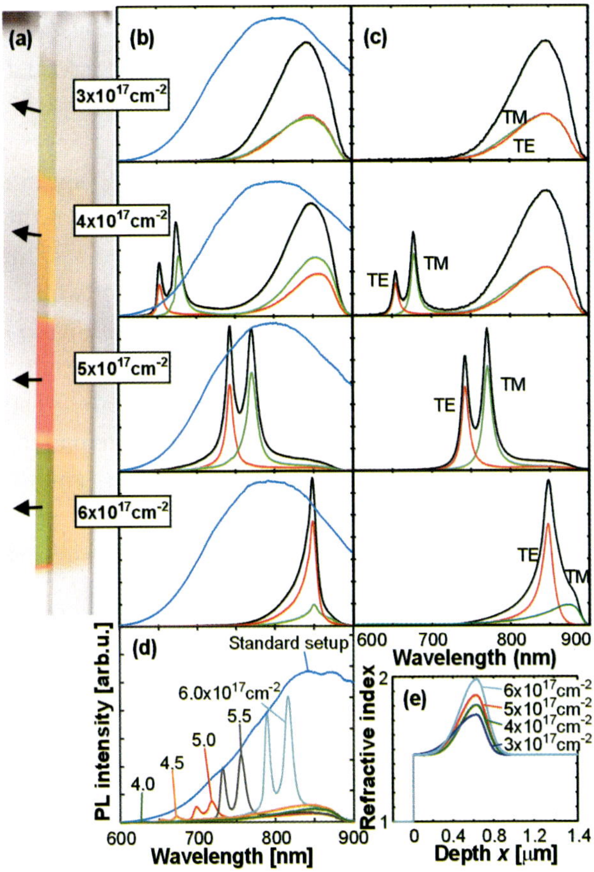

Fig. 2. (a) Photograph of the edge of a set of implanted layers with direction of PL indicated by arrows, the edge is on the left; (b) measured PL from samples implanted to different Si-ion fluences (indicated on the left) in standard (blue line) and waveguiding geometry (black line, red and green lines stand for TE and TM resolved polarizations); (c) theoretically calculated PL spectra; (d) comparison of PL measured in the waveguiding geometry for different series of implant fluences with the PL measured in the standard geometry (broad band); (e) profiles of refractive index of the samples with indicated implant fluences.[17]

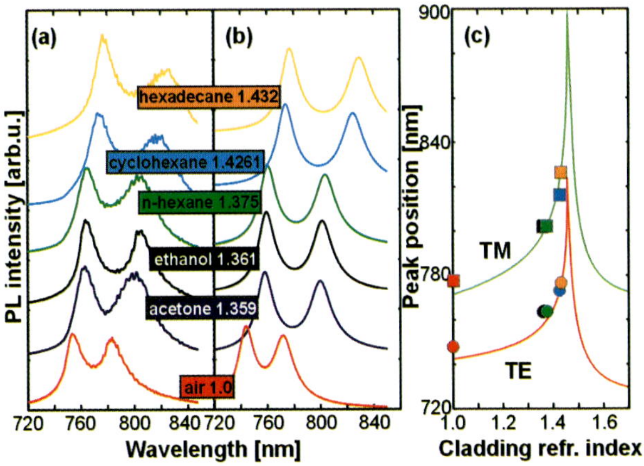

Fig. 3. (a) Measured PL spectra with a drop of various liquids above the excitation spot; (b) theoretical PL spectra; (c) comparison of experimental (points) and theoretically calculated positions of PL maxima for different refractive indices of the liquids. The sample with an implant fluence of $5{\cdot}10^{17}$ cm^{-2}.[24]

In the three-layer waveguides as discussed in this chapter, the TE mode peak is red-shifted with respect to the TM mode peak. The spectral positions of the peaks may merge or interchange in the waveguides with other profile of the refractive index.

We compared the predictions of our model with the PL spectra measured on samples prepared by Si^+ ion implantation (for details of the experimental set-up and sample preparation, see Refs. 2, 8 and 17). With the help of near-IR transmission measurement and the SRIM software,[23] we obtained the refractive index profiles of our waveguides (see Fig. 2e) which were then used in numerical simulations. The samples with higher implant fluences reveal higher peak values of the refractive index. The experimental spectra are shown in Fig. 2b and our theoretical simulations in Fig. 2c. Fig. 2d shows the PL spectra from another series of samples implanted to different fluences where the scaling of the narrow peak intensities can be clearly seen — the peak magnitudes follow well the spectral course of PL of free nanocrystals (PL measured in the standard setup).

The experimental data in Fig. 2b reveal significant modulation of the PL spectra when detecting in the waveguiding geometry. In the standard geometry (blue line), only the broad PL spectrum characteristic of Si nanocrystals is displayed and its shape is almost indcpendent of the density of nanocrystals. Measurements in the waveguiding geometry, on the contrary, reveal the two abovementioned distinct linearly polarized peaks (black line without resolution of polarization, red and green lines represent TE and TM polarization, respectively) which move to longer wavelengths with increasing density of nanocrystals, and a broader peak centred around 850 nm which is unpolarized and its spectral position remains the same in all samples. The broad peak is interpreted as the spectrum of the guided modes: for the lowest concentration of nanocrystals (implant fluence of 3×10^{17} cm^{-1}), our calculations show that the cut-off wavelengths are below 600 nm and above 1.5 μm and thus the substrate modes are missing in the selected spectral region. The interpretation is supported by the fact that the broad peak is not polarized, its shape and magnitude are only weakly dependent on nanocrystal density (Fig. 2d) and its spectrum is similar to the spectrum measured in the standard geometry (absorption cuts the spectrum of the

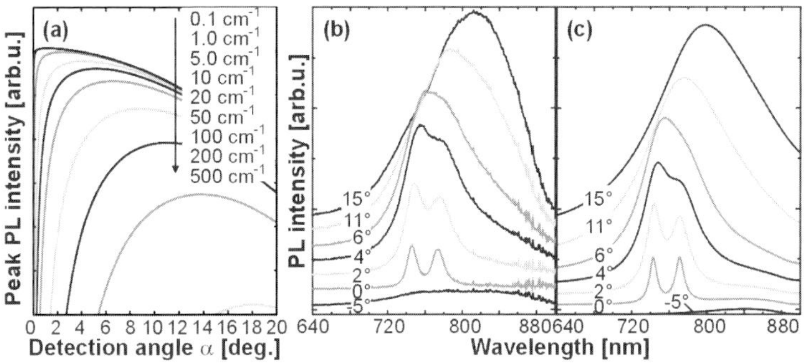

Fig. 4. (a) Theoretical dependence of peak PL intensity on the angle of detection α for various absorption coefficients of the core; (b) measured PL spectra at different detection angles; (c) numerical simulations of the measurements. The sample with an implant fluence of 5×10^{17} cm^{-2}.

guided modes below 750 nm). The substrate modes are not influenced by the absorption since they propagate mostly in the transparent substrate.

According to Eq. (8) and the subsequent discussion, the theoretical spectra represent the sum of the contributions from the two types of the modes. We are not able to determine the spectrum of the guided modes analytically within this model and thus we treat the experimental PL spectrum from the sample with the lowest nanocrystal density as the spectrum of the guided modes in all samples. The spectra of the substrate modes are then calculated according to Eq. (3) and added to the spectrum of the guided modes. Comparison of the resulting theoretical spectra in Fig. 2c to the experimental data in Fig. 2b gives an excellent agreement indicating a good precision of our model.

In experiments, we are able to set the detection angle α or the numerical aperture of the detector, therefore the angular properties of the PL should be discussed. The angle of detection may influence the detected spectra in two distinct ways: (1) magnitude of the photoluminescence and (2) shape of the spectra. The spectrum of the guided modes changes with the angle of detection only slightly,[16] therefore we discuss here only the substrate modes which reveal the most significant changes. The variance of the peak magnitude of the spectra may be evaluated directly from Eq. (3) considering $|E_0(\lambda)| = \text{const.}$ and considering exact resonance, *i.e.* $\mathrm{Im}\{r_{21}r_{23}\exp[i4\pi n_2 d\cos\theta_2/\lambda]\} = 0$ — it is plotted in Fig. 4a for several absorption coefficients of the core material taken at the wavelength of 500 nm. The peak position of the PL magnitude is not constant and shifts towards the larger values of the detection angle with increasing absorption coefficient. Besides the overall intensity, the PL spectra change also in shape when the angle α is varied as seen in our experimental results depicted in Fig. 4b. With increasing angle α, the peaks broaden and shift towards longer wavelengths. The interpretation is straightforward within our model: increase of the angle α implies decrease of the angle θ_2 and therefore decrease of the reflection coefficient r_{23} and, in addition, change of the phase of the reflection coefficient r_{21}. The former is responsible for the peak broadening while the latter causes a change of the phase matching condition and thus the spectral shift of the peaks. Our theoretical simulations in Fig. 4c agree with the experimental results in Fig. 4b which indicates correct interpretation and modelling. We feel that these experiments with detection at different angles can be hardly interpreted in terms of delocalized guided modes since there is no reason for the guided modes to reveal markedly different spectra in different diffraction directions.

3.4. *Differentiation of the substrate modes from the guided modes*

In order to support experimentally our model, we performed a novel experiment published in Refs. 17 and 24. The refractive index of the cladding layer may be easily modulated for example by dropping a small volume of liquid onto the sample. If the refractive index of the whole layer is changed, the PL spectra should change due to the obvious change of the phase of the reflection coefficient r_{21}. The measured spectra for the sample implanted to the fluence of 5×10^{17} cm^{-2} are depicted in Fig. 3a and our numerical simulations in Fig. 3b. The theoretically estimated dependence of the peak positions and their comparison to the experiments are the plotted in Fig. 3c. The guided and the substrate modes may be then unambiguously resolved if a small drop of the liquid is placed only between the excitation spot and the sample edge. The modes with their characteristic spectra form within a short distance from the excitation spot where the waveguide is not modified by the liquid. The guided modes then pass through a modified waveguide where the cut-off frequency and thus the resonance condition for the delocalized modes are shifted. If the model based on delocalized guided modes were valid, we would thus expect a shift of the spectral positions of the peaks. The substrate modes, on the contrary, are already decoupled from the core at the position of the liquid and propagate freely in the substrate and therefore the core/cladding boundary cannot influence them. The PL spectra are then, in the case of validity of our model of substrate modes, expected to remain unchanged when compared to the original spectra without any liquid above the sample. The experimental data[17] prove the correctness of the concept of the substrate modes since the spectra do not modify when dropping a liquid between the excitation spot and the edge.

Obviously the active waveguides may be used as sensitive detectors of refractive index changes (we estimate their sensitivity up to 10^{-5}) which may be tuned for particular applications via tailoring the preparation parameters as shown in Ref. 25. The spectral region in which the detector may work is not limited only to the PL band of nanocrystalline silicon as shown experimentally in Ref. 14. The data presented therein show that light with a broad spectrum may be coupled to the substrate modes from outside. The modes which fulfil the resonance condition are then absorbed in the core and the reflection reveal dips at the theoretical spectral positions of the PL peaks, which is in agreement with our model.

4. Wave Propagation in Waveguides

In this section, we introduce an advanced model capable of describing the wave propagation in the direction along the waveguide layer. Within this theory, we are able to interpret the mechanism of the mode formation and we may evaluate the influence of the net optical gain or losses on the PL intensity. Although the model is formulated in terms of the ray optics (and an approximative description of the wave propagation), it gives good qualitative predictions and it is helpful when interpreting the experiments. A more subtle but with difficulty interpretable model usable for quantitative predictions is introduced in section 5.

Since we are interested here in wave propagation, we must take into consideration the fact that nanocrystals emit generally spherical waves and thus the signal recorded by a spatially limited detector decreases with the distance from the source as already discussed in Ref. 26 for a special case. The real detector collects the signal from a nonzero spatial angle as well as the input aperture has a nonzero spatial area and therefore there are two limiting cases for which we develop our theory: (1) A small detector with a large numerical aperture (NA), typically an optical fibre capable of high spatial resolution and small angular resolution, and (2) A detector with a small NA thus with a high angular resolution. Since the energy density of a spherical wave decreases as $1/r^2$, where r denotes a distance from a source, the intensity detected in the case 1 by a spatially small detector also decreases as $1/r^2$. In the latter case, on the contrary, when the characteristic size of the input lens of the detector is much larger than r, the signal would be independent of the distance from the source. We therefore denote the coupling coefficient as $\rho(\zeta)$ in the following text for spherical waves, *i.e.* for the case 1. The positive variable $\zeta = -z_{\mathrm{N}}$ denotes the distance between the excited spot and the sample edge, z_{N} being the position of the excited spot at the axis z. For detectors with a small NA and high angular resolution, we put $\rho(\zeta)=1$.

We divide the initial discussion into the guided and the substrate modes while the radiation modes are omitted for the sake of simplification (as discussed above, they play only a minor role).

4.1. *Guided modes*

It was stated above in section 3 that the PL from Si nanocrystals couples to the guided modes in the whole spectral region and that the spectrum of the guided modes is nearly equal to the spectrum of the PL of the unperturbed nanocrystals.[d] A wave emitted by a nanocrystal at a distance ζ from the sample edge is amplified or attenuated depending on the gain

[d] The spectral intensity of the modes is proportional to the number of the guided modes. Therefore the spectral intensity of the light carried out by the modes may reveal jumps at distinct wavelengths which correspond to the cut-off wavelengths.

and the loss coefficients, and the intensity transmitted through the edge takes the form (according to Eq. (6))

$$I_{G,2}(\lambda,\alpha,\zeta)=I_0(\lambda)|F(\lambda)|^2\rho_G(\zeta)\exp[(g(\lambda)-\gamma(\lambda))\zeta]\times$$
$$\times\int_{\alpha-\Delta\alpha/2}^{\alpha+\Delta\alpha/2} T_G(\lambda,\theta_2)D_G(\lambda,\theta_2,\alpha')\mathrm{d}\alpha', \tag{9}$$

where $g(\lambda)$ and $\gamma(\lambda)$ stand for the gain and the loss coefficient, respectively, $I_0(\lambda)=|E_0(\lambda)|^2$ is the spectral intensity of the PL and $\rho_G(\zeta)$ is the coupling coefficient. The dependence of the gain and the loss coefficients on wavelength reflects only properties of the material (the role of mode delocalization is neglected since our calculations show it is small) and we consider in the following that the major portion of the energy of guided modes is localized inside the core.

The coupling coefficient cannot be expressed simply as $1/\zeta$ since the detector is outside the waveguide core and the wave refraction at the sample edge, illustrated in Fig. 1c, is responsible for a slight modification. Considering only rays near the normal incidence in Fig. 1c, the intensity radiated to the angle ϕ_2 is $I_0\phi_2/2\pi$. After refraction, this energy further propagates in the yz plane with divergence given by an angle $\phi_0=n_2\phi_2$ and at the position of the detector (its distance from the sample edge being D), the energy is spread over the arc with length $\phi_2(n_2D+\zeta)$. In the xz plane, the wave decoupled from the waveguide is also divergent, however the distance D is fixed and the divergence contributes to the coupling coefficient by a constant term $1/D$. It is then easy to express the coupling coefficient for the guided modes as

$$\rho_G(\zeta)=\frac{1}{n_2D+\zeta}. \tag{10}$$

Note that this formula is valid only for the spatially small detector, for a large detector with small NA we should write $\rho_G(\zeta)=1$.

4.2. *Substrate modes*

Unlike the guided modes, where we usually cannot resolve the particular guided modes by the angle of detection and the PL spectrum is similar to the PL spectrum of unperturbed nanocrystals, the substrate modes reveal maxima and minima in the spectra according to the constructive or destructive interference, respectively. The shape of the spectra depends on the number of internal reflections in the core and thus on the distance which the mode propagates. To express the shape of the spectrum mathematically, we then modify Eq. (3) in order to account for the coupling and Eq. (2) must be adapted to the finite number of reflections. The intensity emitted from the substrate then reads

$$I_{S,3}(\lambda,\alpha,\zeta)=\rho_S(\theta_2,\zeta)\int_{\alpha-\Delta\alpha/2}^{\alpha+\Delta\alpha/2}\left|E_{\text{internal}}(\lambda,\theta_2,\zeta)\right|^2 T_{23}(\theta_2)T_S(\alpha')\frac{d\theta_2}{d\alpha'}d\alpha',\tag{11}$$

where

$$\left|E_{\text{internal}}(\lambda,\theta_2,\zeta)\right|^2=\left|\frac{1-[P(\lambda,\theta_2)]^{N(\theta_2,\zeta)+1}}{1-P(\lambda,\theta_2)}\right|^2 I_0(\lambda),\tag{12}$$

$$P(\lambda,\theta_2)=r_{21}(\theta_2)r_{23}(\theta_2)\exp[(g(\lambda)-\gamma(\lambda))z_R/2]\exp[i4\pi n_2 d\cos\theta_2/\lambda].\tag{13}$$

Here $N(\theta_2,\zeta)$ is the number of reflections of the wave inside the core and z_R is the distance between two points of reflection on the core/substrate boundary (see Fig. 1b). The intensity of radiation which remains in the core may be expressed by a formula similar to Eq. (9), however we account for the wave leaking by the effective attenuation coefficient γ_{eff}:[18]

$$I_{S,2}(\lambda,\alpha,\zeta)=I_0(\lambda)\rho_G(\zeta)\exp[(g(\lambda)-\gamma(\lambda)-2\gamma_{\text{eff}}(\lambda,\theta_2))\zeta]\times \int_{\alpha-\Delta\alpha/2}^{\alpha+\Delta\alpha/2}T_G(\lambda,\theta_2)D_G(\lambda,\theta_2,\alpha')d\alpha'.\tag{14}$$

In order to evaluate N, we consider that after N reflections of a ray on the core/substrate boundary (see Fig. 1b) the intensity in the core with $g=\gamma=0$ may be expressed as $I_{S,2}(N,\zeta)=I_0\left|r_{21}r_{23}\right|^{2N}=I_0\exp[-2\gamma_{\text{eff}}\zeta]$. The number of reflections may be then evaluated in the following way:

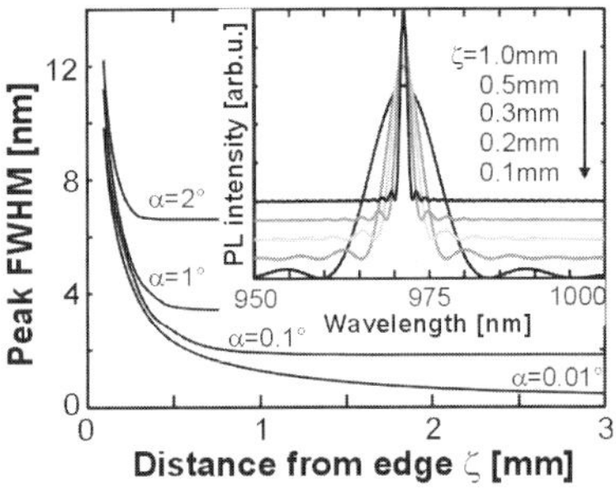

Fig. 5. Calculated dependence of the PL peak width on the length of propagation of the substrate mode for several angles of detection; inset: spectral profiles of PL for α=1°.[25]

$$N = -\gamma_{\mathrm{eff}}\zeta / \ln|r_{21}r_{23}|, \tag{15}$$

$$z_{\mathrm{R}} = -\ln|r_{21}r_{23}| / \gamma_{\mathrm{eff}}. \tag{16}$$

The effective attenuation coefficient may be calculated from the Poynting vector.[18] In the case of an asymmetric waveguide with the step-like profile of the refractive index, the coefficient may be expressed for TE waves as

$$\gamma_{\mathrm{eff}} = -\mathrm{i}\frac{\alpha_{2\beta}{}^{2}}{\beta d}\frac{\alpha_{3\beta}}{(\alpha_{2\beta}+\alpha_{3\beta})^{2}}, \tag{17}$$

with $\alpha_{2,3\beta}$ defined in section 2 and $\beta=2\pi n_2 \sin\theta_2/\lambda$. In structures with more complicated profile of the refractive index, the attenuation coefficient and the reflection coefficients must be evaluated numerically, however Eqs. (13) to (16) remain valid. The coupling coefficient may be, after simplifications and considering θ_2 near the critical angle θ_C, evaluated from the geometry of the system as

$$\rho_S(\theta_2,\zeta) = \left[\sqrt{2\frac{\theta_C-\theta_2}{n_3}\sqrt{n_2^2-n_3^2}\,(\zeta+n_2D)(\zeta+n_3D)}\right]^{-1}. \tag{18}$$

The PL spectrum of the modes radiated to the substrate obviously reveals narrow peaks for sufficiently large number of reflections, *i.e.* after a sufficiently long propagation of the mode. The modes which are excited near the sample edge undergo only several reflections and their spectrum is almost flat. With the increasing propagation length, the resonances become better resolved and the spectrum gets narrower as shown in

Fig. 5 where we plot the results of the numerical evaluation of Eqs. (11)-(13) for the special case of a step-like profile of the refractive index. For different profiles, the modes may form at distances differing by orders of magnitude as seen when comparing *e.g.* Refs. 10 and 27.

4.3. *Optical gain*

Our aim in this paragraph is to discover the role of the optical gain in waveguiding samples. Let us consider the VSL measurement of the net optical gain: a stripe of the length ℓ is optically excited and we detect the total light intensity radiated from the sample in the waveguiding setup. The PL comes from all excited nanocrystals in the stripe and thus the total ASE intensity may be expressed as an integral of Eqs. (9), (11) and (14):

$$I_{\mathrm{ASE}}(\lambda,\alpha)=\int_0^{\ell}\left[I_{\mathrm{G},2}(\lambda,\alpha,\zeta)+I_{\mathrm{S},2}(\lambda,\alpha,\zeta)+I_{\mathrm{S},3}(\lambda,\alpha,\zeta)\right]\mathrm{d}\zeta\,. \quad (19)$$

The above formula is composed of the respective contributions from the guided modes, the substrate modes diffracted from the core and the substrate modes refracted from the substrate. The three contributions have generally different magnitudes because of different mechanisms of decoupling of the associated waves at the sample edge. Since the efficiencies of the wave decoupling and the enhancement factor of the guided modes (see Eq. (7)) are unknown and depend on the particular sample geometry, we separate phenomenologically our discussion into two general cases: 1. majority of the guided modes (substrate modes cannot be distinguished in PL spectra) and 2. majority of the substrate modes.

1. If the *guided modes are dominant*, the energy of the radiating dipole in a silicon nanocrystal is coupled mainly to the guided modes. We may therefore neglect the contributions from the substrate modes and evaluate only the integral of Eq. (9). For the detector with a very small numerical aperture (*i.e.* when detecting plane waves, $\rho=1$), the integration yields the well known formula[15]

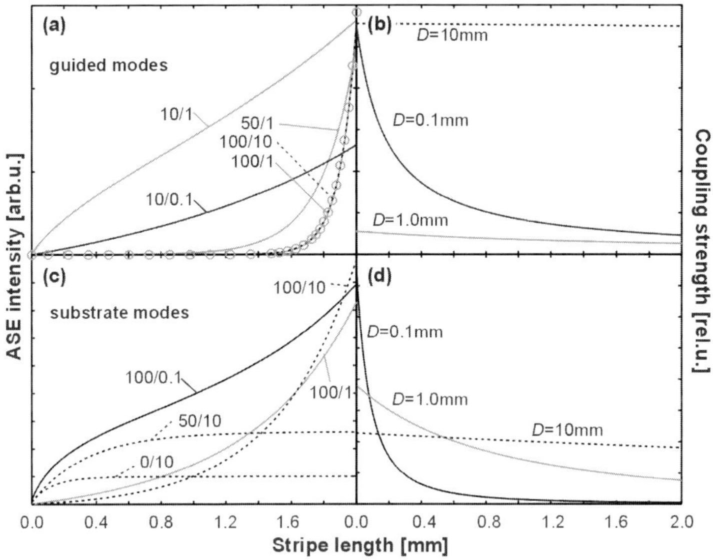

Fig. 6. Calculated ASE intensity from Eq. (19) as a function of the excited stripe length for the guided (a) and the substrate (c) modes; a small detector with a large NA is supposed. Coupling coefficients are plotted for comparison for the guided (b) and the substrate (d) modes. Scales of the curves are set for better comparison of their behavior. Numbers in (a) and (c) mean (gain coefficient in cm^{-1})/(distance D between the sample and the detector in mm). Circles in (a) denote ASE intensity according to Eq. (20) with $(g-\gamma)=100\ cm^{-1}$.

$$I_{\mathrm{ASE}}(\ell) \propto \frac{\exp[(g-\gamma)\ell]-1}{g-\gamma}, \tag{20}$$

where $(g-\gamma)$ denotes the net gain. The result for a spatially small detector with a large NA obviously differs, however the result is not analytically integrable. We thus give only the numerical results in Fig. 6a — we compare evaluated Eq. (20) with Eq. (19) for the guided modes and the small detector for several values of the gain coefficient and the position of the detector. In order to give a clear interpretation of the graph, we plot in Fig. 6b the strength of coupling $\rho(\zeta)$ of the excited elements to the detector (the strength of coupling may be directly measured in the Shifting Excitation Spot[2] experiment (SES)). It is obvious from Figures 6a and 6b that Eq. (20) is valid only in special cases of the detection of plane waves by a spatially large detector or by a small detector at the distance from the sample edge much larger than the length of the excited stripe. In the latter case, however, we may expect experimental complications because of the low PL intensity.

2. If the *substrate modes are dominant*, we consider that we detect ASE emission at the wavelength of one of the modes because at other wavelengths the contribution due to the substrate modes is negligible as discussed above and then the ASE intensity is modelled according to the discussion in the previous paragraph. The dominance of the substrate modes is caused by inefficient decoupling of the waves from the core at the sample edge and thus the major contribution comes only from the third term in Eq. (19). We plot in Figs. 6c-d the ASE intensity and the magnitudes of the coupling coefficient for several values of the gain coefficient and of the distance of the detector from the sample edge. The curves obviously show that Eq. (20) cannot be used for evaluation of the net gain in this case. For clear resolution of the net gain, one should use the SES method.

The theory developed in this section fits very well our experimental observations in VSL experiments. By choosing proper wavelength, we are able to select whether we detect guided or substrate modes and then we can directly compare their characteristics in one sample. An instructive result published in Ref. 28 means that the guided modes are more effectively amplified in the presence of a positive net optical gain than the substrate modes, as follows from the theory. We are also able to explain the unusual behaviour of the ASE and SES curves presented in Figs. 1e-f. The SES curve in Fig. 1f for the guided mode at 825 nm decreases monotoneously as the coupling coefficient decreases. The curves for the substrate TE and TM modes, however, initially increase and for longer distance of the excited spot from the edge they start to decrease due to the decrease of the coupling coefficient. The initial increase follows from the initial refraction of the waves into the substrate — since energy is radiated to the core and we detect the waves from the substrate, there is a nonzero propagation length necessary for leakage of the energy from the core. The position at which the SES curve reaches its peak depends on the detection angle (numerical aperture of the detector) as shown experimentally in Ref. 8 and as may be shown using our model.

5. Numerical Analysis of the Modes

Both models presented above are based on assumptions which simplify either the real geometry of the problem (consideration of the infinite waveguide in section 3) or the level of physical description (simple use of rays in section 4). In this section, we present a rigorous model which correctly describes propagation of the optical waves along the waveguide. The calculations in this section are performed numerically because it is impossible to derive the final expressions analytically.

The structure under description has three layers, two of them being infinite (the cladding and the substrate). According to the discussion in Ref. 20, the mode spectrum is then composed of discrete guided modes and the continuum of the substrate and the radiation modes. Presence of the continuum represents an unwanted feature in the numerical evaluation and therefore we overcome this problem by considering that all the layers are finite and the top and the bottom of the whole guiding structure is covered by a medium with reflectivity r=1. The waves propagate in the z direction and we consider the edge at the coordinate z=0. We assume the infinite structure in the y direction and we consider detection of the signal in the limit of Fraunhoffer diffraction (propagation of "plane" waves). The thickness of the cladding and the substrate is set to 0.6 mm, thus the angular resolution of the pattern diffracted from the edge is resolved with precision better than 0.15°.

The spatial profile of the intensity of electric field is given by Eq. (1) for TE modes, *i.e.* for the waves whose vector of electric intensity lies in the y direction — the numerical analysis is performed only for the TE waves since the mode splitting is well explained by the simplified models and TM modes reveal similar behaviour to TE ones in other characteristics. Using the theory of Ref. 20, the calculations have been performed in the way briefly summarized in the following. First, the allowed propagation modes (*i.e.* modes with Re β>0) were found numerically — let us assume that the 2×2 transfer matrix for the whole structure without mirrors is T_β and the relation between the field coefficients at the two opposite boundaries of the structure is[e]

[e] The medium with index 0 is an infinitesimally thin layer at the top of the cladding with refractive index n_0=n_1,

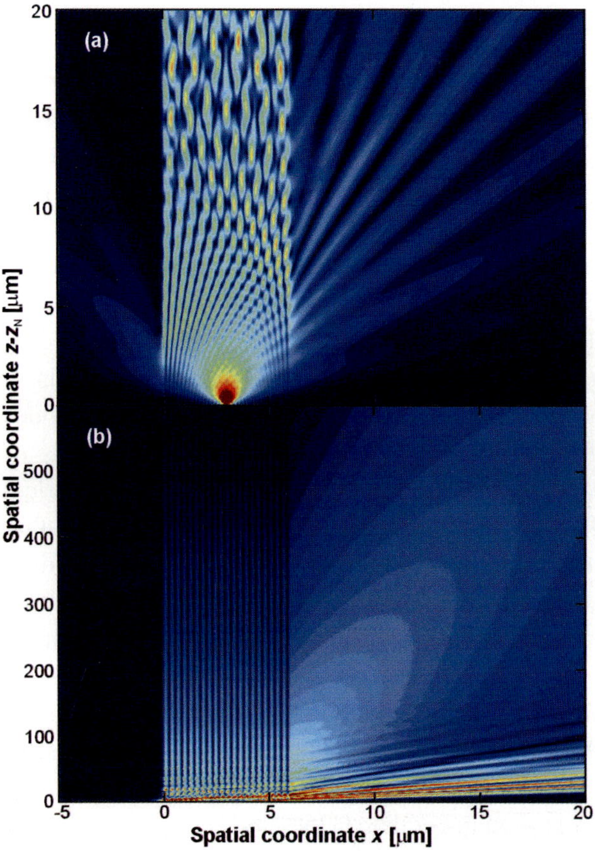

Fig. 7. (a) Distribution of the light intensity at wavelength 624.4 nm in the sample with 6 μm thick core; all modes are plotted. The coordinate x means depth below the sample surface. (b) Detailed view of decoupling of the lowest–order substrate mode at long propagation distances; only substrate modes are plotted for clarity. Dark blue colour denotes minimum intensity, maximum intensity is dark red.

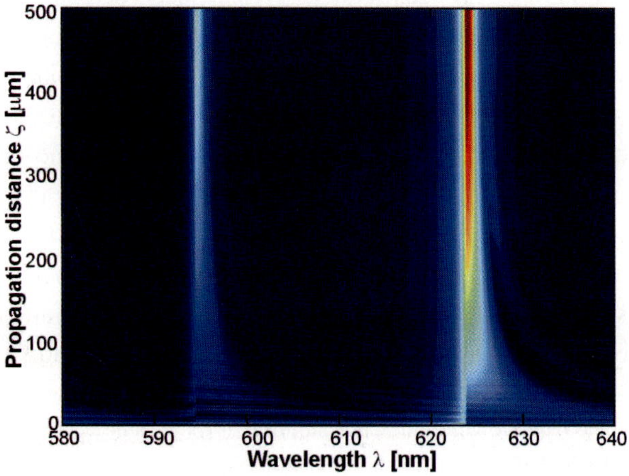

Fig. 8. Dependence of the light intensity on wavelength (horizontal axis) and propagation distance of the substrate modes (vertical axis). The waveguide has a core 6 μm thick.

$$\begin{pmatrix} A_{3\beta} \\ B_{3\beta} \end{pmatrix} = T_{\beta} \begin{pmatrix} A_{0\beta} \\ B_{0\beta} \end{pmatrix}. \tag{21}$$

Boundary conditions imply $A_{0\beta}=B_{0\beta}$ and $A_{3\beta}=B_{3\beta}$ and the (generally complex) coefficients β which fulfil the above constraints were found iteratively. The second step consisted in calculation of the field distribution at the sample edge. We consider a nanocrystal at the coordinates $(x_{\mathrm{N}}, z_{\mathrm{N}} = -\zeta)$ which is coupled to the waveguide modes. The field $E_{\mathrm{rad}}(x,z_{\mathrm{N}})$ radiated by the nanocrystal may be decomposed to the modes as follows:

$$E_{\mathrm{rad}}(x, z_{\mathrm{N}}) = \sum_{m} \frac{[E_{\beta_m}(x', z_{\mathrm{N}}), \delta(x' - x_{\mathrm{N}})]}{\sqrt{[E_{\beta_m}(x', z_{\mathrm{N}}), E_{\beta_m}(x', z_{\mathrm{N}})]}} E_{\beta_m}(x, z_{\mathrm{N}}), \tag{22}$$

where $[\cdot,\cdot]$ stands for a scalar product (integral over the x coordinate) and the summation goes over all allowed propagating modes. The field distribution at the sample edge is then, according to the propagating factor in Eq. (1), expressed as

$$E_{\mathrm{rad}}(x, z=0) = \sum_{m} \frac{[E_{\beta_m}(x', z_{\mathrm{N}}), \delta(x' - x_{\mathrm{N}})]}{\sqrt{[E_{\beta_m}(x', z_{\mathrm{N}}), E_{\beta_m}(x', z_{\mathrm{N}})]}} E_{\beta_m}(x, z_{\mathrm{N}}) \exp[-i\beta_m z_{\mathrm{N}}]. \tag{23}$$

In the third step, we calculated the field diffracted from the sample edge in the Fraunhoffer limit by performing the Fourier transform on the function $E_{\mathrm{rad}}(x,z=0)$ and considering field refraction according to Snell's law and Fresnel formulae for simplification.

In the numerical results presented below, we consider a model system with the cladding with refractive index $n_1=1$ and width $w_1 = 600$ μm, the substrate with refractive index $n_3 = 1.46$ and width $w_3 = 600$ μm and the core has refractive index $n_2 = 1.8$ and its width is $d = 6$ μm. Formation and propagation of the modes is illustrated in Figures 7a and 7b where we plotted the distribution of the light intensity in the structure (the coordinate $x = 0$ refers to the core/cladding boundary). The beams which leave the core to the right are the substrate modes — we may assign an angle of propagation to each of them. It is interesting that obviously they are not radiated from the core at one distinct point but they decouple over long distances. This property is better visualized in the detailed view in Fig. 7b where we may identify the substrate modes which are rapidly

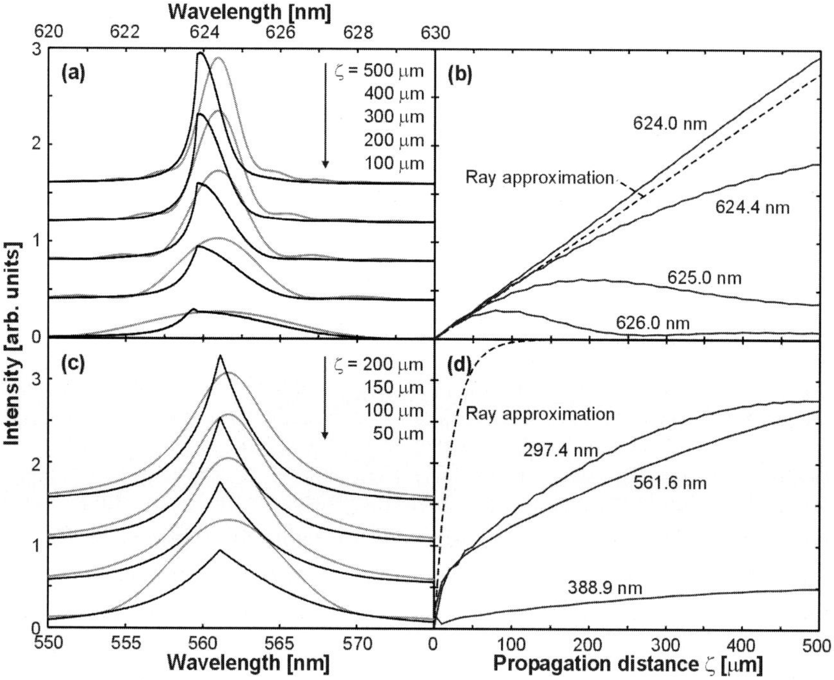

Fig. 9. (a), (c) Narrowing of the substrate mode during propagation along the waveguide. Spectral profiles of the modes at several distances from the excited spot calculated by the numerical model (black lines) and results of the ray model for comparison (gray lines). (b), (d) Calculated energy decoupled from the core as a function of the propagation distance from the excited spot (solid) and results of the ray model for comparison (dashed line). Parameters of the curves denote wavelengths in the vicinity of the resonance of the mode (b) and wavelengths of several resonant modes (d). The core thicknes is 6 μm (a), (c) and 0.6 μm (b), (d), respectively.

decoupled from the core and one mode which decouples at much longer distances comparable to 0.5 mm.

The results depicted in Fig. 7 directly support the model from section 4 based on the assumption of the gradual decoupling of the waves from the core to the substrate. As we have shown, this fact is responsible for gradual narrowing of the PL spectra when increasing $|z_N|=\zeta$, *i.e.* when the excited spot is moved further from the sample edge. We performed calculations in order to confirm this hypothesis and the results are plotted in Fig. 8 — the dependence of the PL intensity measured at the angle α=2.5° on wavelength λ and distance $|z_N|=\zeta$ between the nanocrystal and

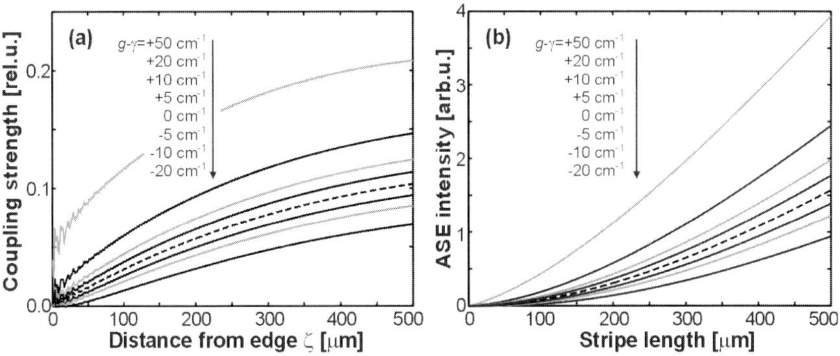

Fig. 10. (a) Calculated coupling strength of the excited spot at the distance ζ from the sample edge considering a sample with the core thickness of 6 μm for several values of gain coefficient at the resonant wavelength 624.4 nm of the substrate mode (*cf.* Figs. 7, 8); (b) calculated ASE intensity recorded from the sample with same parameters as in (a).

the sample edge. The cuts of this graph in the directions of both axes ζ, λ are plotted in Figures 9a and 9b. In Fig. 9a, the mode narrowing is obvious and we compare these numerical results with the results of the model from section 4 (plotted as grey lines). We may conclude that the approximate model gives accurate results and is therefore usable as an estimate for practical use in evaluation of experiments. The rate of energy decoupling from the core is depicted in Fig. 9b for several wavelengths around the resonance together with the estimate performed according to Ref. 18. Decoupling rate is obviously well estimated by the theoretical curve for the resonant mode and thus the ray model is usable also in this case. Graphs similar to those in Figures 9a and 9b are plotted in Figures 9c and 9d for the waveguide core with thickness of 0.6 μm together with comparison with the simplified model. The energy flow from the core to the substrate is more pronounced when compared to the waveguide with the wider core. We may interpret this feature in terms of ray propagation: as seen from Eqs. (16) and (17), the distance z_R between the two points of reflection of the ray is proportional to the waveguide thickness and therefore the energy loss due to transmission to the substrate is faster in thin waveguides. Although qualitative discussion well reflects the situation, quantitative comparison in Fig. 9d fails. The limitations of the ray model from section 4 are thus clearly seen — it gives good numerical predictions only when $d \gg \lambda$, *i.e.* when wave propagation is well described by ray optics. The narrowing of the modes, nevertheless, may be roughly estimated by using the ray model also for structures with $d \sim \lambda$ as seen in Fig. 9c.

We may include small optical losses or optical gain (up to tens cm^{-1}) into the numerical model in the form of a complex refractive index. We present in Fig. 10a the dependence of the intensity of light emitted by a nanocrystal and transmitted through the structure with gain or losses on the distance of propagation. The detection angle is $\alpha = 2.5°$, wavelength 624.4 nm and other parameters equal to those in Fig. 7. By integration of the curves in Fig. 10a, we get the ASE intensity depicted in Fig. 10b. According to section 4, all the ASE curves seem to have exponential behaviour near the origin $\zeta = |z_N| = 0$, however this is not a consequence of the optical gain but the result of the mode leaking.

We may conclude this section by stating that the approximate models are capable of a qualitative description of the mode propagation and leaking in the waveguiding structures. They give good quantitative results if the waveguide core is wide enough ($d \gg \lambda$) but one should use them only as rough estimates in other cases where numerical calculation should be applied in order to get quantitatively correct results.

6. Conclusions and Acknowledgements

The results presented in this chapter show that the substrate modes, usually considered as an unwanted artificial feature, have properties very different from the guided modes of the waveguide. Because of their limited lifetime inside the core, they cannot be effectively absorbed or amplified and therefore they are hardly controllable. In spectroscopic measurements, it appears almost impossible to evaluate the magnitude of the net optical gain from the VSL measurements on the substrate modes and therefore it is desirable to avoid (by properly designing the core parameters) the presence of the substrate modes in most applications including optical spectroscopy. The spatial separation of the substrate and the guided modes can be exploited in order to filter out the substrate modes, *e.g.* by using a detection with high spatial resolution or using techniques for decoupling of the guided waves from the core (by placing a prism on the top plane of the sample *etc.*).

The substrate modes may be, however, of importance in nano-silicon active optoelectronic devices where light is emitted, guided or amplified. Because of usual small optical gain in Si nanocrystals, the intensity of the substrate modes may be comparable to or even larger than the intensity of the guided modes and they may be then a source of noise. Devices must be then constructed in order to avoid wave propagation in the substrate — one may use for example absorbing substrate or a special geometry of the device in order to physically block the waves.

Although the substrate modes represent in most cases a parasitic feature in waveguiding devices, it appears, however, possible to use them in a wide range of applications in nano–optics. They may be used for optical sensing profiting from their sensitivity to the refractive index of the cladding. Another promising application may be the use of active waveguides as simple tunable sources of narrow–line directional optical emission and as tunable optical filters. However, further research in this direction is needed.

Financial support through research projects GAAVCR IAA1010316, GACR 202/01/D030, GACR 202/07/0818, Centrum LC510 and the institutional research plan AV0Z 10100521 (Institute of Physics) is greatly appreciated. This work is also a part of the research plan MSM 0021620834 that is financed by the Ministry of Education of the Czech Republic. Thanks are also due to Prof. R.G. Elliman (ANU Canberra) for kindly supplying samples used throughout this study.

References

1. L. Pavesi, L. Dal Negro, C. Mazzoleni, G. Franzo and F. Priolo, *Nature* **408**, 440 (2000).
2. J. Valenta, I. Pelant and J. Linnros, *Appl. Phys. Lett.* **81**, 1396 (2002).
3. L. Dal Negro, M. Cazzanelli, L. Pavesi, D. Pacifici, G. Franzo, F. Priolo and F. Iacona, *Appl. Phys. Lett.* **82**, 4636 (2003).
4. L. Khriachtchev, M. Räsänen, S. Novikov and J. Sinkkonen, *Appl. Phys. Lett.* **79**, 1249 (2001).
5. J. Ruan, P.M. Fauchet, L. Dal Negro, M. Cazzanelli and L. Pavesi, *Appl. Phys. Lett.* **83**, 5479 (2003).
6. M. Cazzanelli, D. Navarro-Urrios, F. Riboli, N. Daldosso, L. Pavesi, J. Heitmann, L.X. Yi, R. Scholz, M. Zacharias and U. Gösele, *J. Appl. Phys.* **96**, 3164 (2004).

7. L. Khriachtchev, M. Räsänen and S. Novikov, *Appl. Phys. Lett.* **83**, 3018 (2003).
8. J. Valenta, T. Ostatnický, I. Pelant, R.G. Elliman, J. Linnros and B. Hönerlage, *J. Appl. Phys.* **96**, 5222 (2004).
9. L. Khriachtchev, M. Räsänen, S. Novikov and J. Lahtinen, *J. Appl. Phys.* **95**, 7592 (2004).
10. L. Khriachtchev, S. Novikov, J. Lahtinen and M. Räsänen, *J. Phys.: Condens. Matter* **16**, 3219 (2004).
11. D. Navarro-Urrios, F. Riboli, Massimo Cazzanelli, A. Chiasera, N. Daldosso, L. Pavesi, C.J. Oton, J. Heitmann, L.X. Yi, R. Scholz and M. Zacharias, *Opt. Mater.* **27**, 763 (2005).
12. R. Ferre, B. Garrido, P. Pellegrino, M. Perálvarez, C. García, J. A. Moreno, J. Carreras and J. R. Morante, *J. Appl. Phys.* **98**, 084319 (2005).
13. J. Valenta, I. Pelant, K. Luterová, R. Tomasiunas, S. Cheylan, R.G. Elliman, J. Linnros and B. Hönerlage, *Appl. Phys. Lett.* **82**, 955 (2003).
14. P. Janda, J. Valenta, T. Ostatnický, E. Skopalová, I. Pelant, R.G. Elliman and R. Tomasiunas, *J. Lumin.* **121**, 267 (2006).
15. K. L. Shaklee, R. F. Leheny and R. E. Nahory, *Phys. Rev. Lett.* **26**, 888 (1971).
16. T. Ostatnický, J. Valenta, I. Pelant, K. Luterová, R.G. Elliman, S. Cheylan and B. Hönerlage, *Opt. Mater.* **27**, 782 (2005).
17. I. Pelant, T. Ostatnický, J. Valenta, K. Luterová, E. Skopalová, T. Mates and R.G. Elliman, *Appl. Phys. B* **83**, 87 (2006).
18. D. Marcuse, *Theory of Dielectric Optical Waveguides* (Academic Press, Boston, 1991).
19. A.W. Snyder and J.D. Love, *Optical Waveguide Theory* (Chapman&Hall, London, 1991).
20. R. E. Smith and S. N. Houde-Walter, *J. Opt. Soc. A* **12**, 715 (1995).
21. H. Kogelnik and V. Ramaswamy, *Appl. Opt.* **13**, 1867 (1974).
22. M. Born and E. Wolf, *Principles of Optics: Electromagnetic Theory of Propagation, Interference and Diffraction of Light* (Pergamon Press, Oxford, 1993).
23. The software package SRIM developed by J.F. Ziegler *et al.*, http://www.srim.org.
24. K. Luterová, E. Skopalová, I. Pelant, M. Rejman, T. Ostatnický and J. Valenta, *J. Appl. Phys.* **100**, 074307 (2006).
25. T. Ostatnický, P. Janda, J. Valenta and I. Pelant, *Proc. SPIE* **6609**, 66090F (2007).
26. L. Dal-Negro, P. Bettotti, M. Cazzanelli, D. Pacifici and L. Pavesi, *Opt. Commun.* **229**, 337 (2004).
27. K. Luterová, D. Navarro, M. Cazzanelli, T. Ostatnický, J. Valenta, S. Cheylan, I. Pelant and L. Pavesi, *phys. stat. sol. (c)* **9**, 3429 (2005).
28. K. Luterová, M. Cazzanelli, J.-P. Likforman, D. Navarro, J. Valenta, T. Ostatnický, K. Dohnalová, S. Cheylan, P. Gilliot, B. Hönerlage, L. Pavesi and I. Pelant, *Opt. Mater.* **27**, 750 (2005).

CHAPTER 11

SILICON NANOCRYSTALS IN SILICA: OPTICAL PROPERTIES AND LASER-INDUCED THERMAL EFFECTS

Leonid Khriachtchev

Laboratory of Physical Chemistry, University of Helsinki
P. O. Box 55, FIN-00014 Helsinki, Finland
E-mail: leonid.khriachtchev@helsinki.fi

Silicon nanocrystals embedded in silica matrix are studied by optical spectroscopy. Correlations between the optical and structural properties suggest the photoluminescence mechanism involving oxygen-related defects. Wavelength-selective optical waveguiding by silica films containing silicon nanocrystals is described. This effect allows to estimate the optical properties of the film material. In particular, birefringence of silicon nanocrystals in silica matrix is found for some samples. Laser-induced thermal effects on structural and optical properties of free-standing silica films containing silicon nanocrystals are described in detail. The obtained results show very efficient Si-SiO_2 phase separation by intense laser light (laser annealing). Laser-induced stress of silicon nanocrystals is evidenced by Raman spectroscopy. The laser manipulations with stress of silicon nanocrystals offer an approach to optical memory with an extremely long retention time.

1. Introduction

Promising applications and fundamental value of silicon nanoscale structures require detailed studies of a variety of their properties. Silica layers containing silicon nanocrystals (Si-nc) is an interesting material with optoelectronic perspectives.[1,2] Studies of Si-nc embedded in silica have been stimulated by reports on optical gain in this material,[3-9] giving chances for a Si-nc laser.[10] The mechanism of the broad photoluminescence (PL) around 1.5-eV is still uncertain, and several models have been under discussion.[11-27] The PL spectrum of Si-nc embedded in silica matrix is quite independent of the crystallite size, which supports a "surface-state model".[12] Based on analogies with oxidized porous silicon,[14,17,20] the light emitting centers probably involve Si=O covalent bonds and locate at the interface between Si-nc and host oxide. On the other hand, the 1.5-eV PL does not correlate straightforwardly with the concentration of Si-nc embedded in the oxide host so that the emission can originate from less ordered areas.[23,27]

Waveguiding of light is an important photonic phenomenon, which has been extensively studied. However, waveguiding in Si-nc/SiO_2 layers exhibits unusual properties due to substantial absorption of the core material. Efficient wavelength-selective waveguiding in Si-nc/SiO_2 layers was found,[4,28] and tunable spectral filtering of light in Si-nc/SiO_2 optical wedges was demonstrated.[29] The effect of spectral filtering has been reported for various samples (silicon-rich silicon oxide, Si-implanted silica, and Si/SiO_2 superlattices).[4,28-34] The observed spectra were interpreted using the known dependence of mode localization on the generalized frequency in asymmetrical waveguides[28-30] and within the model of leaky modes.[33-35] This effect was used to study optical properties of silica layers containing Si-nc.[28-30,36]

Laser light can increase temperature in the illuminated volume. Thermal effects were found previously for free-standing films of porous silicon.[37-40] It has been observed that intense laser light increases absorption and leads to an extremely non-linear rise of light emission in the visible and near-IR spectral regions. Koyama and Fauchet explained these effects by laser-induced heat and presented supporting simulations.[39,40] We expanded those studies on free-standing silica films containing Si-nc and found that laser-induced thermal effects are strong for this material as well.[41,42] In addition, continuous-wave laser control of Si-nc stress was achieved in a free-standing Si/SiO_2 superlattice.[43] In the proposed qualitative model, laser radiation melts Si-nc in a silica matrix, and silicon crystallization from the liquid phase produces strong compressive stress. This stress can be tuned in a wide range using laser annealing below the melting point of silicon and then recovered by laser annealing above the melting temperature.

In this chapter, we describe the spectroscopic studies on Si-nc/SiO_2 materials recently performed in the University of Helsinki. Silicon-rich silicon oxide films and Si/SiO_2 superlattices are studied. Probable mechanisms of light emission are briefly discussed. Spectral filtering of light in these waveguiding systems is described. For free-standing films, laser-induced heat leads to structural reorganization of the material and changes its optical and light-emission properties. In particular, laser-controlled stress of Si-nc in silica matrix is evidenced by Raman spectroscopy, which offers an approach to Si-nc optical memory.

2. Experimental Details

Silicon-rich silicon oxide (SiO_x, x<2) layers and Si/SiO_2 superlattices were prepared by molecular beam deposition in the Electron Physics Laboratory (Helsinki University of Technology).[21,23] Si/SiO_2 superlattices are constituted of alternating ultra-thin layers of silicon and SiO_2. Crystalline silicon wafers and silica plates were used as substrates. The samples were typically annealed up to 1200°C for 1 hour in nitrogen atmosphere in a furnace, which promoted the formation of silicon nanocrystals. In some experiments, samples were heated by using rapid thermal annealing technique.[21] In order to prepare free-standing films, a sample was annealed in a furnace and silicon dioxide patterns were made on the silicon substrate backside using photolithography.[41-43] Then the silicon substrate between the patterns was etched with tetramethyl ammonium hydroxide solution producing free-standing Si-nc/SiO_2 film areas with dimensions up to several millimeters.

Two silicon-rich silicon oxide samples were prepared with a gradient of silicon concentration in one direction along the surface.[27,29] The silicon concentration gradient was achieved by tilting the silicon beam with respect to the substrate. In this situation, the distance from the silicon source to the deposition surface changes across the sample, leading to a variation of about ±8% in the silicon concentration. The silicon concentrations were estimated from the deposition conditions yielding x = 1.70 and 1.65 at the sample centers. Some changes of the layer thickness along the samples were also found. In fact, the gradient of the optical thickness of the deposited layer was seen as interference patterns clearly visible by the naked eye. These samples are called Si-nc/SiO_2 optical wedges. The Si concentration gradient was confirmed by X-ray photoelectron spectroscopy, which also excluded any significant incorporation of contaminants during the sample processing.[29]

The PL and Raman spectra were typically recorded with an Ar^+ laser (488 nm, Omnichrome 543-AP), a single-stage spectrometer (Acton SpectraPro 500I), and a CCD camera (Andor InstaSpec IV). The PL light was collected either perpendicularly to the sample surface (transverse geometry) or from the sample edge in a parallel direction to the sample surface (waveguiding geometry).[28] The PL spectra measured in the transverse geometry are smoothed and presented spectrally corrected. The PL spectra measured in the waveguiding geometry are presented without smoothing, and spectral correction is not done in this case. The absorption coefficients at 488 nm were obtained by measuring transmission of the laser beam, taking into account interference of light.[44] The visible transmission spectra were recorded with a fiber-optics spectrometer (SD2000, Ocean Optics) and a broadband light source (DH-2000, Top Sensor Systems).

3. Structural and Optical Properties

3.1. *Raman and photoluminescence spectra*

Raman spectroscopy is a powerful tool to study structural properties of amorphous and crystalline solids, and it has been extensively applied for silicon materials. The as-grown SiO_x films with a substantial excess of silicon have clear amorphous silicon inclusions evidenced by the broad Raman band around 470 cm^{-1} (see Figure 1a).[45] This result shows that silicon excess in the SiO_x material is not homogeneously distributed but it forms amorphous silicon agglomerates, which was noticed long ago.[46] A similar spectrum is obtained for as-grown Si/SiO_2 superlattices, which highlights the analogy of SiO_x films with Si/SiO_2 superlattices where silicon excess concentrates in amorphous layers. For small silicon excess in SiO_x films ($x \sim 1.9$) and very thin silicon layers (1 nm) in Si/SiO_2 superlattices, the characteristic 470 cm^{-1} phonon band is not seen in the Raman spectra.[16,19,23] These results suggest efficient oxidation of the Si

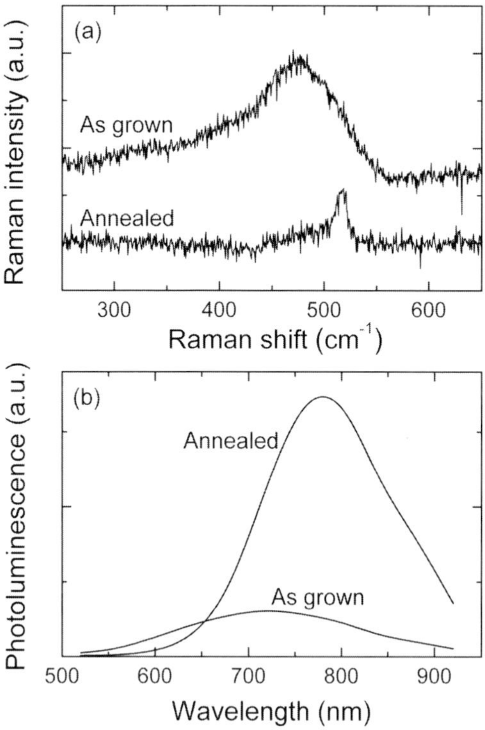

Fig. 1. (a) Raman spectra and (b) PL spectra of a Si/SiO_2 superlattice constituted by 600 repeats of Si (1.5 nm) and SiO_2 (2 nm) layers.[23] Shown are the spectra of as grown and annealed (1150°C) samples.

excess, which is consistent with the optical properties of the same samples.[19] The refractive index and extinction coefficient of ultra-thin Si layers was found to differ substantially from those of bulk silicon. These differences were interpreted in terms of interaction of silicon layers with the oxide surrounding and formation of suboxide interfaces with optical properties different from amorphous Si or SiO_2. In the proposed model, Si and SiO_2 amorphous layers in a superlattice are separated by 0.8-nm-thick SiO interfaces with $n = 2$ and low absorption. If the SiO interfaces are taken into account, the refractive index and extinction coefficient in the middle of ultra-thin silicon layers become quite close to those of amorphous silicon.[19]

For the as-grown amorphous samples, the PL spectrum typically has a maximum at 1.7 - 1.8 eV (see Figure 1b). The PL intensity often decreases under laser irradiation, which supports its interface origin.[19] In accord, the 1.9-eV PL in as-grown SiO_x films was assigned by Kenyon *et al.* to non-bridging oxygen hole centers.[47] Zyubin *et al.* also assigned the red PL in ultra-thin (~1 nm) SiO_x layers to oxygen-related defects.[48] In these models, the PL originates from molecular-like emitters which can be thermodynamically unstable. Light emission of amorphous (as-grown) Si/SiO_2 superlattices was also connected by some authors with the quantum confinement effect, based on the dependence of the PL position on the silicon layer thickness.[49] However, other interpretations of the size effect are possible. For example, the microstructure of the interfaces most probably changes with the layer thickness, and this might change the light-emitting properties. The PL spectra can be also influenced by non-radiative recombination at the amorphous Si/SiO_2 interface.[50]

Thermal annealing above 1000°C leads to crystallization of silicon inclusions and increases the red PL intensity as repeatedly reported for various related materials. The narrowing and up-shift of Raman bands indicate silicon crystallization (see Figure 1).[51] The relatively narrow Raman bands at 515-517 cm^{-1} forming upon annealing correspond to Si-nc with diameters of 3-4 nm.[52,53] No clear evidence of smaller crystallites (≤ 2 nm) is provided by Raman spectroscopy for these samples. It is possible that the oxide surrounding prevents crystallization of amorphous Si inclusions with very small diameters in agreement with the results of Zaharias *et al.*[54] Very small silicon grains are thermodynamically less stable than more extensive structures with hundreds of atoms.[55] Furthermore, the absence of Raman fingerprints of ultra-small silicon grains (< 2 nm) does not prove their absence. It is probable that very small silicon grains embedded in a SiO_2 matrix are invisible in Raman spectra due to their specific structural organization. The concept of a crystal lattice is not straightforward in the case of small silicon grains embedded in silica matrix. The proportion of properly coordinated atoms substantially decreases for smaller grains. The ordering of small silicon crystallites is known to depend on bonding of the surface atoms, and oxidized crystallites show stronger disorder.[56] In all cases, the integrated Raman intensity of Si-nc after thermal annealing is much smaller than

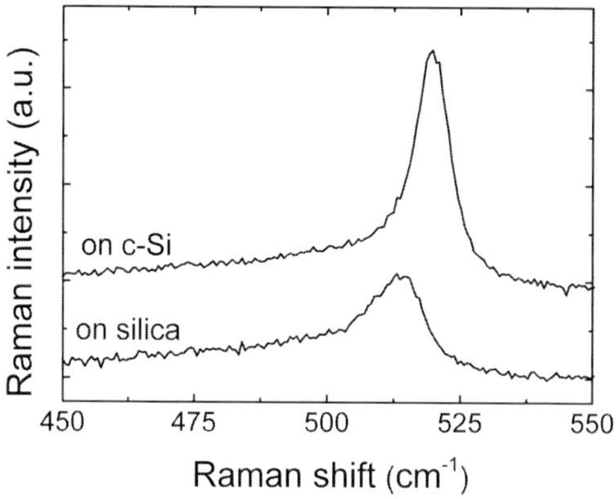

Fig. 2. Effect of substrate material on Raman spectra. The samples are Si/SiO_2 superlattice annealed at 1200°C (rapid thermal annealing) on crystalline-silicon (c-Si) and silica substrates.[21]

the expected value derived from the silicon excess in the material, and this remarkable observation will be addressed later.

Raman spectra of Si-nc/SiO_2 samples were found to depend on substrate material. In Ref. 21, we comparatively studied Si/SiO_2 superlattices and single silicon films grown on crystalline silicon and amorphous silica substrates. After rapid thermal annealing at 1200°C, all samples exhibit some crystallization as indicated by narrowing and up-shift of the Raman bands. As seen in Figure 2, the samples on silicon wafers have up-shifted and narrower Raman bands compared with the samples on silica plates. This difference was connected with annealing-induced stress due to different thermal expansion of the film and substrate. At elevated temperatures, the stresses are relaxed. However, cooling down produces a stress:

$$\sigma = \frac{E}{1-\nu}(\alpha_f - \alpha_s)\Delta T \tag{1}$$

where ΔT is the temperature difference, E is the Young's modulus, ν is the Poisson ratio ($E = 1.9\times10^{11}$ Pa, $\nu = 0.262$ for Si), and α_f and α_s are the thermal expansion coefficients of the film and substrate. For a silicon film on a silica substrate ($\alpha_f = 2.6\times10^{-6}$ and $\alpha_s = 0.4\times10^{-6}$/°C)[57] and $\Delta T =$ 1200°C, this simple model yields a tensile stress of ~700 MPa, which

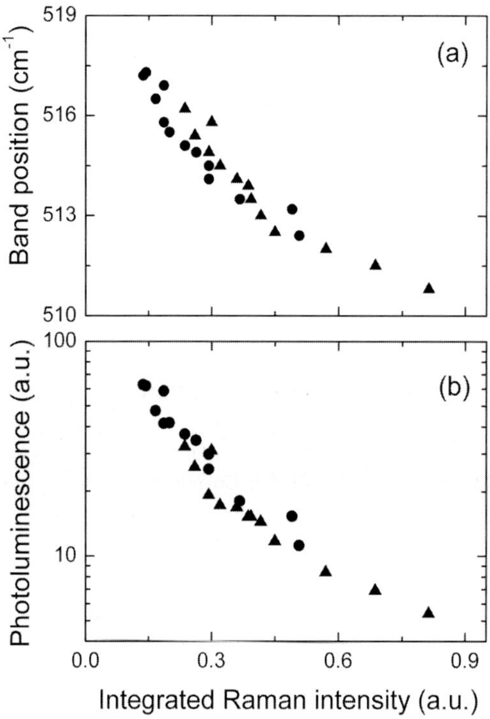

Fig. 3. (a) Correlation between Raman band position and integrated Raman intensity of Si-nc. (b) Correlation between photoluminescence at 800 nm and integrated Raman intensity of Si-nc. The data points correspond to different positions on the Si-nc/SiO_2 optical wedges.[27]

decreases the Raman frequency by ~3 cm^{-1}.[58,59] A silicon film on a silicon substrate is not stressed after annealing. These estimates agree with experiment for both Si/SiO_2 superlattices and single silicon films. It seems that the ultra-thin SiO_2 layers in the superlattices have a minor effect on the formation of stress. The broader crystalline bands for the samples on silica substrates might be connected with a distribution of stresses. It is also possible that a stable silica substrate tears to some extent the contracting silicon layer to smaller grains upon cooling down, which increases the structural disorder.

The superlattices on silicon wafers produce visible PL (~2.1 eV) after RTA at 1200°C.[21] It can be interpreted that RTA forms some specific oxygen-related defects in the lattice. The known simulation for Si=O bonds stabilized at small Si crystallites yielded 2.1 eV for recombination of trapped electrons and holes, and the 2.1-eV PL observed in oxygen-passivated porous silicon was assigned to these double bonds.[17] Similar visible PL was connected to oxygen-related defects in silicon-rich silicon oxide[47] and defective silicon oxide in porous silicon.[60] The *ab initio* simulations have supported the contribution of Si=O bonds to visible light emission.[20] The visible light emission enhanced by RTA was found to bleach under laser irradiation, similarly to the results on as-deposited amorphous Si/SiO_2 superlattices, which shows thermodynamic instability of the formed light-emitting centers. The PL rising upon RTA also depends on substrate material. The 2.1-eV PL is weaker and the 1.5-eV PL is stronger for the samples on silica plates than for the samples on silicon wafers. It follows that tensile stresses found in the samples on silica substrates can destroy unstable centers responsible for the 2.1-eV PL. The 1.5-eV light-emitting centers are obviously more thermodynamically stable.

The 1.5-eV PL intensity of various Si/SiO_2 materials usually increases after furnace annealing at 1000-1200°C. The Si-nc/SiO_2 optical wedges annealed at 1100°C were used to investigate Raman spectra as a function of Si concentration. It was found that the Raman bands systematically appeared at lower energies for higher Si concentrations. In Figure 3a, the Raman band position is plotted versus the integrated Raman intensity that is proportional to the Si concentration. The change of the Raman band position was explained by stresses in the lattice. Stress is known to shift Raman bands, the shift being 1 cm^{-1} per ~250 MPa for crystalline silicon.[58,59] The phonon confinement model and the available experimental data suggest that unstressed 4-nm spherical Si-nc scatter at 517-518 cm^{-1} with a bandwidth of ~12 cm^{-1}.[52,53] Thus, the observed Raman band width of 12 cm^{-1} relates to Si-nc diameters of ~4 nm. This value is a lower limit of the Si-nc size due to possible stress and size distributions. The Raman bands at 517-518 cm^{-1} observed for the lowest Si concentration suggest very small stresses in this case. The decrease of the Raman shift observed for the higher Si content could not be due to size effect and it was explained by tensile stress.[27] The appearance of tensile stresses in annealed Si-nc/SiO_2 materials originates from the larger thermal expansion of silicon compared to silica as discussed above. The formation of the tensile stress upon cooling down seems to evidence bonding between the silicon grains and oxide surrounding.

The PL spectra were also measured at various points on the thermally annealed optical wedges and compared with the Raman spectra.[27] The 1.5-eV PL intensity decreases when the Raman band of Si-nc increases (see Fig. 3b). This observation is remarkable because Si-nc (including the Si/SiO_2 interface) have been most often considered as a source of this emission: it is hence expected that the PL intensity is rather proportional to the Raman intensity. The present data show that this expectation is not necessarily valid. It appears that silicon nanocrystals observed in the Raman spectra are not the direct emitting phase. Alternatively, the model of molecular-like emitting centers localized at the Si/SiO_2 interface can explain light emission from the annealed samples. In this model, the quantum confinement participates in the stage of photon absorption, and the excitation energy is then transferred to a radiative oxygen-related defect. By using the analogy with oxidized porous silicon,[14,17,20] light-emitting centers containing non-bridged oxygen can grow in the network upon annealing. The similarity between oxidized porous silicon and silicon nanocrystals in silica matrix was discussed by Prokes *et al.*[15] In the surface state model, the PL energy is rather independent of the Si-nc size, which agrees with early experiments.[12] We also observed no significant difference in PL peak positions for different Si concentrations.[27] Garoufalis and Zdetsis have shown that the oxygen related states determine the optical gap and make it substantially lower as compared to oxygen-free silicon nanocrystals.[25]

As seen in Figure 3b, the PL intensity shows a clear anti-correlation with the Raman band intensity. The absorption coefficient at 488 nm increases with the concentration of silicon, and it is a nearly linear function of the Raman signal. The absorption coefficient for these samples increases by a factor of 6 as the PL intensity decreased by a factor of 20. It follows that the PL quantum yield is 120 times greater for the lowest than for the highest concentration of silicon. Several reasons for this behavior can be suggested. First, non-radiative processes can become more efficient with increasing Si concentration. Second, the mean size of silicon nanostructures may increase to some extent with the

Si concentration, and this can decrease the PL intensity because lower PL quantum yields are found for larger silicon nanocrystals (> 4 nm).[61] Third, it was also suggested that the emitting center might be located between silicon nanocrystals detected by Raman spectroscopy.[27] In this case, the light-emitting area would become smaller when the silicon concentration increases. The light-emitting centers can be also stabilized at ultra-small silicon inclusions invisible by Raman spectroscopy. Finally, the light-emitting centers can be influenced by annealing-induced stresses in the Si-nc/SiO_2 network. Wolkin *et al.* have suggested that local stresses are capable of changing coordination of interfaces that are thermodynamically less stable compared with the bulk structures due to the lattice mismatch.[17] A network with the Si=O bonds (probable light-emitting centers) is metastable with respect to the Si-O-Si structures.[20] In this situation, the effect of stress is an interesting possibility to consider because the decrease of the 1.5-eV PL intensity obviously correlates with the decrease of Raman shift (see Figure 3). As this spectral change is due to tensile stresses, these data makes probable that the stress influences the light-emitting centers by destroying the emitting centers such as Si=O interface bonds. In accord, stresses introduced during the thinning-oxidation step in SiO_2/Si/SiO_2 structures were found to influence the red PL.[62]

3.2. *Effect of spectral filtering and optical properties*

When the PL spectrum is measured in the waveguiding detection geometry shown in Figure 4, it exhibits a set of narrow and polarized peaks; thus, the as-emitted light is "filtered". This effect of spectral filtering was interpreted in terms of delocalization of waveguide modes near the cutoff frequency[28-30] or by considering leaky (substrate) modes.[33,34] In both interpretations, the observed effect is due to the lower absorption of the substrate material compared to the core layer. It seems that no serious contradiction exists between these two models. In this chapter, we consider the interpretation based on delocalized light.

The mode localization is a function of the generalized frequency parameter written for the asymmetrical waveguide in the form:

$$V = 2\pi(n_1^2 - n_2^2)^{1/2} d / \lambda \tag{2}$$

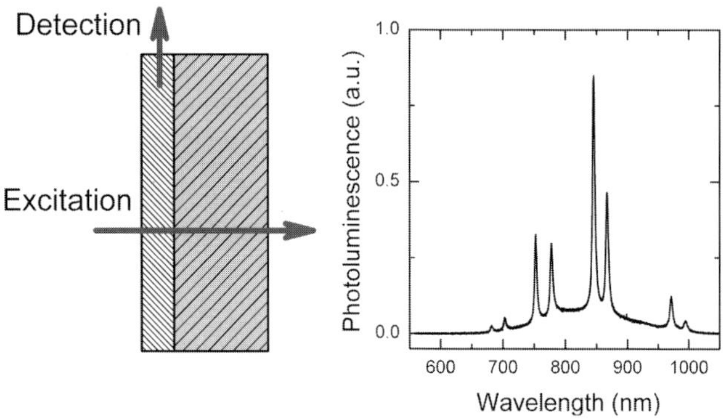

Fig. 4. Waveguiding geometry (on the left) and measured PL spectrum (on the right). The sample is a 4.3-μm thick SiO_x (x = 1.85) layer annealed at 1150°C for 1 h in nitrogen atmosphere.[28] The bands are of the TM (on the left in pairs) and TE polarizations.

where n_1 and n_2 are the refractive indexes of the waveguide layer and the substrate ($n_1 > n_2$), d is the waveguide layer thickness, and λ is the wavelength of light.[63] The guided light becomes delocalized on the cutoff conditions $V = (2m+1)\pi/2$ when it travels mainly outside the waveguide layer. In this case, the waveguiding attenuation is equal to the bulk loss α_2 of the substrate material (or air for free-standing films). Far above the cutoff frequency, the mode is well localized, and the attenuation approaches the bulk loss α_1 of the waveguiding layer. The Si-nc/SiO_2 layer is more absorbing than silica substrate ($\alpha_1 n_1 > \alpha_2 n_2$), hence the more delocalized modes have a lower attenuation. It follows that minimal losses occur for the mode cutoff condition, which corresponds to the observed narrow spectral bands. Thus, the PL spectra measured from the sample edge show the waveguide cutoff frequencies.

The spectral peaks obtained in the waveguiding geometry have TE and TM polarizations. The TE peaks (polarization parallel to the sample surface) usually appear at lower energy (i.e. at longer wavelengths) compared with the TM peaks (polarization perpendicular to the sample surface).[28,30] This situation is consistent with the classical theory of an isotropic asymmetrical waveguide. In the terminology of Kogelnik and Ramaswamy, this order of the polarized peaks is due to a difference of the asymmetry measure that is larger for the TM modes by a factor of $(n_1)^4$.[64]

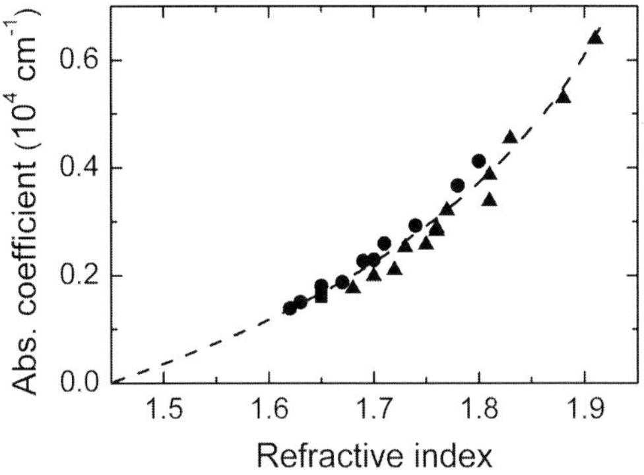

Fig. 5. Correlation between absorption coefficient and refractive index for annealed SiO_x samples with different silicon concentrations. Data shown by circles and triangles are obtained at various positions on two optical wedges.[29] The data point shown by a square is for is a 4.3-μm thick SiO_x (x = 1.85) layer.[28] The line demonstrates the extrapolation to the optical properties of silica.

However, the opposite spectral order between the TE and TM peaks was surprisingly found for some samples: the TE peaks appeared at shorter wavelengths (higher energy) than the TM peaks.[29] The reversed order of the TE and TM peaks was explained by assuming optical birefringence of the material.[36] The larger difference between the cutoff frequencies was proposed for the TE modes than for the TM modes. In this situation, the TE peaks could appear at shorter wavelength (higher energy) than the TM peaks for waveguide mode order $m >> 1$. It follows that the refractive index is smaller for the TE light (ordinary refractive index) than it is for the TM light, i.e. the birefringence is positive. The estimates show that the birefringence increases with Si concentration, and it is ~5% for the highest studied Si concentration (x ~ 1.5 as estimated from deposition condition). The birefringence obtained in these measurements applies to the critical angle in the core layer.

Analysis of the results shows that optical birefringence mainly occurs in samples with high Si concentrations and annealed at relatively low temperatures.[36] It was suggested that the optical birefringence could originate from non-spherical silicon nanocrystals. It is possible that the birefringent nanocrystals are formed upon annealing at relatively low temperatures (1100°C), and annealing at a higher temperature (1200°C) makes them more spherical or/and randomize their orientations. Non-spherical silicon nanocrystals in silica were found by Yurtsever *et al.* using combination of electron topography with plasmon-filtered microscopy.[65] In agreement with our model, their sample had a high Si concentration of 47 at.% and was annealed at 1100°C. Their silicon nanocrystals had a mean volume of 64.5 nm^3, which corresponds to a sphere with a diameter 4.2 nm, which is similar to our samples.

The effect of spectral filtering allows us to estimate the waveguide optical properties. By comparing the transmission and cutoff spectra, we could find the refractive index and layer thickness.[28-30] The obtained layer thickness is used to estimate the absorption coefficient from the transmission measurements. Figure 5 presents the correlation between the absorption coefficient and refractive index for a number of samples. The presented data demonstrate that the optical properties of silicon-rich silicon oxide can be adjusted by changing the silicon concentration. It should be mentioned that Eq. (2) was derived using approximation $(n_1 - n_2) << 1$,[63] which limits its accuracy in practical cases. In principle, the waveguide equations can be treated numerically with high accuracy. As other approximations, the dispersion of the material was neglected in the present estimates and the step-like profile of refractive index was assumed. Despite these obvious limitations, the obtained results are reasonable. Moreover, these results were confirmed by m-line measurements, which yielded very similar optical properties (including optical birefringence) of the same samples.[36]

Absorption coefficients were compared for two free-standing Si/SiO_2 superlattices with different SiO_2 layer thickness (2 and 6 nm) whereas the silicon layer thickness was 2 nm for both samples. The samples were first crystallized by laser irradiation (see later). The absorption coefficient of the sample with 2-nm SiO_2 layer thickness is $(4.6 \pm 0.3)\times10^3$ cm^{-1} at 488 nm, and it is $(0.9 \pm 0.2)\times10^3$ cm^{-1} for the sample with 6-nm SiO_2 layer thickness. These values show no proportionality with respect to the Si excess in the materials. It follows that a large proportion of silicon in the sample with 6-nm SiO_2 layer thickness does not absorb the light; hence it is most probably oxidized. The absorption coefficient of crystalline silicon is 1.7×10^4 cm^{-1} at 488 nm at room temperature,[66] which suggests a substantial oxidation effect already for

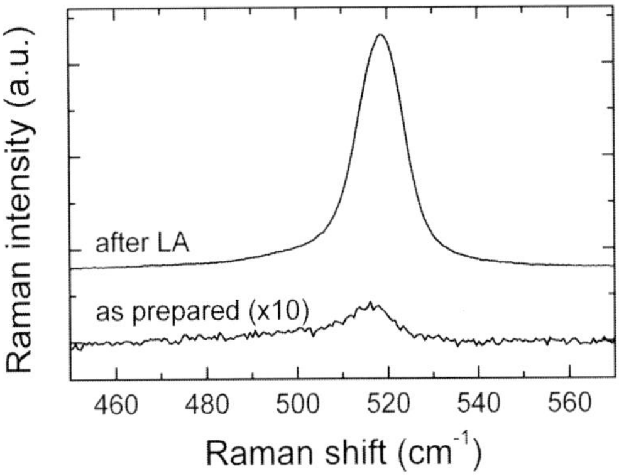

Fig. 6. Raman spectrum of silicon nanocrystals in silica for as prepared SiO_x free-standing film (x = 1.7) and for the same sample after laser annealing (LA).[41] For laser annealing, 488-nm laser light (100 mW) was focused to a 40-μm spot. The Raman measurements were done with a weak probing beam.

the sample with similar amounts of Si and SiO_2 materials. A similar SiO_x interface effect on optical properties was found for the as-grown Si/SiO_2 superlattices as discussed above.[19] The value of 4.6×10^3 cm^{-1} obtained for the laser-annealed sample with 2-nm thick Si and SiO_2 layers is much smaller than the absorption coefficient of 3.3×10^4 cm^{-1} measured for an as-grown Si/SiO_2 superlattice with the same layer thicknesses,[19] which agrees with a much higher absorption coefficient of amorphous silicon compared to that of crystalline silicon.

4. Laser-Induced Thermal Effects

4.1. *Laser annealing*

It is clear that intense laser light can increase temperature in absorbing materials. Pulsed laser heating (annealing) of silicon materials is well known.[67,68] For continuous-wave laser light, the thermal effects are weaker, and the laser-induced heat is very often neglected. However, the laser heat becomes substantially stronger for free-standing films that allow high steady-state temperatures due to the absence of thermal flow to substrate.[40] We have studied thermal effects produced by an argon-ion laser in free-standing Si-nc/SiO_2 films. The free-standing films were

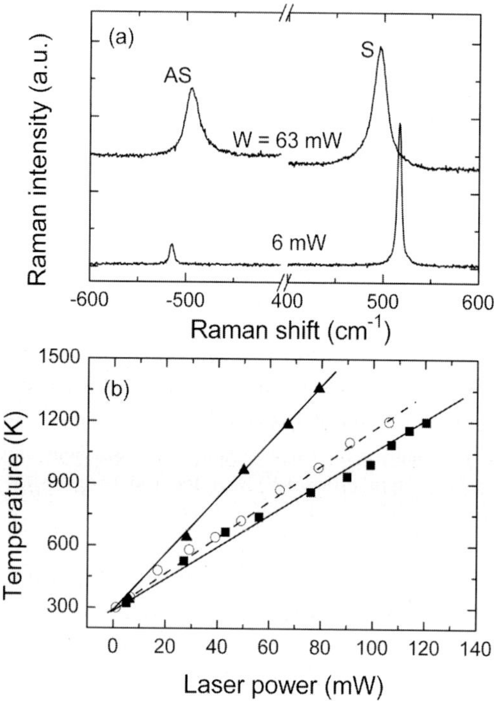

Fig. 7. (a) Stokes (S) and anti-Stokes (AS) Raman bands of a free-standing Si/SiO_2 superlattice measured with two laser powers (6 and 63 mW). (b) Temperature as a function of laser power. Solid symbols show the data for Si/SiO_2 superlattices with the SiO_2 layer thickness of 2 nm (triangles) and 6 nm (squares).[75] The thickness of Si layers is 2 nm for both superlattices. Open circles show the data for a SiO_x (x = 1.7) free-standing film.[42] The 488 nm laser spot diameter is ~40 μm. The temperature was obtained by Raman spectroscopy using Eq. (3). The samples were crystallized by laser light prior to the measurements. The lines guide the eye.

thermally annealed during the fabrication procedure, and the presence of silicon nanocrystals was evidenced by Raman spectroscopy. A short exposure of an as-prepared free-standing film to intense laser light leads to irreversible structural changes. As a result of the laser annealing, the Raman intensity of Si-nc strongly increases (see Figure 6). These irreversible changes produced by laser light were connected with high temperatures achieved in the irradiated volume.[41] Furnace annealing at up to 1200°C does not produce such strong Raman bands of Si-nc.

The laser-induced change of Raman spectra shows that furnace annealing at ~1150°C does not lead to complete Si-SiO_2 phase separation, in agreement with the results by Fazio *et al.*[69] This conclusion has been recently confirmed by Boninelly *et al.* using energy filtered transmission electron microscopy. In a recent work,[70] they have corrected their earlier dependence of the crystalline fraction on the annealing temperature.[71] It should be stressed again that it is not fully clear how to define crystallites in the case of ultra-small sizes, and different characterization methods can lead to different estimates.

The laser-induced change of Raman spectra shows that a large proportion of silicon excess in thermally annealed SiO_x materials is undetectable by Raman spectroscopy.[41] The reduced Raman scattering cross-section from ultra-small silicon grains embedded into oxide host can be caused by a large proportion of oxidized interface and disorder of the core. After laser annealing, the Raman intensity greatly increases (see Figure 6). It can be concluded that laser annealing makes the silicon grains better ordered and more isolated from the silica matrix. The integrated Raman intensity of Si-nc after laser annealing is similar (if not higher) to the value expected from the Si excess. This means that practically all Si excess is visible in Raman spectra after laser annealing. The observed laser-annealing effect makes a case of permanent optical memory readable by means of Raman spectroscopy.

The temperatures in the laser-irradiated volume can be measured using the anti-Stokes and Stokes Raman bands. This analysis can be easily done due to the exceptionally strong Raman bands after laser annealing. The intensity ratio between the anti-Stokes and Stokes Raman bands (I_{AS}/I_S) increases with the laser power (see Figure 7a). The temperature upon laser irradiation can be extracted using the measured I_{AS}/I_S intensity ratio:

$$\frac{I_{AS}}{I_S} = \exp\left(-\frac{E_R}{kT}\right) \tag{3}$$

where E_R is the phonon energy and T is the temperature. The experimental correlation between the I_{AS}/I_S ratio and temperature for crystalline silicon was verified by Balkanski *et al.*[72]

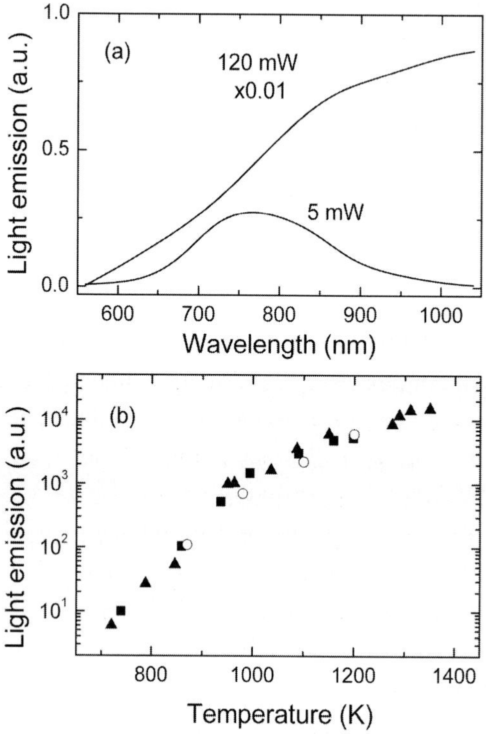

Fig. 8. (a) Light-emission spectra excited at 488 nm. The spectra are presented for a Si/SiO_2 superlattice with Si and SiO_2 layer thickness of 2 and 6 nm, respectively.[75] The spectra are measured with 5 and 120 mW excitation as marked in the plot (notice the multiplication factor). The 488 nm laser spot diameter is ~40 μm. (b) Light emission at 475 nm as a function of temperature. Solid symbols show the data for Si/SiO_2 superlattices with the SiO_2 layer thickness of 2 nm (triangles) and 6 nm (squares). The Si layer thickness is 2 nm for both superlattices. Open circles correspond to a SiO_x ($x = 1.7$) free-standing film.[42] The temperature was measured by Raman spectroscopy. The probed areas were firstly crystallized by laser light.

Figure 7b shows the temperature as a function of laser power obtained for three different samples. The temperature of ~1300 K is the highest temperature which can be practically measured using Raman spectroscopy because of very strong thermal emission background at higher temperatures. These results demonstrate that strong heat can be produced by a continuous-wave laser in free-standing films. The experimental temperature is a nearly linear function of laser power. The different curve slopes can be qualitatively explained by different absorption coefficients and thermal capacities of these materials. The measurements of laser-induced temperatures support our previous discussion on the mechanism of laser crystallization. Indeed, the high temperatures were evidenced by Raman spectroscopy only in the laser-crystallized areas. However, it is reasonable to assume that similar temperatures are obtained in as-prepared (fresh) areas because the material absorption is not changed significantly by laser annealing.

4.2. *Light emission and absorption*

For free-standing Si-nc/SiO_2 films, the PL signal is proportional to laser power at low laser intensities (<<1 kW/cm^2). The PL spectrum has a maximum at ~1.5 eV as described previously (see Figure 8a). At laser intensities of ~1 kW/cm^2, the PL signal saturates and even decreases. At high laser intensities (>>1 kW/cm^2), the light emission increases strongly nonlinearly with laser power.[41,42] A qualitatively similar effect was observed earlier for free-standing films of porous silicon.[37-40]

The non-linear increase of light emission is caused by high laser-induced temperatures, and it essentially originates from thermal radiation. Based on the exponential temperature dependence of blackbody radiation, the experimentally observed increase of the 840-nm emission by a factor of 100 corresponds to a rise of temperature from 1300 K to 2000 K, which is consistent with the temperature estimates from Raman spectra.[41] For high laser intensities (see Figure 8a), the light-emission spectrum has no maximum in the studied spectral region (<1.05 μm). The light-emission spectrum measured at high excitation powers corresponds to a blackbody radiation spectrum. Peculiarities of thermal emission from nc-Si were discussed elsewhere.[73]

The model of thermal radiation from free-standing films is strongly supported by the rise of emission in the anti-Stokes region (λ<488 nm) shown in Figure 8b. The anti-Stokes emission cannot be explained within one-photon PL process. Using the temperatures extracted from the Raman spectra, it is possible to obtain the light emission intensity as a function of sample temperature. In particular, it was found that the temperature dependence measured at 475 nm (anti-Stokes region) was

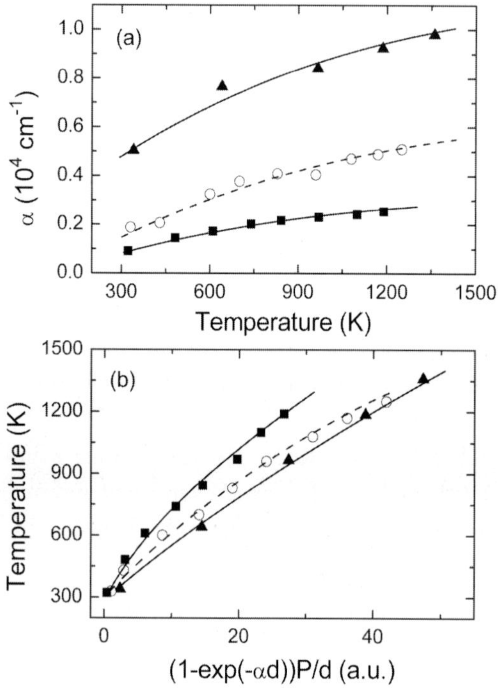

Fig. 9. (a) Absorption coefficient at 488 nm of the free-standing samples as a function of temperature. Solid symbols show the data for Si/SiO_2 superlattices with the SiO_2 layer thickness of 2 nm (triangles) and 6 nm (squares). The Si layer thickness is 2 nm for both superlattices. Open circles correspond to a SiO_x ($x = 1.7$) free-standing film. The 488 nm laser spot diameter is ~40 μm. (b) Laser-induced temperature as a function of the laser power absorbed by a unite volume. The symbols are the same as in plot (a). The temperature was measured by Raman spectroscopy. The probed areas were firstly crystallized by laser light. The lines guide the eye. The data are from Refs. 42 and 75.

weaker than the corresponding dependence for blackbody radiation, which indicates the deviation of the present material from blackbody.[42]

Laser-induced absorption is another thermal effect on free-standing Si-nc/SiO_2 films (see Figure 9a).[41] A similar enhancement of absorption was observed for free-standing films of porous silicon, and the suggested explanation was the thermal shrinking of the band gap.[39,40] A strong argument for the heat-based model is the absence of this effect for films on substrates because the substrate considerably decreases the laser-induced temperature. The obtained temperature dependences of absorption coefficients vary for different samples, and they substantially differ from the data for crystalline silicon.[66] The absorption coefficient of crystalline silicon changes by a factor of 5 from room temperature to 1000 K whereas, in our case, the increase is much smaller. This fact shows that silicon nanocrystals have different thermal optical properties compared to crystalline silicon, which is reasonable taking into account the different band gaps.

The temperature increase is caused by absorption of laser light. The laser power density absorbed in the material can be found as

$$P_{abs} = P_0\left[1-\exp(-\alpha d)\right]/d \qquad (4)$$

where P_0 is the initial laser intensity, α is the absorption coefficient, and d is the sample thickness. Figure 9b presents the temperatures produced by various absorbed laser power densities in three samples, and straightforward correlations take place. Some difference found between these samples is much smaller than the difference of their absorption coefficients. However, the same absorbed power density leads to a somewhat higher temperature for the superlattice with 6-nm oxide layers than for the superlattice with 2-nm oxide layers. This difference can be qualitatively connected with the much lower thermal conductivity of silica compared to crystalline silicon. On the other hand, the situation is not so simple because the thermal conductivity of a superlattice was found to be lower than that of silicon dioxide,[74] and the specific heat capacity of the material is not accurately known either.

4.3. *Laser-induced compressive stress*

Laser-induced compressive stress is a thermal effect which deserves special discussion. These experiments were done on two free-standing Si/SiO_2 superlattices, which have the same (2 nm) silicon layers and different (2 and 6 nm) silicon dioxide layers.[75] The as-prepared samples contain some amount of Si-nc evidenced by relatively narrow Raman bands. The Raman bands of Si-nc are very weak compared to the signal from a silicon wafer in the same experimental arrangement. However, the Raman intensity of Si-nc strongly increases by exposure to intense laser light (high temperature annealing, HTA) indicating Si-SiO_2 phase

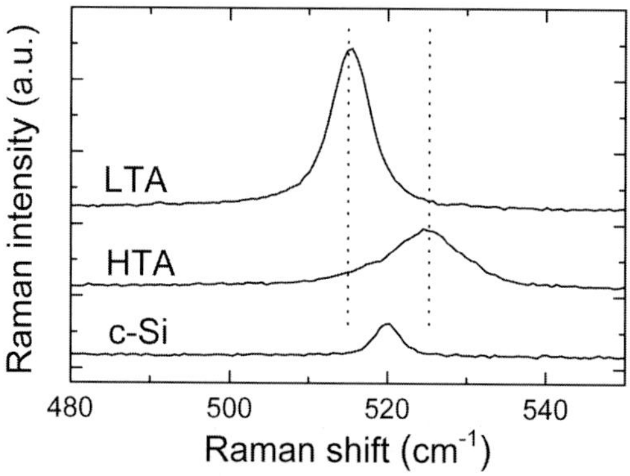

Fig. 10. Laser-induced stress of silicon nanocrystals. Shown are Raman spectra of crystalline silicon (c-Si) and spectra of Si/SiO_2 superlattice after laser annealing above silicon melting temperature (HTA) and below this temperature (LTA). The Si and SiO_2 layers of the superlattice are 2-nm thick.[43]

separation. The Raman band observed after such intense laser annealing is at ~525 cm^{-1}, i.e. above the Raman frequency of crystalline silicon (see Figure 10). Upon subsequent exposure to lower laser power (low-temperature annealing, LTA) the Raman band position gradually shifts to a lower energy, and the smallest Raman frequency achieved in these experiments is ~515 cm^{-1}. The next HTA shifts the Raman peak back to ~525 cm^{-1}. The preparation of the Raman band at 525 cm^{-1} and its shift down to 515 cm^{-1} can be repeated many times, and the Raman intensities show practically no degradation. The structural effect of laser annealing is localized within the laser beam (diameter ~40 μm). The Raman peak positions are very similar for the two samples studied here. The light-induced changes of band positions were explained by Si-nc stress.[43,75] In the case of crystalline silicon, the shift of 10 cm^{-1} is produced by a stress of ~3 GPa.[58,59]

Raman spectroscopy was used to estimate laser-induced temperatures in these experiments. Figure 7b shows the temperature as a function of laser power. The threshold power to produce the Raman shift of 525 cm^{-1} can be measured experimentally. By extrapolating the laser-power dependences of temperature, the corresponding threshold temperatures can be found and they appear to be close to the melting temperature of crystalline silicon (1685 K). Based on this result, we can discuss the mechanism of stress formation. In our qualitative model, laser annealing above the silicon melting temperature (HTA) melts silicon inclusions and relaxes stresses in the vicinity. Crystallization of a melt silicon particle from the liquid phase increases its volume by ~10%,[76] which leads to compressive stress. The induced stress can be estimated in the approximation of a spherical silicon particle with volume V_S inserted into a sphere with volume V_M in silica matrix:[77]

$$P = \frac{V_S - V_M}{V_S / K + 3V_M / 4G} \tag{5}$$

where K is the silicon modulus of compression (~100 GPa), and G is the shear modulus of silica (28 GPa). This simple estimate yields a stress difference of 2.8 GPa between HTA (compressive stress) and LTA (tensile stress), which is close to the experimentally obtained value of 3.0 GPa.[43,75] The present consideration is a simplification due to a complicated morphology of Si-nc embedded in silica matrix and the use of crystalline-silicon parameters for nanocrystals. However, these simple estimates agree well with experiment.

The relaxation of Si-nc stress strongly depends on temperature. No change of Raman spectra in day scale was noticed at room temperature, and the process is activated at elevated temperatures induced by laser light. The stress decay rate was numerically described by using the time required for relaxation of 1 GPa, which approximately corresponds to a shift of the Raman band from 525 to 522 cm^{-1}. Figure 11 presents this retention time as a function of 1/T for the two studied samples. These results suggest a practically infinite lifetime of stress at temperatures below 500 K. The slope of the Arrhenius plot yields an activation energy of (1.6 ± 0.2) eV for both samples. On the other hand, the relaxation of stress is an order of magnitude faster for the sample with thicker SiO_2 layers.

The different stress retention time for the two samples is an interesting observation, which is not easy to interpret. The mechanism of compressive stress relaxation of Ge nanocrystals in silica matrix was studied by Sharp *et al.*,[78] and the diffusion of matrix atoms away from the stressed particle was suggested as a probable mechanism. Our data

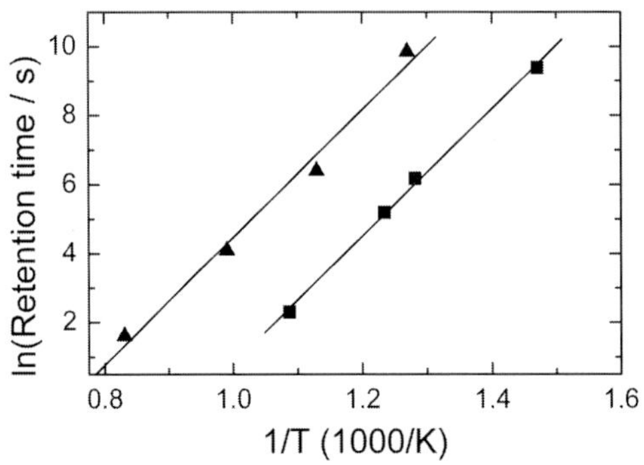

Fig. 11. Retention time of Si-nc compressive stress as a function of 1/T. Shown are the data for Si/SiO_2 superlattices with the SiO_2 layer thickness of 2 nm (triangles) and 6 nm (squares). The Si layer thickness is 2 nm for both superlattices.[75]

are consistent with the diffusion model. The close activation energies obtained for the two superlattices in Ref. 75 suggest the same relaxation mechanism in the two samples. The difference in the relaxation rates can be at least partially explained by different Si-nc sizes in these two samples, which can be caused by efficient merging of silicon layers in the case of thinner SiO_2 layers. We can speculate that the stress-induced diffusion rate is proportional to the Si-nc surface area whereas the stress relaxation needs a volume change. In addition, the structure of the interface region can contribute to the relaxation process because atomic diffusion in the SiO_x ($x < 2$) material presumably depends on Si concentration.

The laser-controlled Si-nc stress makes a case of optical memory. The demonstrated effect allows writing and erasing information and its reading by optical means, which is practically provided by means of micro-Raman mapping. For instance, the data reading can be realized by mapping the Raman intensity at 525 cm^{-1} from stressed Si-nc. The potential data density can be very high because silicon melting with submicron dimensions was demonstrated (see Ref. 79) and Raman mapping has submicron resolution as well. A very long retention time for the Si-nc stress at high elevated temperatures is practically important. All described manipulations with Si-nc stress were done in air meaning that no significant oxidation of Si-nc in silica matrix from the atmosphere occurs, even at high laser-induced temperatures.

5. Concluding Remarks

Raman and photoluminescence studies of silicon rich silicon oxide have been described. After thermal annealing of as-grown materials, silicon nanocrystals are formed in silica matrix and the red photoluminescence is increased. The material structure and photoluminescence are influenced by annealing-induced stress formed upon furnace annealing.[21,27] An attractive model of the light emission is based on the Si=O bonds stabilized on Si-nc/SiO_2 interface or even in disordered SiO_x areas.

Wavelength-selective guiding of light was found for Si-nc/SiO_2 layers on silica substrates and as free-standing films and explained by delocalization of guided light near the cutoff frequencies.[28-30] This effect of spectral filtering allows narrowband and polarized tunable light to be obtained and optical properties of the Si-nc/SiO_2 material to be studied. Birefringence of Si-nc/SiO_2 materials was found for samples with relatively high Si concentration, which was tentatively connected with a non-spherical shape of the silicon nanocrystals.[36]

In free-standing Si-nc/SiO_2 films, continuous-wave laser light is capable of producing high temperatures.[41] The intense laser irradiation leads to efficient Si-SiO_2 phase separation, and the laser-induced structural effect presumably originates from melting of silicon inclusions in silica matrix. It appears that a large part of silicon excess is invisible by Raman spectroscopy after furnace annealing and it becomes detectable after laser annealing. These results show that the furnace annealing at up to 1200°C does not complete structural reorganization of silicon nanostructures and a large part of silicon excess is oxidized and disordered. The laser induced heat increases absorption of Si-nc in silica matrix and leads to a very strong rise of thermal emission, which was earlier observed for porous silicon.

One thermal effect caused by laser light is of special interest. It was found that melting of Si-nc in silica matrix leads to strong compressive stress evidenced by the Raman spectra.[43,75] The compressive stress can be relaxed using laser annealing at temperatures below the melting temperature of silicon. This effect provides a potential approach to Si-nc memory with very long retention time and high data density. The data can be written, erased, and read by optical means, which is realized within conventional silicon technology.

Acknowledgments

This work was supported by the Academy of Finland partially through the Finnish Center of Excellence "Computational Molecular Science" and the FinnNano Program "Optical and Surface Properties of Nanoparticles". The author thanks all colleagues who have contributed to the results described in this chapter, especially, Sergei Novikov, Olli Kilpelä, Jouko Lahtinen, Markku Räsänen, Juha Sinkkonen, Lorenzo Pavesi, Daniel Navarro-Urrios, and Claudio Oton.

References

1. *Silicon photonics*, L. Pavesi and D. J. Lockwood (Eds.), Springer, 2004.
2. L. Pavesi, *J. Phys.-Condens. Matter.* **15**, R1169 (2003).
3. L. Pavesi, L. Dal Negro, C. Mazzoleni, G. Franzo and F. Priolo, *Nature (London)* **408**, 440 (2000).
4. L. Khriachtchev, M. Räsänen, S. Novikov and J. Sinkkonen, *Appl. Phys. Lett.* **79**, 1249 (2001).
5. K. Luterova, I. Pelant, I. Mikulskas, R. Tomasiunas, D. Muller, J.-J. Grob, J.-L. Rehspringer and B. Hönerlage, *J. Appl. Phys.* **91**, 2896 (2002).
6. L. Dal Negro, M. Cazzanelli, N. Daldosso, Z. Gaburro, L. Pavesi, F. Priolo, D. Pacifici, G. Franzo and F. Iacona, *Physica* **E16**, 297 (2003).
7. L. Dal Negro, M, L. Pavesi, S. Ossicini, D. Pacifici, G. Franzo, F. Priolo and F. Iacona, *Appl. Phys. Lett.* **82**, 4636 (2003).
8. J. Ruan, P. M. Fauchet, L. Dal Negro, M. Cazzanelli and L. Pavesi, *Appl. Phys. Lett.* **83**, 5479 (2003).
9. M. Cazzanelli, D. Kovalev, L. Dal Negro, Z. Gaburro and L. Pavesi, *Phys. Rev. Lett.* **93**, 207402 (2004)
10. S. L. Jaiswal, J. T. Simpson, S. P. Withrow, C. W. White and P. M. Norris, *Appl. Phys.* **A77**, 57 (2003).
11. Y. Kanemitsu, H. Uto, Y. Masumoto, T. Masumoto, T. Futagi and H. Mimura, *Phys. Rev.* **B48**, 002827 (1993)
12. M. Rückschloss, B. Landkammer and S. Veprek, *Appl. Phys. Lett.* **63**, 1474 (1993).
13. B. T. Sullivan, D. J. Lockwood, H. J. Labbe and Z.-H. Lu, *Appl. Phys. Lett.* **69**, 3149 (1996).
14. J. L. Gole, F. P. Dudel, D. Grantier and D. A. Dixon, *Phys. Rev.* **B56**, 2137 (1997).

15. S. M. Prokes, W. E. Carlos, S. Veprek and Ch. Ossadnik, *Phys. Rev.* **B58**, 15632 (1998).
16. L. Khriachtchev, M. Räsänen, S. Novikov, O. Kilpelä and J. Sinkkonen, *J. Appl. Phys.* **86**, 5601 (1999).
17. M. V. Wolkin, J. Jorne, P. M. Fauchet, G. Allan and C. Delerue, *Phys. Rev. Lett.* **82**, 197 (1999).
18. F. Iacona, G. Franzo and C. Spinella, *J. Appl. Phys.* **87**, 1295 (2000).
19. L. Khriachtchev, S. Novikov and O. Kilpelä, *J. Appl. Phys.* **87**, 7805 (2000).
20. F. Zhou and J. D. Head, *J. Phys. Chem.* **B104**, 9981 (2000).
21. L. Khriachtchev, O. Kilpelä, S. Karirinne, J. Keränen and T. Lepistö. *Appl. Phys. Lett.* **78**, 323 (2001).
22. P. Carrier, L. J. Lewis and M. W. C. Dharma-Wardana, *Phys. Rev.* **B65**, 165339 (2002).
23. L. Khriachtchev, S. Novikov and J. Lahtinen, *J. Appl. Phys.* **92**, 5856 (2002)
24. P. Photopoulos and A. G. Nassiopoulou, *J. Phys.-Condens. Matter* **15**, 3641 (2003).
25. C. S. Garoufalis, and A. D. Zdetsis, *Phys. Chem. Chem. Phys.* **8**, 808 (2006).
26. N. Daldosso, M. Luppi, S. Ossicini, E. Degoli, R. Magri, G. Dalba, P. Fornasini, R. Grisenti, F. Rocca, L. Pavesi, S. Boninelli, F. Priolo, C. Spinella and F. Iacona, *Phys. Rev.* **B68**, 085327 (2003).
27. L. Khriachtchev, M. Räsänen, S. Novikov and L. Pavesi, *Appl. Phys. Lett.* **85**, 1511 (2004).
28. L. Khriachtchev, M. Räsänen and S. Novikov, *Appl. Phys. Lett.* **83**, 3018 (2003).
29. L. Khriachtchev, M. Räsänen, S. Novikov and J. Lahtinen, *J. Appl. Phys.* **95**, 7592 (2004).
30. L. Khriachtchev, S. Novikov, J. Lahtinen and M. Räsänen, *J. Phys.-Condens. Matter.* **16**, 3219 (2004).
31. J. Valenta, I. Pelant and J. Linnros, *Appl. Phys. Lett.* **81**, 1396 (2002).
32. J. Valenta, I. Pelant, K. Luterova, R. Tomasiunas, S. Cheylan, R. G. Elliman, J. Linnros and B. Hönerlage, *Appl. Phys. Lett.* **82**, 955 (2003).
33. J. Valenta, T. Ostatnický, I. Pelant, R. G. Elliman, J. Linnros and B. Hönerlage, *J. Appl. Phys.* **96**, 5222 (2004).
34. I. Pelant, T. Ostatnicky, J. Valenta, K. Luterova, E. Skopalova, T. Mates and R. G. Elliman, *Appl. Phys.* **B83,** 87 (2006).
35. J. L. Shen, J. Y. Chang, H. C. Liu, W. C. Chou, Y. F. Chen, T. Jung and M. C. Wu, *Solid State Commun.* **116**, 431 (2000).
36. L. Khriachtchev, D. Navarro-Urrios, L. Pavesi, C. J. Oton, N. E. Capuj and S. Novikov, *J. Appl. Phys.* **101**, 044310 (2007).
37. T. Matsumoto, M. Daimon, H. Mimura, Y. Kanemitsu and N. Koshida, *J. Electrochem. Soc.* **142**, 3528 (1995).
38. H. Koyama, L. Tsybeskov and P. M. Fauchet, *J. Luminescence* **80**, 99 (1998).
39. H. Koyama and P. M. Fauchet, *Appl. Phys. Lett.* **73**, 3259 (1998).
40. H. Koyama and P. M. Fauchet, *J. Appl. Phys.* **87**, 1788 (2000).

41. L. Khriachtchev, M. Räsänen and S. Novikov, *Appl. Phys. Lett.* **86**, 141911 (2005).
42. L. Khriachtchev, M. Räsänen and S. Novikov, *J. Appl. Phys.* **100**, 053502 (2006).
43. L. Khriachtchev, M. Räsänen and S. Novikov, *Appl. Phys. Lett.* **88**, 013102 (2006).
44. L. Khriachtchev, *Appl. Phys, Lett.* **81**, 1357 (2002).
45. N. Maley, D. Beeman and J. S. Lannin, Phys. Rev. **B38**, 10611 (1988).
46. A. Hartstein, J. C. Tsang, D. J. DiMaria and D. W. Dong, *Appl. Phys. Lett.* **36**, 836 (1980).
47. A. J. Kenyon, P. F. Trwoga, C. W. Pitt and G. Rehm, *J. Appl. Phys.* **79**, 9291 (1996).
48. A. S. Zyubin, Yu. D. Glinka, A. M. Mebel, S. H. Lin, L. P. Hwang and Y. T. Chen, *J. Chem. Phys.* **116**, 281 (2002).
49. Z. H. Lu, D. J. Lockwood and J. M. Baribeau, *Nature* **378**, 258 (1995).
50. Y. Kanemitsu, M. Iiboshi and T. Kushida, *Appl. Phys. Lett.* **76**, 2200 (2000).
51. H. Richter, Z. P. Wang and L. Ley, *Solid State Commun.* **39,** 625 (1981).
52. I. H. Campbell and P. M. Fauchet, *Solid State Commun.* **58**, 739 (1986).
53. Ch. Ossadnik, S. Veprek and I. Gregora, *Thin Solid Films* **337**, 148 (1999).
54. M. Zaharias, J. Bläsing, P. Veit, L. Tsybeskov, K. Hirschman and P. M. Fauchet, *Appl. Phys. Lett.* **74**, 2614 (1999).
55. T. Bachels and R. Schäfer, *Chem. Phys. Lett.* **324**, 365 (2000).
56. J. C. Tsang, M. A. Tischler and R. T. Collins, *Appl. Phys. Lett.* **60**, 2279 (1992).
57. M. Madou, *Fundamentals of Microfabrication,* CRC Press, New York, 1997.
58. J. Macia, E. Martin, A. Perez-Rodriguez, J. Jimenez, J. R. Morante, B. Aspar and J. Margail, *J. Appl. Phys.* **82**, 3730 (1997).
59. V. Domnich, and Y. Gogotsi, *Rev. Adv. Mater. Sci.* **3**, 1 (2002).
60. I. Coulthard, W. J. Antel, Jr., J. W. Freeland, T. K. Sham, S. J. Naftel and P. Zhang, *Appl. Phys. Lett.* **77**, 498 (2000).
61. G. Ledoux, J. Gong, F. Huisken, O. Guillois and C. Reynaud, *Appl. Phys. Lett.* **80**, 4834 (2002).
62. P. Photopoulos and A. G. Nassiopulou, *J. Phys.- Condens. Matter.* **15**, 3641 (2003).
63. H. G. Unger, *Planar Optical Waveguides and Fibers,* Clarendon Press, Oxford, 1977.
64. H. Kogelnik and V. Ramaswamy, *Appl. Opt.* **13**, 1857 (1974).
65. A. Yurtsever, M. Weyland and D. A. Muller, *Appl. Phys. Lett.* **89**, 151920 (2006).
66. G. E. Jellison, Jr. and F. A. Modine, *Appl. Phys. Lett.* **41**, 180 (1982).
67. P. Baeri, S. U. Campisano, G. Foti and E. Rimini, *J. Appl. Phys.* **5**0, 788 (1979)
68. M. O. Thompson, G. J. Galvin, J. W. Mayer, P. S. Peercy, J. M. Poate, D. C. Jacobson, A. G. Cullis,and N. G. Chew, *Phys. Rev. Lett.* **52**, 2360 (1984).
69. A. Fazio, M. Vulpio, C. Gerardi, Y. Liao, I. Crupi, S. Lombardo, S. Trusso and F. Neri, *J. Electrochem. Soc.* **149**, G376 (2002).
70. S. Boninelli, F. Iacona, G. Franzo, C. Bongiorno, C. Spinella and F. Priolo, *J. Phys.: Condens. Matter* **19**, 225003 (2007).
71. F. Iacona, C. Bongiorno, C. Spinella, S. Boninelli and F. Priolo, *J. Appl. Phys.* **95**, 3723 (2004).

72. M. Balkanski, R. F. Wallis and E. Haro, *Phys. Rev.* **B28**, 1928 (1983).
73. P. Roura and J. Costa, *Eur. J. Phys.* **23**, 191 (2002).
74. A. V. Kamenev, H. Grebel, L. Tsybeskov, *Appl. Phys. Lett.* **88**, 143117 (2006).
75. L. Khriachtchev and S. Novikov, *Appl. Phys.* **A87**, 761 (2007).
76. T. Ujihara, G. Sazaki, K. Fujiwara, N. Usami and K. Nakajima, *J. Appl. Phys.* **90**, 750 (2001).
77. V. V. Voronkov and R. Falster, *J. Appl. Phys.* **89**, 5965 (2001).
78. I. D. Sharp, D. O. Yi, Q. Xu, C. Y. Liao, J. W. Beeman, Z. Liliental-Weber, K. M. Yu, D. N. Zakharov, J. W. Ager, D. C. Chrzan and E. E. Haller, *Appl. Phys. Lett.* **86**, 063107 (2005).
79. S. Kouteva-Arguirova, T. Arguirov, D. Wolfframm and J. Reif, *J. Appl. Phys.* **94**, 4946 (2003).

CHAPTER 12

LIGHT EMISSION FROM SILICON-RICH NITRIDE NANOSTRUCTURES

Luca Dal Negro, Rui Li, Joseph Warga and Selcuk Yerci

Department of Electrical and Computer Engineering, Boston University
8 Saint Mary's street, Boston, MA, 02215-2421, USA
E-mail: dalnegro@bu.edu

Soumendra N. Basu

Department of Manufacturing Engineering, Boston University
15 Saint Mary's Street, Brookline, MA 02446, USA

Sebastien Hamel

Physical Sciences, Lawrence Livermore National Laboratory
7000 East Ave. L-415, Livermore, CA ,USA

Giulia Galli

Department of Chemistry, University of California at Davis,
Davis, CA 95616, USA

In this chapter, we discuss our main results on the fabrication, optical characterization and theoretical modeling of silicon nanocrystals (Si-nc) embedded in silicon-rich nitride (SRN) matrices by direct magnetron co-sputtering. In particular, we study the structural and optical emission properties of Si-nc dispersed in matrices of varying stoichiometry and annealed at different temperatures, and we discuss their implications with respect to the nature of light emission in co-sputtered SRN films. In addition, we discuss Erbium emission and energy sensitization in Si-nc films embedded in SRN matrices. Combining high-resolution transmission electron microscopy with absorption and emission spectroscopy we demonstrate efficient room-temperature light emission from dense arrays of Si-nc with 2nm average diameter. These Si-nc systems show weakly tunable, near-infrared emission spectra with large absorption/emission Stokes shifts and multi-component, nanosecond recombination dynamics. Our results are interpreted using first-principles simulations and indicate that nitrogen atoms bonded to the surface of nanometer-size silicon clusters play a crucial role in the emission mechanism of small Si-nc in SRN systems. These findings suggest that Si-nc embedded in SRN matrices have a large potential for the fabrication of optically active photonic devices based on Si technology.

1. Introduction

Silicon-based nanophotonics has the potential to become a disruptive technology by enabling full-scale on-chip integration of optical functions within mainstream microelectronics.[1,2] However, progress towards this goal is curtailed by the lack of efficient light emission from silicon. Photon emission from silicon originates from low-probability phonon-mediated transitions that compete unfavourably with fast non-radiative de-excitation paths, such as Auger and free carrier recombinations.

In view of these limitations, several strategies have been recently developed to engineer Si into a more efficient light-emitting material.[2-7] In particular, quantum confinement has led to a dramatic improvement of light emission efficiency in Si nanostructures.[2-5,8,9] Emission efficiencies as high as ~60% in the 700-800 nm range have been reported for Si-nc embedded in Si-rich silicon oxide films (SRO) and sizable optical gain under optical excitation has been demonstrated by several groups.[10–18] Alternatively, Erbium (Er) doping of Si-nc systems provides a very valuable approach for 1.54 μm light-emitting devices, which could be easily integrated within a Si platform. In particular, it has been discovered that phonon-mediated non-radiative energy transfer from Si-nc to Er ions can significantly sensitize Er emission,[19-22] leading to broad-band Er excitation with up to three orders of magnitude increased excitation cross section.

However, in the last few years, it has become progressively clearer that the optical emission and energy transfer in Si nanostructures can be strongly affected by the complex interplay of two main contributions: the system's dimensionality (quantum confinement) and the structural/chemical properties of the Si-nc interfaces.[23-25] In particular, it has been shown that the presence of Si=O double bonds at the surface of small Si crystallites embedded in SiO_2 plays a crucial role in determining the mechanism of light emission and optical gain.[16,23-25]

The demonstration of efficient light emission and optical gain from surface-localized exciton transitions highlight the importance to achieve a more robust understanding of the Si-nc interface structure and composition, which can be strongly affected by the nature of the embedding matrix. In addition, the ever-expanding need to develop stable and efficient electroluminescent devices based on Si-nc systems has led to the investigation of Si-nc as an alternative to CMOS dielectrics with smaller bandgaps than SiO_2, which would facilitate electrical pumping. Recently, visible and near-infrared light-emission with nanosecond–fast relaxation dynamics have been demonstrated[26-28] for Si-nc embedded in silicon nitride matrices fabricated by different techniques. In addition, it has been shown[29,30] that nitrogen-related surface groups can introduce strongly-localized gap states which play a crucial role, akin to Si=O double bond, in the optical emission properties of Si-nc smaller than approximately 3 nm. However, a general picture for the detailed understanding of light emission in nitride-based Si nanostructures is significantly complicated by the enormous structural complexity of Si-rich nitride matrices, which sustain a large number of structural configurations and defect centers. Therefore, more investigations are needed to clarify the exact nature of light emission in nitride-based Si systems. In this chapter we will discuss our main experimental and theoretical results on the optical emission and sensitization properties of Si-nc fabricated by RF magnetron co-sputtering in silicon nitride matrices.

The rest of the chapter is organized as follows: in section 2 we describe our approach for the fabrication of Si-nc in Silicon-Rich Nitride (SRN) and Er-doped structures. In section 3 we discuss Transmission Electron Microscopy (TEM) and infrared absorption characterization of

Si-nc. In section 4, we discuss the optical emission properties of Si-nc structures via steady-state and time-resolved photoluminescence spectroscopy. In section 5, we discuss Er sensitization via Si-nc in silicon nitride and in section 6 we discuss computational models for the structural and emission properties of Si-nc embedded in silicon nitride.

2. Fabrication of Silicon Nanostructures via Magnetron Co-Sputtering

In order to fabricate Si-nc and investigate their optical properties, we have deposited thin (300nm) SRN films followed by thermal annealing. The annealing process induces a precipitation transformation where the meta-stable Si-rich dielectric film decomposes into two stable phases: Si clusters and a matrix which, depending on Si supersaturation and annealing temperature/time, can be closer in composition to the equilibrium (stoichiometric) composition.

We fabricated Si-nc and Er-doped Si-nc embedded into amorphous Si_3N_4 through Radio Frequency (RF) magnetron co-sputtering from Si, Si_3N_4 and metallic Er targets. In order to induce the precipitation of small crystalline Si clusters, all the SRN films were annealed using a standard rapid Thermal Annealing (RTA) furnace in N_2/H_2 forming gas (5% hydrogen) for 10 minutes at temperatures ranging between room temperature (as-grown) and 900°C. The sputtering was performed, without substrate heating, in a Denton Discovery 18 confocal-targets sputtering system. The atomic concentrations of Si, N an Er in the deposited films are measured, within 0.5% accuracy, with Energy Dispersive X-ray analysis (EDX, Oxford ISIS), while the refractive indices of the SRN films are measured by ellipsometry (Rudolph, AutoEL-II) using a wavelength of 830 nm.

Figure 1 shows the effective refractive indices of SRN films deposited by magnetron sputtering. The effective index is mainly determined by the Si atomic fraction, which can be conveniently controlled by varying the plasma power applied to the Si target. It is important to notice that the refractive indices plotted in Figure 1 have been measured after annealing (700°C for 10 minutes). In fact, we have found that the refractive index changes upon the annealing process as

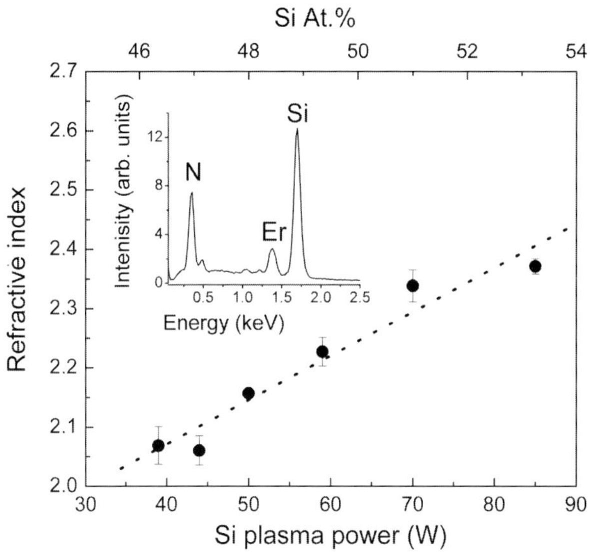

Fig. 1. Refractive indices of SRN films deposited with different plasma powers on the Si target. The silicon nitride cathode power was kept constant at 350 W. The deposition pressure was kept constant at 0.8 mTorr. The Si atomic percent is measured by EDX analysis with 4 keV beam energy. Inset: Typical EDX spectrum of an Er:SRN sample.

result of compositional changes and phase separation in the SRN matrix, as we will discuss in section 3.

The inset in Figure 1 shows a typical EDX spectrum for Er-doped SRN films (Er:SRN) obtained by direct co-sputtering. The peaks corresponding to Si, Er and N elements are clearly visible and are labeled in the picture. The atomic fraction of Er ions in the Er:SRN films can be linearly controlled by the plasma power applied to the metal Er target, while keeping the power on the other two targets (Si and Si_3N_4) constant.

3. Structural Characterization of Si-nc Films

As mentioned in the previous section, magnetron co-sputtering of Si-rich dielectrics is a particularly convenient approach for the fabrication of Si nanostructures. In the case of SRN, we have found that, depending on the Si concentration, a relatively short (10 minutes) thermal annealing step

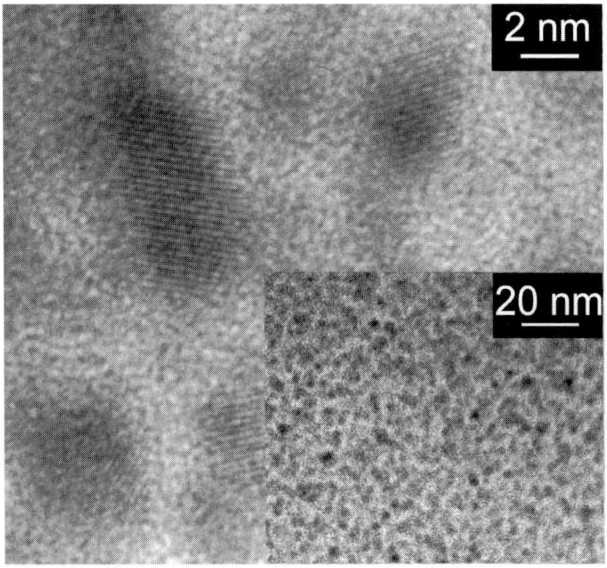

Fig. 2. High resolution TEM cross section of a typical Si-nc sample with 48% Si. The sample has been annealed at 700°C for 10 minutes. Inset: bright field image of Si-nc.

performed in RTA within the 600-800°C temperature range results in a high density (~5×10^{18} cm^{-3}) of small Si-nc. In order to image the Si-nc, electron transparent, cross-sectional samples of the thin films were prepared for TEM observation.

Figure 2 shows a high-resolution TEM cross section of a co-sputtered SRN film with 48% Si and annealed for 10 minutes at 700°C. The inset in Figure 2 shows a TEM bright-field image of the Si quantum dots, taken using a JEOL 2010 TEM, operated at 200 kV. The lattice fringes in the image clearly indicate that the nano-dots are crystalline in nature, and have an average diameter of approximately 2 nm. These Si-nc samples lead to intense light emission in the 750-800 nm range at room temperature, as will be discussed in section 4. We notice that the size of Si-nc is not the only parameter that controls their light emission properties. The optoelectronic properties of Si-nc embedded in silicon nitride matrices are in fact significantly complicated by the large variety of structural phases and atomic configurations that are typical of

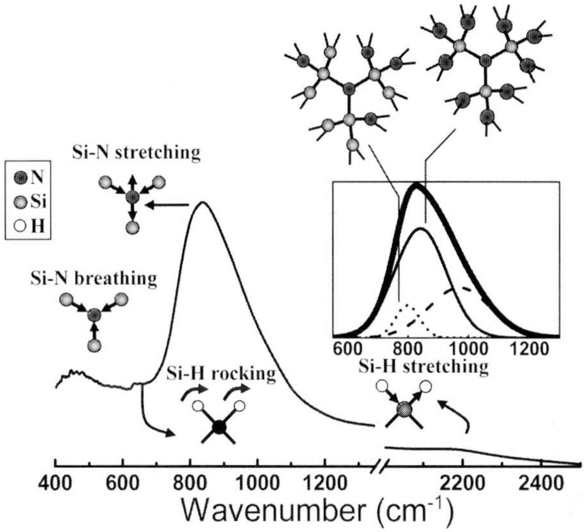

Fig. 3. FTIR spectrum is of the sample with 48% Si. Schematic cartoons representing the assignment of the peaks to the vibrational bands. Inset: The deconvolution of Si-N asymmetric stretching peak.

amorphous silicon nitride. Therefore, an understanding of the structural/chemical bonding properties of the embedding matrix becomes necessary. The chemical bonding structures can be conveniently examined by means of vibrational absorption.

In the following, we will discuss the effect of thermal annealing on phase separation and chemical bond modifications in the matrix as studied by Fourier Transform Infrared Spectroscopy (FTIR). FTIR measurements were performed using a Bruker IFS 66/S instrument. Figure 3 shows the main vibrational contributions to the infrared absorption of our light-emitting SRN samples (48% Si). The peaks observed at 465, 635, 837 and 2185 cm^{-1} can be assigned to Si-N breathing, Si-H bending, Si-N asymmetric stretching and Si-H stretching vibrational modes, respectively.[31-34] The dominant absorption peak is given by the Si-N asymmetric stretching mode, which has been widely studied in the literature of glow discharge[31-33] of SiH_4-NH_3 and PECVD silicon nitride.[32] The energy position of this peak is typically observed in the range 835 cm^{-1} and 885 cm^{-1}, depending on the fabrication conditions. This peak can shift to lower or higher wavenumbers mainly as a result of the presence of excess Si or N atoms as next-nearest neighbors, respectively.[31] Therefore, structural disorder and next-neighbors coordination in amorphous silicon nitride play a crucial role in determining the position and the width of the in homogeneously-broadened Si-N stretching band. However, it is possible to describe the Si-N stretching mode by considering the convolution of different Gaussian peaks, despite the exact assignment of both the peak position and their chemical interpretation are still controversial.[34]

The inset of Figure 3 shows the Gaussian deconvolution of the Si-N stretching band into three main components, which depend on the local bonding geometry near the Si-N bonds. In particular, based on a modified random bonding model,[31] it is possible to assign the peaks located around 790, 840, and 970 cm^{-1} to different chemical environments surrounding the Si-N bonds. The 790 cm^{-1} band results from first and second-neighbor coordination with excess Si atoms present in the SRN matrix. On the other hand, the assignments of the 840 cm^{-1} and 970 cm^{-1} bands are still controversial. Some authors associate the 840 cm^{-1} band to stoichiometric $N(Si_3)$ bonds in samples with low hydrogen content,[32] while others suggest that the presence of at least one hydrogen atom as a second neighbor plays a crucial role in determining the peak positions of both the 840 cm^{-1} and 970 cm^{-1} bands.[33] An alternative picture[35] is to assume that the 840 cm^{-1} and 970 cm^{-1} bands are due to Si_3-SiN and N_3-SiN bonds, respectively. It is worth mentioning that a shift in Si-N peak towards higher wavenumbers has also been reported due to oxygen incorporation, which is not present in co-sputtered films.[36]

Based on the previous picture, it is interesting to investigate the effect of the annealing temperature and phase separation on the infrared absorption of SRN samples. Fig. 4a shows the evolution of the main Si-N peak with annealing temperature, in the case of samples with 48% Si. We have chosen to describe the behavior of this specific set of samples since they give rise to intense room-temperature photoluminescence, while showing very little amount of ambient hydrogen incorporation. As mentioned in section 2, all the samples are annealed in a Rapid Thermal

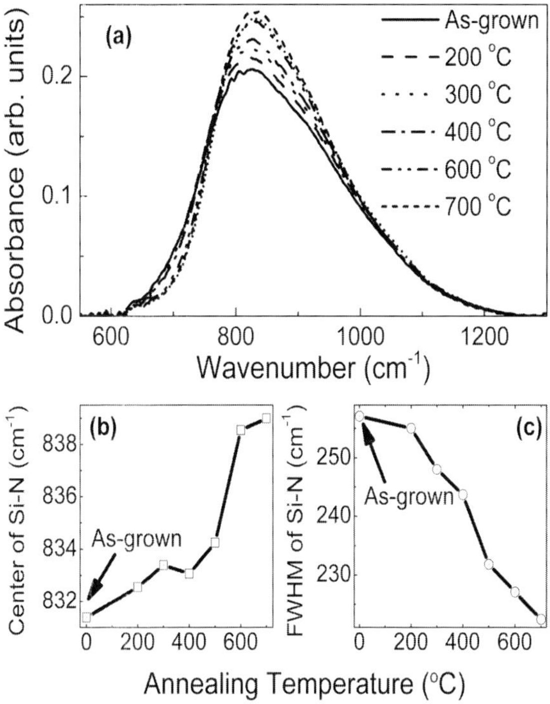

Fig. 4. (a) FTIR spectra of the samples with a Si concentration of 47.2% annealed at temperatures between 200 and 700°C for 10 minutes in RTA furnace under forming gas. (b) The shift in the center of the peaks and (c) the change in the FWHM of the Si-N asymmetric stretching peak with annealing temperature.

Annealing (RTA) furnace in nitrogen forming gas (95% N_2, 5% H_2) atmosphere in a temperature range between 200 and 700°C for 10 minutes. Figures 4b and 4c show that, by increasing the annealing temperatures, the Si-N bond peak position shifts to higher biding energies while the Full Width Half Maximum (FWHM) of the Si-N band decreases. We interpret these findings as the result of thermally-induced phase separation, which leads to a more stoichiometric silicon nitride matrix phase surrounding the Si-nc.

Larger Si-N peak shifts are observed for the samples annealed at 600°C and 700°C, which correspond to the nucleation of 2-nm-diameter Si-nc shown in Figure 2. Interestingly, Fig. 4b demonstrates that by

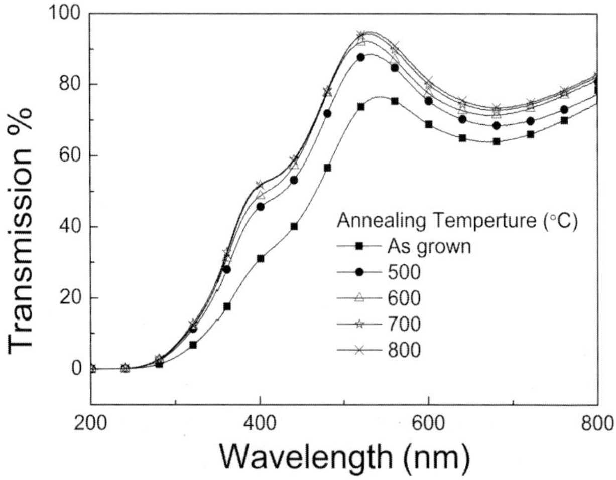

Fig. 5. Transmission spectra of samples with 46% Si concentration and annealed at temperatures between 500 and 800°C for 10 minutes in RTA furnace under forming gas. The as-grown (un-annealed) sample is also included.

increasing the annealing temperature above 500°C, the stoichiometric quality of the silicon nitride matrix can be substantially improved, since the Si-N peak position approaches the energy of the equilibrium phase. Furthermore, an increase in the annealing temperature results in the reduction of the asymmetric Si-N band FWHM. This is consistent with the increased atomic order in the stoichiometric silicon nitride matrix. Remarkably, the improved quality of the silicon nitride matrix upon annealing also results in more transparent films in the optical regime.

In Figure 5, we show the dramatic effect of the annealing temperature on the optical transmission spectra of SRN matrices. We observe that, by increasing the annealing temperature, the transmission spectra shift to shorter wavelengths while their peak transmission values increase. A simple analysis reveals that the horizontal shift of the transmission spectra with annealing temperature results from a shift in the interference fringes induced by a reduction (up to 10%) of the film refractive index, as measured by ellipsometry. However, the vertical shift in the transmission observed at high annealing temperatures cannot be explained by the change in the refractive index of the films. We correlate this vertical shift (higher transmission) with the improved stoichiometric quality and dangling bond passivation of thermally annealed SRN matrices. In fact, our FTIR absorption analysis suggests an increase in the ordering of the amorphous matrix at temperature in excess of 500°C, which initiate the compositional changes leading to a more stoichiometric phase. Our results demonstrate that during the annealing process, the relatively strong Si-N bonds (Si-N 439.0 kJ/mol) replace the weaker Si-Si bonds (Si-Si 326.8 kJ/mol) and Si-H bonds (Si-H 323.0 kJ/mol) to form a silicon nitride phase with improved stoichiometry. At the same time, the excess Si present in the Si-rich (non-stoichiometric) matrix precipitates into small Si clusters embedded in a stoichiometric silicon nitride matrix with improved optical transmission.

4. Optical Characterization of Si-nc in Silicon Nitride

The optimization of the light emission properties of Si-nc dispersed in SRN amorphous matrices can have important implications for the fabrication of Si-based light sources. However, the large variability of structural defects observed in SRN can severely reduce the emission efficiency of Si-nc. In fact, the Si-nc PL efficiency is affected by the complex interplay between the Si-nc dimensionality, their surface structure and defect passivation as well as by the optical quality of the surrounding silicon nitride matrix, as discussed in the previous section. Therefore, in order to gain a robust understanding of this complex system, it is imperative to investigate the optical properties of Si-nc in SRN matrices fabricated under a wide range of deposition conditions. In this section we will discuss the optical emission properties of Si-nc in silicon nitride, and we will discuss their implications with respect to the nature of light emission in co-sputtered SRN films.

The optical emission of Si-nc has been studied by steady-state and time-resolved photoluminescence as well as absorption spectroscopy. Room temperature photoluminescence (PL) experiments were performed using a 488-nm Ar^+ pump laser (Spectra Physics, 177-602) and a standard photomultiplier tube. All the PL spectra have been accurately corrected by the response of the experimental set-up. Optical

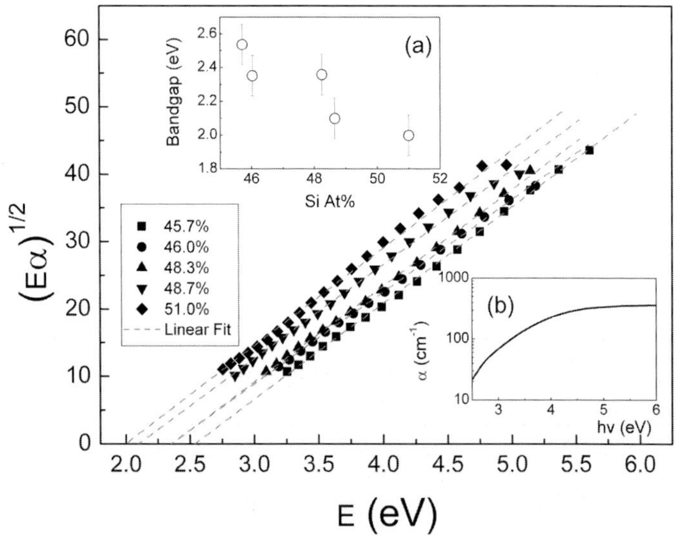

Fig. 6. Tauc plot of the absorption of samples with different Si concentrations and annealed at 700°C for 10 minutes. (a) Extrapolated band-gaps for samples with different Si concentration derived from the Tauc plot. Error bars take into account the error propagation in the subtraction of the reflectance from the optical transmission spectra. (b) Typical semi-logarithm plot of the absorption coefficient of Si-nc in SRN.

transmission experiments were performed using a commercial Cary 5000 double beam spectrophotometer, equipped with a variable-angle reflectance module. Figure 6 shows a Tauc plot[37,38] of annealed (700°C, 10 minutes) SRN films with different Si concentrations. The absorption spectra are smooth, and show quadratic energy dependence like the absorption edge of indirect band gap semiconductors (see Fig. 6, Inset b). In addition, the semi-logarithm absorption plot (see Fig. 6, Inset b) indicates an additional absorption transition occurring at about 5 eV. This is consistent with direct absorption in the stoichiometric silicon nitride matrix surrounding the Si-nc.

From Figure 6, an average value of the Si-nc band gap can be obtained from the intersection of the Tauc plot of $\sqrt{\alpha E}$ with the E axis. Following this procedure, we have found that the Si-nc band gap depends critically on Si concentration (see Fig. 6, Inset a). In particular, the band

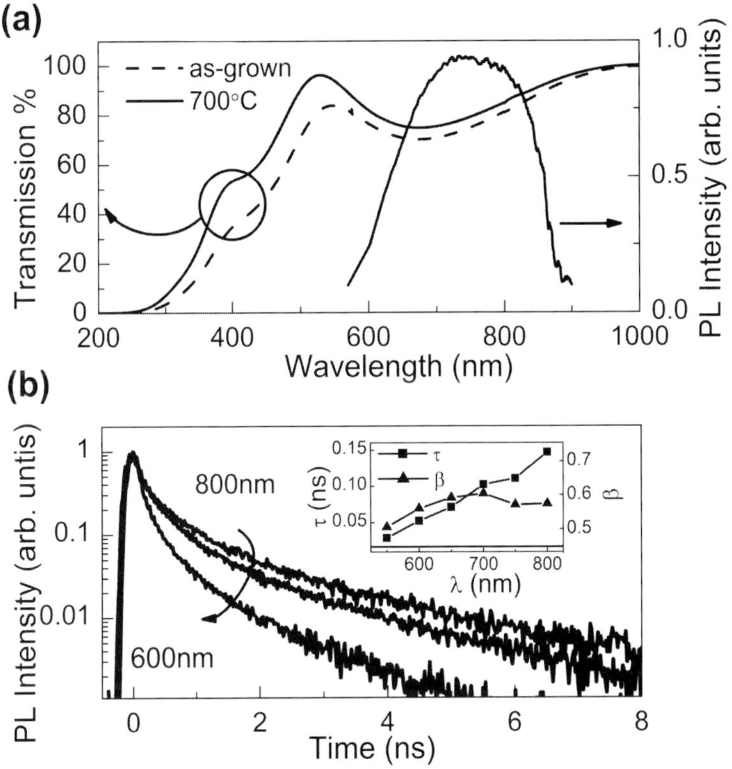

Fig. 7. Summary of the linear optical properties of SRN sample. (a) Comparison of transmission spectra for as-grown and annealed samples at 700°C for 10 minutes. The plot shows also the typical PL spectrum of Si-nc in SRN excited at 488 nm with 5mW pump power. (b) PL decay measured at 600, 700 and 800 nm and excited with 100 fs pulses at 430nm. The pump fluence was 3.7 μJ/cm^2. The inset shows the lifetime (square) and stretching parameter β (triangle) as derived from a stretched exponential fit.

gap increases for smaller Si concentrations, consistently with the nucleation of smaller Si-nc.

Time-resolved photoluminescence experiments have been performed using a frequency doubled Ti:Sa laser (Mai Tai HP, Spectra Physics) with 100 fs pulses at 430 nm. The emitted light was dispersed through a double grating spectrometer (Acton Spectra Pro. 2300i) and detected using a single-photon counting streak camera with 10 ps time-resolution (Hamamatsu, C4770).

Figure 7 summarizes the optical properties of our best emitting sample, which has been annealed at 700°C for 10 minutes. The PL spectrum (Fig. 7a) consists of an intense and broad room-temperature emission centered at about 750 nm, which is similar to typical emission bands observed in Si-nc embedded in SiO_2. However, a significant Stokes-shift (in excess than 1eV) can be extracted by comparing the peak emission energy and the measured band-gaps from Figure 6. These large Stokes-shift values suggest the strongly localized nature of the radiative transitions in Si-nc embedded in silicon nitride. Additionally, as discussed in the previous section, Figure 7a shows that the optical transmission of annealed SRN films is increased with respect to un-annealed samples of identical thickness. This effect reflects the improvement of the SRN optical quality after annealing, and can certainly play in important role in determining the balance between radiative and non-radiative transitions in Si-nc embedded in silicon nitride matrices. Figure 7b shows typical PL decay curves measured at different wavelengths for the best emitting SRN sample. Wc have found that for sputtered SRN samples the recombination dynamics consists of an initial sub-nanosecond component followed by a much longer decay component (10-100 ns time scale). In addition, the fast decay, which can be modeled using a stretched exponential, strongly depends on wavelength. On the other hand, the ns-long decay is only weakly wavelength dependent. The inset of Figure 7 (b) shows the decay time and stretching constant dispersion associated to the fast recombination process. *Ab-initio* microscopic models of Si-nc, which will be discussed in section 6, indicate that light emission in Si-nc embedded in silicon nitride results from localized transitions related to different nitrogen groups at the surface of Si-nc smaller than 3nm. According to this picture, the wavelength-dependent 50-200-ps fast recombination dynamics can be interpreted as due to non-radiative trapping on a distribution of surface states or non-radiative centers present in the silicon nitride matrix. On the other hand, the longer (1-20 ns) decay components, which are almost independent by the observation wavelength, directly result from surface-state recombination (both radiative and non-radiative). Another unique aspect related to the light emission properties of small Si-nc in silicon nitride matrices is the almost

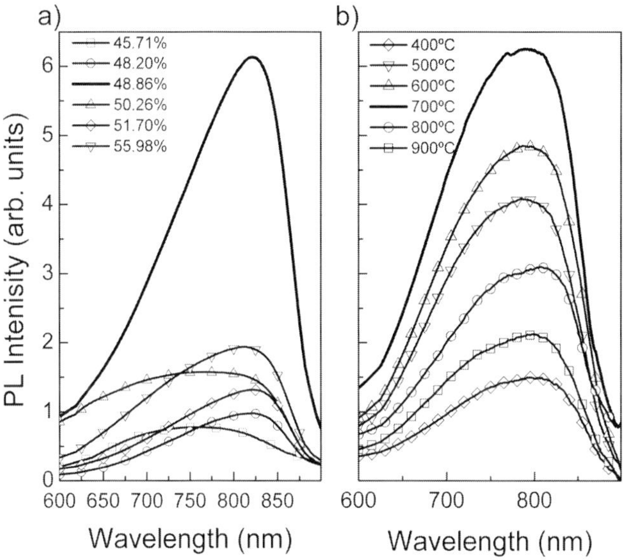

Fig. 8. (a) PL spectra for samples with different Si concentrations and annealed at 700°C for 10 minutes. (b) PL spectra for samples with the same Si concentration (51.7%) and annealed from 400°C to 900°C for 10 minutes. The PL are excited at 488 nm with 5 mW optical power.

complete absence of emission tunability. Figure 8a shows the evolution of the room-temperature PL spectra for samples fabricated with varying Si concentration, but annealed under identical conditions (700°C, for 10 minutes). We notice that the weak PL emission corresponding to the smallest (45%) Si concentration is significantly blue-shifted into the visible range and it can be explained by a broad-band defect emission of the silicon nitride matrix.

On the other hand, when the Si concentration is increased above 45% an intense PL band appears in the 750-800 nm wavelength range, resulting from Si-nc surface states. When the Si concentration is increased beyond approximately 48%, the nucleation of Si-nc approaches a diffusion-limited maximum density, and the PL reaches its maximum value before it quenches at the highest Si concentrations (56%). This PL behavior suggests that the nucleation of Si-nc in silicon nitride is a diffusion-limited process which initiates from a high density of optically active defect centers in the silicon nitride matrix.

Figure 8b shows the evolution of the room-temperature PL spectra for samples annealed at different temperatures but with a fixed Si concentration (51.7%) It is remarkable that, by comparing the near-infrared Si-nc PL bands shown in Figure 8, almost no emission tunability can be observed.

Therefore we believe that the combined observations of wavelength-insensitive nanosecond recombinations, the large absorption/emission Stokes-shifts (1.0-1.5 eV) and the absence of PL tunability observed in nitride-based Si-nc systems indicate that light emission originates from strongly localized gap states introduced by N-groups at the surface of small Si-nc. A more detailed discussion of the optical properties of nitrogen-related surface states in nitride-based Si-nc systems will be presented in section 6.

5. Energy Transfer to Erbium Ions

It has been recently discovered that Si-nc act as efficient light emission sensitizers for rare-earth ions, particularly Er, enabling broad-band pumping with almost three orders of magnitudes enhancement in the pumping efficiency.[19-22] Several studies have been devoted to the understanding and the optimization of energy sensitization in SRO systems. However, little is known on the energy transfer in silicon nitride-based systems. In this section we discuss the effect of annealing temperature on the energy transfer in Er:SRN systems fabricated by direct co-sputtering. In addition, we will discuss the efficiency of the energy transfer in SRN-based Si-nc systems. Figure 9 shows the PL spectra of Er in silicon nitride matrices annealed at different temperatures. All the PL spectra are obtained under non-resonant (457 nm) optical pumping and demonstrate that the Er excitation is mediated by the Si-nc in the silicon nitride matrix. In addition, Figure 9 shows that the sensitized Er emission peaks at approximately 700°C, which is the temperature that maximizes the Si-nc emission (at 800 nm) in the absence of Er ions (see Fig. 8b). Furthermore, we notice that this

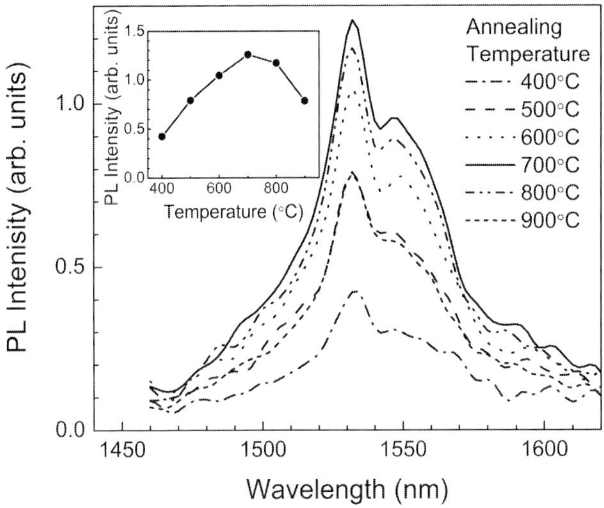

Fig. 9. Typical PL spectra of Er: SRN samples annealed at different temperatures for 10 minutes. The PL has been excited at 457 nm with 5 mW. The inset shows the evolution of the PL peak intensity at 1.54 μm for different annealing temperatures.

temperature behavior of the Er PL has been observed in differently prepared samples, irrespective of their Si concentrations.

The direct correlation between the Si-nc PL and the sensitized Er emission is more clearly displayed in Fig. 10a, where we compare the sensitized 1.54-μm Er emission and the 800 nm Si-nc emission for samples annealed at different temperatures. Unfortunately, due to the limitations of EDX analysis, we can only measure an upper limit of 0.5% for the Er concentration in these samples. On the other hand, the samples with/without Er are display identical Si concentration, which in this specific case is 47%. Figure 10a shows that the decrease in the 800nm Si-nc emission intensity, which is due to energy transfer to Er ions, depends dramatically on the annealing temperature. In particular, the quenching of the Si-nc emission is stronger at 700°C. On the other hand, the 800nm Si-nc emission can be completely recovered at higher annealing temperatures, where we expect the nucleation of larger Si-nc to take place. Therefore, our data suggest that the smallest Si-nc, which nucleate at temperatures below 700°C, are directly involved in the energy transfer to Er ions in silicon nitride. At temperatures below 700°C, the 1.54 μm sensitized Er emission (full triangles) follows the same temperature trend than the 800 nm emission of the Si-nc sample without Er (open squares). Therefore these results strongly indicate that energy transfer in silicon nitride occurs via sequential coupling with small Si-nc that form at temperatures below 700°C. Therefore, simultaneous information on the modifications of the Si-nc decay dynamics in the presence and absence of Er ions is needed in order to unambiguously attribute the energy transfer to non-radiative resonant processes.

In order to quantify the energy transfer to Er ions we have measured the effective excitation cross section of Er from the linear dependence of the Er PL rise time versus the pumping photon flux, as shown in Figure 10b. However, before discussing our experiments further, we need to introduce some important relations which will help to better understand the implications of our results. We first notice that, within a simple rate equation picture of Förster dipole coupling,[39] and only considering two-level systems (Er and Si-nc ground and first excited states), the measured effective Er excitation cross section σ_{eff} can be related to the direct Er absorption cross section σ_{abs} by the simple expression:

$$\sigma_{eff} = \sigma_{abs} + \gamma \bar{n}_b / \phi_P \tag{1}$$

where we have defined the steady-state population of excited Si-nc $\bar{n}_b$ and the energy transfer coupling coefficient γ, which measures the coupling strength between one Si-nc and one Er ion embedded in a given matrix. Moreover, in the weak pumping regime the Si-nc excited state population is directly proportional to the pumping rate and equation (1) can be reduced to the simpler expression:

$$\sigma_{eff} = \sigma_{abs} + \gamma \tau \sigma_P \rho_{NC} \tag{2}$$

where τ is the Si-nc decay time, σ_P is the Si-nc absorption cross section at the pump wavelength and ρ_{NC} is the Si-nc volume-density. This simple expression allows the direct determination of the fundamental coupling constant γ via steady-state measurements of the effective excitation cross section, provided that the Si-nc decay time, pump absorption cross

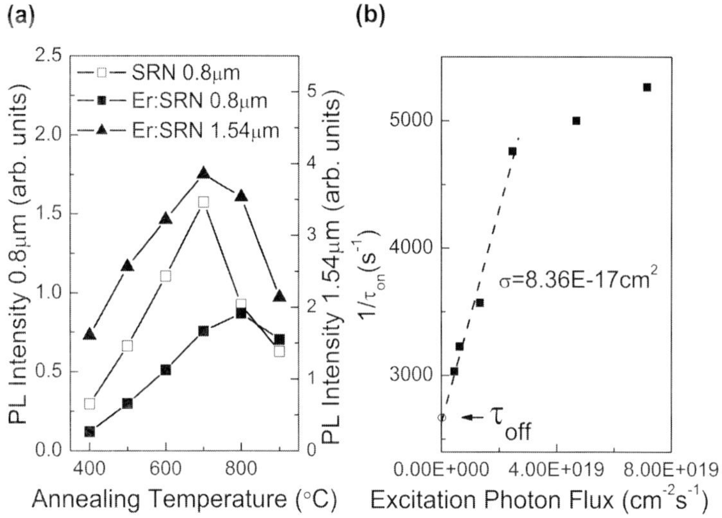

Fig. 10. (a) PL peak intensity at 0.8 μm from SRN (open square) and Er:SRN (close square), and PL peak intensity at 1.54 μm from Er:SRN (close triangle) annealed at temperatures from 400 to 900°C. The PL spectra are excited at 457 nm with 5 mW. (b) Reciprocal of the rise time of the PL at 1.54 μm excited at 457 nm with different excitation photon fluxes.

section and Si-nc density are known. It is worth noticing that by substituting into equation (2) the published values for the coupling constant $\gamma \approx 10^{-15}$ cm^3/s and the typical parameters for Si-nc in silicon oxides ($\sigma_p \approx 10^{-16}$ cm^2, $\tau = 50\,\mu$s, $\rho_{NC} \approx 10^{19}$ cm^{-3}) we can predict an effective Er cross section $\sigma_{eff} \approx 10^{-16}$ cm^2, which is in excellent agreement with the published experimental literature.[39] On the other hand, the coupling constant γ describing the Si-nc/Er coupling strength in a silicon nitride matrix is not known. However, following our simple model for the sensitized Er cross section, the coupling constant γ can easily be determined using equation (2) and the measured value for the Er effective excitation cross section in silicon nitride.

A linear fit of the data in Figure (b) yields approximately $\sigma_{eff} = 8\times10^{-17}$ cm^2 for the Er effective excitation cross section of our best-emitting Si-nc samples. The Er decay time, extrapolated from the linear regime in Figure 10, is approximately 0.5 ms, in good agreement with the directly measured Er lifetime values. The saturation in the Er emission risetime observed at high pumping rates is attributed to the onset of sensitized up-conversion.

We notice that the effective Er cross section that we measured in nitride-based Si-nc is of the same order of the cross sections obtained in silicon oxide. However, since the decay time of Si-nc in nitride is almost three orders of magnitudes faster than in oxide, our data strongly indicate that the Si-nc coupling constant (γ) to Er ions must be considerably larger in silicon nitride. The application of equation (2), which is based on general rate equations within the weak pumping approximation, leads to a value of $\gamma \approx 10^{-13}$-10^{-12} cm^3/s. These values are two to three orders of magnitude larger than the ones reported in Si-nc dispersed in SiO_2.

It is possible to conjecture that the high radiative rates (nanosecond radiative lifetimes) of surface-localized exciton transitions in N-passivated Si-nc, which will be discussed in the next section, play an important role in the enhanced energy transfer rate observed in silicon nitride, potentially enabling more efficient Er sensitization schemes.

6. *Ab-Initio* Modeling of Si-nc in Silicon Nitride

While the characterization of thin film SRN Si-nc presented in sections 3 and 4 provides extensive information on the size, shape and opto-electronic properties of Si-nc, current limitations in imaging resolution makes it difficult to directly correlate the properties of Si-nc with the details of their atomistic structure. *Ab-initio* computational modeling allows us to bridge the gap between structures and properties by predicting in detail the influence of the atomistic structure/surface of Si-nc on the optical properties. In particular, different types of bonds at the interface between the Si cluster and the surrounding amorphous silicon nitride matrix are shown to have very different impacts on the optical properties of Si-nc.

6.1. *Structural models of SRN Si-nc*

We first consider different models of the amorphous silicon nitride matrix. Based on the experimental evidence presented in section 3, we target a nearly stochiometric matrix and a crystalline Si cluster with a

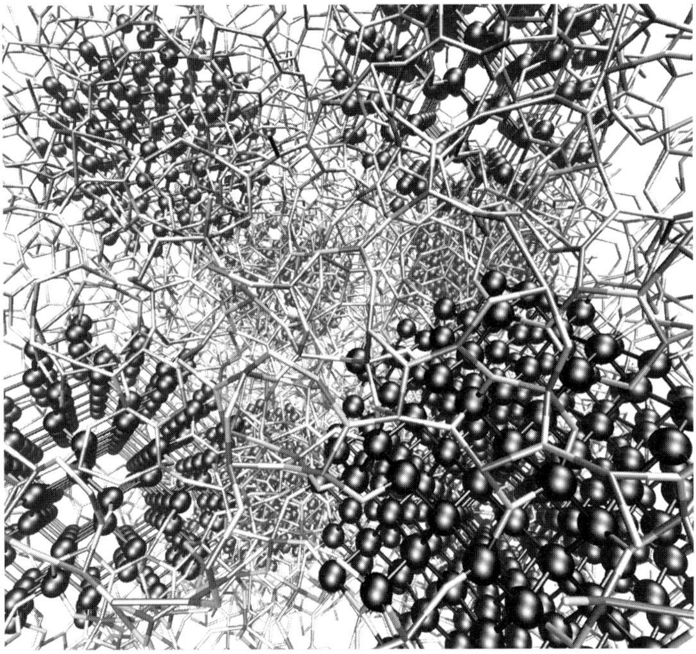

Fig. 11. Structural model of a 2 nm Si-nc embedded in amorphous Si_3N_4 matrix. The silicon atoms in the Si-nc are black spheres; the nitrogen atoms are light grey and the Si atoms in the matrix are dark grey.

2 nm diameter. Amorphous models can be obtained through procedures such as melting and rapidly cooling a superlattice of crystalline Si_3N_4.[41] Another method proposed by Valladares *et al.*[40] consists in starting from the Si diamond structure and replacing silicon atoms by nitrogen atoms until the right stochiometry is obtained. The system will amorphize when the structure is relaxed. Bond switching (WWW) is yet another method used to generate amorphous models of silicon nitride.[43,44]

Clearly, having a good model of the amorphous silicon nitride matrix is important, but for the purpose of studying the electronic structure of the Si-nc it is key to get a good model of the interface between the amorphous matrix and the Si-nc.

As the timescales involved in segregating the excess silicon and nucleating a Si-nc are well out of the reach of current atomistic methods (quantum or classical Molecular Dynamics (MD)), we have explored different schemes to generate models of the interface between the Si-nc and the matrix. One simple approach is to carve a cavity in an amorphous embedding matrix, insert a Si-nc and then anneal the interface, keeping the Si-nc atoms fixed. Another approach is to have the Si-nc present from the start during the amorphization of the matrix and prevent the Si atoms from diffusing during the annealing cycles. Although not very precise, these empirical or semi-empirical potential[42] based models of the interface between the Si-nc and the matrix give insight into the variety of configurations of silicon-nitrogen bonding that are likely to be present in such complex interfaces. Figure 11 shows a structural model of a 2 nm Si-nc embedded in stoichiometric amorphous silicon nitride. Several bonding configurations between the Si-nc and nitrogen atoms in the matrix are chosen and studied with *ab initio* methods in the next section.

6.2. *Electronic structure of H-, O-, and N- terminated Si-nc*

To investigate how the structure of the Si-nc and the bonds at the interface may influence the optical properties of SRN materials, we performed a series of first-principles density functional theory (DFT) calculations,[a] of the structural, electronic and optical properties of silicon nanoclusters with different numbers of nitrogen and oxygen atoms bonded to their surface.[b] The configurations were chosen based on models of the interface between the Si-nc and the amorphous silicon nitride embedding matrix obtained in classical MD simulations.

---

[a] The calculations use norm conserving, Troullier-Martins pseudo-potentials and a plane-wave basis with a 90 Ry cutoff. The simulation super-cell sizes were chosen to place >10 Å of vacuum between the periodic replica of the Si-nc, so as to remove any interactions. The atomic relaxations were performed until the residual force on all atoms was lower than $3\cdot10^{-3}$ eV/Å. The QBOX[49] code was used for the structural relaxations and the PWSCF[50] code was used to evaluate gaps and obtain the excited states. All calculations are done with the PBE exchange-correlation functional.[53]

[b] One can envision a pair of N interstitials getting trapped in the core of the cluster but our calculations have shown that a similar configuration at the surface is energetically favored by 0.7 eV. This is due to a minimization of the stress on the crystalline core when the di-interstitial pair of N is at the surface.[30]

Previous calculations[45,46] have shown that it is not possible to directly compare DFT calculated gaps of doped silicon nanoclusters with experimental gaps for two reasons; (i) DFT calculated gaps of Si nanoclusters may underestimate the true optical gap, (ii) gaps calculated with a single surface dopant, typically overestimate the gap compared to clusters covered with multiple surface dopants. However, Quantum Monte Carlo (QMC) studies of oxygen terminated Si nanoclusters[46] showed that the qualitative changes in the gap of nanoclusters with different surface dopant structures are well reproduced by DFT calculations. In Figure 6, the onset of absorption, as determined by a Tauc plot, is approximately in the range 2-2.5 eV, depending on the Si concentration. This is lower than the theoretical 3 eV absorption gap of a 2 nm Si-nc which was estimated based on quantum Monte-Carlo calculations.[45] However, DFT calculations on small Si-nc have shown that the addition of oxygen or nitrogen significantly lowers the calculated gaps[c] and the smallest gaps are found when nitrogen is added on the surface of the cluster.[46,47,52,29,30] This suggests that absorption and emission may originate at the interface between the Si-nc and the SRN matrix rather than from the core of the Si-nc. These oxygen or nitrogen dopants introduce surface states and, in the presence of a few (1 to 6) dopants, the HOMO-LUMO band gap is practically independent of the size of the Si-nc and, for most configurations, of the number of dopants. An exception is provided by the bridged NH group (Fig 12, structure f) with a DFT gap that can be lowered below 2.0 eV with 6 dopants; other exceptions are the di-interstitial N pairs whose gap can be lowered to 2.1 eV with 6 pairs. In the limit of complete coverage with NH_2 group on a small Si-nc, although NH_2 groups still introduce surface states, these states are not as localized as in the case of other surfactants and the DFT

[c] In Figure 12, the inset plots show the detail of the bonding of the nitrogen containing groups on the surface of the clusters. Our reference H-terminated 1 nm Si-nc, structure (a), consists of a diamond crystalline core of 35 Si atoms where dangling bonds are capped with 36 H atoms. In our reference crystalline cluster both the HOMO and LUMO orbitals are delocalized throughout the core of the cluster. Upon addition of nitrogen, the HOMO orbital becomes localized around the nitrogen atom and exhibits a *p* character. Analysis of the individual HOMO and LUMO eigenvalues shows that the gap of these clusters is reduced because the nitrogen localized HOMO states are higher in energy than the HOMO of the reference hydrogen terminated clusters.

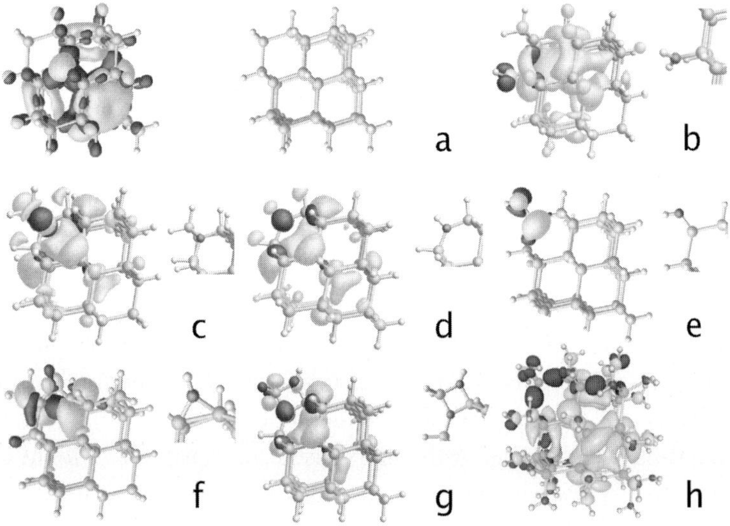

Fig. 12. Isosurface plots of the density of the highest occupied molecular orbital (HOMO) in dark grey and lowest unoccupied molecular orbital (LUMO) in light grey. Silicon and hydrogen atoms are light grey, nitrogen atoms are dark grey. The isosurfaces contain 30% of an electron charge. For the reference structures (a) the full structure is shown. For structures (b)-(g) the inset plots show the detail of the bonding of the nitrogen containing groups on the surfaces of the clusters.

gap is 1.6 eV. This limit may actually be a more realistic model of the interface with the SRN matrix than just assuming the presence of a few dopants. Considering the known DFT underestimate of the experimental band gap, this result is consistent with the experimental value of 2.0 eV. Although the experimental results on PL decay rates point to still significant non-radiative decay mechanisms associated to the imperfect quality of the SRN matrix, it is still instructive to look at the radiative lifetime of surface states, especially at the difference between nitrogen and oxygen dopants in similar configurations.

Table 1 shows our calculated radiative lifetimes, τ, evaluated using the following question:[48]

$$\frac{1}{\tau}=\frac{1}{2\pi m_e}\frac{e^2}{\varepsilon\varepsilon_0}\left(\frac{n}{c}\right)^3 f_{HL}\omega_{HL}^2 \tag{3}$$

Table 1. Calculated DFT-PBE optical gaps and radiative lifetimes for different Si nanoclusters. The lifetimes are calculated with a 2.23 index of refraction for the SRN.

Structure	DFT-PBE Gap (eV)	Lifetimes (ns)
(a) $Si_{35}H_{36}$ – reference crystalline structure	3.5	10^2
(b) $Si_{35}H_{35}NH_2$ – NH_2 substituting H	3.0	10^3
(c) $Si_{34}H_{35}N$ – N substituting SiH	2.9	10^2
(d) $Si_{34}H_{34}NH$ – NH substituting SiH_2	2.8,2.9	forbid.*, 10^3
(e) $Si_{35}H_{34}NH$ – NH substituting 2H	2.5	10^2
(f) $Si_{35}H_{34}NH$ – Si-NH-Si bridge	2.3	10^1
(g) $Si_{35}H_{36}N_2$ – di-interstitial N on the surface of the cluster	2.7	10^2
(h) $Si_{35}N_{36}H_{72}$	1.6	10^2
(i) $Si_{35}H_{34}O$ – O substituting 2H	2.3	10^6
(j) $Si_{35}H_{35}OH$- OH substituting H	3.3	10^2

where

$$\frac{e^2}{\varepsilon\varepsilon_0}$$

is the screened interaction of the exciton for $Si_{35}H_{36}$,[51] c is the velocity of light, n the index of refraction ($n = 2.23$ for our SRN samples with 49% Si), m_e the mass of the electron, ω_{HL} is the gap and f_{HL} is the HOMO-LUMO dipole oscillator strength. This oscillator strength depends on the symmetry and the degree of overlap of the orbitals, which in turn is sensitive to the details of the atomistic structure.

Structures (b), (d), and (e) in Figure 12 have large radiative lifetimes for their lowest transitions. In structure (d), the HOMO-LUMO transition is forbidden by symmetry and the lowest allowed transition is the HOMO to LUMO+1. For structures (b) and (e), the high atomic symmetry of the nitrogen groups is weakly broken by structural relaxations, such as the rotation of an N-H bond, yielding only a weak oscillator strength.

Structure (f) has the shortest lifetime, an order of magnitude lower than the reference $Si_{35}H_{36}$ cluster and at least one order of magnitude lower than any other structure. We also calculated the radiative lifetimes for oxygen defects considered in Ref. 46. The double bonded O in structure (i) has a very long lifetime due to an almost symmetry forbidden first transition. A Si-nc with a single bonded OH surface group such as structure (j) has a shorter lifetime than the single bonded N defect in $Si_{35}H_{35}NH_2$ (structure b), but an order of magnitude longer lifetime than the bridged nitrogen structure (f). The completely covered Si-nc (h), with 36 NH_2 group replacing the 36 hydrogen atoms of the reference Si-nc shows an order of magnitude smaller lifetime than the single NH_2 doped structure (b) but still larger than the bridged nitrogen (f). We note that these radiative lifetimes have been calculated in the frozen lattice approximation (0 K in calculation) and that phonon-assisted radiative recombination could lower the lifetimes; in addition to radiative recombination of carriers, we also expect significant contributions from non-radiative carrier decay. Therefore the radiative lifetime should be considered an upper bound to the measured lifetime.

The structures considered here and even more complex combination of structures (e.g., NH_2 groups with a few NH bridges and/or di-interstitial N pairs) point towards a distribution of possible surface states with different gaps and radiative lifetimes. This is consistent with the interpretation put forth in section 4 about the wavelength-dependent fast recombination dynamics as non-radiative trapping on a distribution of surface/matrix-related states; the longer decay (1-20 ns) components would result from recombination (both radiative and non-radiative) originating from this distribution of surface states.

6.3. *Calculated Stokes shifts of H-, O-, and N- terminated Si-nc*

For a subset of Si-nc with different surface groups, we have calculated the structural Stokes shift between the onset of absorption and emission by constraining the cluster in an excited triplet state and calculating the structural relaxation induced by this electronic excitation. The magnitude of the Stokes shift depends on the degree to which the nitrogen atoms on the surface of the clusters relax their geometry in an excited electronic state. In structures (b) and (e) the NH groups are relatively free to relax and the clusters exhibit large Stokes shifts between 1.5 and 2.0 eV. In structures (c) and (d) the nitrogen is constrained within the silicon network producing a Stokes shift of only 0.2 eV, smaller than the Stokes shift of the reference $Si_{35}H_{36}$ cluster (1.0 eV). In Ref. 47, it was found that double bonded O also induces Stokes shifts larger than 1 eV. The experimental Stokes shifts presented in section 4 are consistent with the calculated Stokes shifts for Si-nc with localized surface states.

7. Conclusions and Outlook

One of the key challenges to developing an efficient silicon based light source is to synthesize nanoscale silicon with well characterized structural and optical properties. Here we have presented the results of fabrication, characterization and modeling of Si-nc embedded in silicon nitride matrices. We demonstrate that thin film co-sputtering deposition of SRN can produce high quality, nanometer sized, crystalline silicon nanostructures. High resolution TEM imaging of these Si-nc shows an average diameter of about 2 nm. As the Si-nc have dimensions of only a few nanometers, they have a large surface to volume ratio and their properties are dominated by the structure and chemistry of their surfaces, and the interface between the Si-nc and the surrounding matrix. First-principles models provide an ideal bridge between the measured structural and optical properties of these Si-nc and the atomistic details of their structures. In particular, the predicted properties of Si-nc with different surface structures and surface chemistries can be used to discriminate between different candidates structures. Optical characterization of the thin film SRN structures indicates that the material has a large Stokes shift (in excess than 1 eV) between absorption and emission, an intense near-infrared emission at room temperature and a wavelength-insensitive, nanosecond-long emission lifetime. By comparing the predicted optical properties of a range of model Si-nc with different surface structures we find that the absorption gap, radiative lifetime and Stokes shift are all sensitive to the structure, strain and chemistry of the surface of the Si-nc. We identify structures with surface nitrogen groups exhibiting both short (10-100 ns) radiative lifetimes and an onset of absorption consistent with the measured optical absorption and lifetime of our SRN thin film samples.

In addition, we have studied energy transfer to Er ions and demonstrated optimum transfer efficiency for SRN samples annealed at 700°C. We have directly measured the effective excitation cross section of Er ions in nitride-based Si-nc, showing that the coupling coefficient γ to Er ions can be significantly larger than in the widely investigated Si-nc/oxide materials systems. These findings suggest that Si-nc embedded in silicon nitride matrices have a large potential for the fabrication of optically active photonic devices based on Si technology.

Acknowledgments

This work was partially supported by the U.S. Air Force MURI program on "Electrically-Pumped Silicon-Based Lasers for Chip-Scale Nanophotonic Systems" supervised by Dr. Gernot Pomrenke. Part of this work was also supported by the DOE under grant #DE-FG02-06ER46262 and the U.S. Department of Energy by Lawrence Livermore National Laboratory, partly under contracts W-7405-Eng-48 and contracts DE-AC52-07NA27344.

References

1. N. Savage, *Light from Silicon: Holy Grail*, IEEE Spectrum, **59** (2004).
2. *Silicon Photonics*, edited by L. Pavesi, D. J. Lockwood, Springer-Verlag, Berlin (2004).
3. L. T. Canham, *Appl. Phys. Lett.* **57**, 1046 (1990).
4. K. D. Hirschman, L. Tsybeskov, S. P. Duttagupta and P. M. Fauchet, *Nature* **384**, 338 (1996).
5. R. J. Walters, J. Kalkman, A. Polman, H. A. Atwater and M. J. A. de Dood, *Phys. Rev.* **B73**, 132302 (2006).
6. G. T. Reed and A. P. Knights, *Silicon Photonics* (John Wiley, 2004).
7. L. Dal Negro, J. H. Yi, L. C. Kimerling, S. Hamel, A. Williamson and G. Galli, *Appl. Phys. Lett.* **88**, 183103 (2006)
8. D. Kovalev, H. Heckler, M. Ben-Chorin, G. Polisski, M. Swartzkopff and F. Koch, *Phys. Rev. Lett.* **81**, 2803 (1998).
9. B. Gelloz, A. Kojima and N. Koshida, *Appl. Phys. Lett.* **87**, 031107 (2005).
10. L. Pavesi, L. Dal Negro, C. Mazzoleni, G. Franzò and F. Priolo, *Nature* **408**, 440 (2000).
11. L. Khriachtchev, M. Rasanen, S. Novikov and J. Sinkkonen, *Appl. Phys. Lett.* **79**, 1249 (2001).
12. M. Nayfeh, S. Rao, N. Barry, A. Smith and S. Chaieb, *Appl. Phys. Lett.* **80**, 121 (2002).

13. K. Luterova, I. Pelant, I. Mikulskas, R. Tomasiunas, D. Muller, J. J. Grob, J. L. Rehspringer and B. Hönerlage., *J. Appl. Phys.* **91**, 2896 (2002).
14. L. Dal Negro, M. Cazzanelli, L. Pavesi, S. Ossicini, D. Pacifici, G. Franzò, F. Priolo and F. Iacona, *Appl. Phys. Lett.* **82**, 4636 (2003).
15. J. Ruan, P. M. Fauchet, L. Dal Negro, M. Cazzanelli and L. Pavesi, *Appl. Phys. Lett.* **83**, 5479 (2003).
16. L. Pavesi, S. Gaponenko and L. Dal Negro, *Towards the first silicon laser*, NATO Advance Studies Institute, series 11, Kluwer Academic, Dordrecht **93** (2003).
17. L. Dal Negro, M. Cazzanelli, B. Danese, L. Pavesi, F. Iacona, G. Franzò and F. Priolo, *J. Appl. Phys.* **96**, 5747 (2004).
18. M. Cazzanelli, D. Kovalev, L. Dal Negro, Z. Gaurro and L. Pavesi, *Phys. Rev. Lett.* **93**, 207402 (2004).
19. M. Fujii, M. Yoshida, Y. Kanzawa, S. Hayashi and K. Yamamoto, *Appl. Phys. Lett.* **71**, 1198 (1997).
20. G. Franzo, V. Vinciguerra and F. Priolo, *Appl. Phys.* **A69**, 3 (1999).
21. M. Fujii, K. Imakita, K. Watanabe and S. Hayashi, *J. Appl. Phys.* **95**, 272 (2004).
22. K. Imakita, M. Fujii and S. Hayashi, *Phys. Rev.* **B71**, 193301 (2005).
23. M. V. Wolkin, J. Jorne, P. M. Fauchet, G. Allan and C. Delerue, *Phys. Rev. Lett.* **82**, 197 (1999).
24. A. Puzder, A. J. Williamson, F. A. Reboredo and G. Galli, *Phys. Rev. Lett.* **91**, 157405 (2003).
25. E. Degoli, G. Cantele, E. Luppi, R. Magri, D. Ninno, O. Bisi and S. Ossicini, *Phys. Rev.* **B69**, 155411 (2004).
26. T. Y. Kim, N. M. Park, K. H. Kim, G. Y. Sung, Y. W. Ok, T. Y. Seong and C. J. Choi, *Appl. Phys. Lett.* **85**, 5355 (2004).
27. N. M. Park, C. J. Choi, T. Y. Seong and S. J. Park, *Phys. Rev. Lett.* **86**, 1355 (2001).
28. L. Dal Negro, J. H. Yi, J. Michel, L. C. Kimerling, T.-W. F. Chang, V. Sukhovatkin and E. H. Sargent, *Appl. Phys. Lett.* **88**, 233109 (2006).
29. L. Dal Negro, J. H. Yi, L. C. Kimerling, S. Hamel, A. J. Williamson and G. Galli, *Appl. Phys. Lett.* **88**, 183103 (2006).
30. Luca Dal Negro, Sebastien Hamel, Natalia Zaitseva, Jae Hyung Yi, Andrew Williamson, Michael Stolfi, Jurgen Michel, Giulia Galli and Lionel C. Kimerling, *IEEE J. Select. Topics Quantum Electron.* **12**, 1151 (2006).
31. S. Hasegawa, L. He, Y. Amano and T. Inokumu, *Phys. Rev.* **B48**, 5315 (1993).
32. G. N. Parsons, J. H. Souk and J. Batey, *J. Appl. Phys.* **70**, 1553 (1991).
33. G. Lucovsky, J. Yang, S. Chao, J. E. Tyler and W. Czubatyj, *Phys. Rev.* **B28**, 3234 (1983).
34. M. Kessels, J. Hong, F. J. H. van Assche, J. D. Moschne, T. Lauinger, W. J. Soppe, A. W. Weeber, D. C. Schram and M. C. M. van de Sanden, *J. Vac. Sci. Technol.* **A20**, 1704 (2002).
35. D. Della Sala, C. Coluzza, G. Fortunato and F. Evangelisti, *J. Non-Cryst. Solids* **77-78**, 933 (1985).

36. M. Lattemann, E. Nold, S. Ulrich, H. Leiste and H. Holleck, *Surf. Technol.* **174-175**, 365 (2003).
37. N. E. Cusack, *The physics of structurally disordered matter*, Adam Hilger IOP Publishing Ltd, Bristol, UK (1987).
38. B. Garrido, M. Lopez, O. Gonzalez, A. Perez-Rofriguez, J. R. Morante and C. Bonafos, *Appl. Phys. Lett.* **77**, 3143 (2003).
39. D. Pacifici, G. Franzò, F. Priolo, F. Iacona and L. Dal Negro, *Phys. Rev.* **B67**, 245301 (2003).
40. F. Alvarez and A. A. Valladares, *Phys. Rev.* **B68**, 205203 (2003).
41. J. F. Justo, M. Z. Bazant, E. Kaxiras, V. V. Bulatov and S. Yip, *Phys. Rev.* **B58**, 2539 (1998).
42. F. De Brito Mota, J. F. Justo and A. Fazzio, *Phys. Rev.* **B58**, 8323 (1998).
43. F. Wooten, K. Winer and D. Weaire, *Phys. Rev. Lett.* **54**, 1392 (1985).
44. P. Kroll, *J. Non-Cryst. Solids* **293-295**, 238 (2001).
45. A. Puzder, A. J. Williamson, J. C. Grossman and G. Galli, *Phys. Rev. Lett.* **88**, 097401 (2002).
46. A. Puzder, A. J. Williamson, J. C. Grossman and G. Galli, *J. Chem. Phys.* **117**, 6721 (2002).
47. A. Puzder, A. J. Williamson, J. C. Grossman and G. Galli, *J. Am. Chem. Soc.* **125**, 2788 (2003).
48. A. Yariv, *Quantum Electronics*, 3rd ed. (Wiley, New York, 1989).
49. F. Gygi, Lawrence Livermore National Laboratory.
50. S. Baroni, A. Dal Corso, S. de Gironcoli, P. Giannozzi, C. Cavazzoni, G. Ballabio, S. Scandolo, G. Chiarotti, P. Focher, A. Pasquarello, K. Laasonen, A. Trave, R. Car, N. Marzari and A. Kokalj, http://www.pwscf.org/.
51. The value of ε is taken from A. Tsolakidis, D. Sanchez-Portal and R. M. Martin, *Phys. Rev.* **B66**, 235416 (2002).
52. E. W. Draeger, J. C. Grossman, A. J. Williamson and G. Galli, *Phys. Rev. Lett.* **90**, 167402 (2003).
53. J. P. Perdew, K. Burke and M. Ernzerhof, *Phys. Rev. Lett.* **77**, 3865 (1996); *Phys. Rev. Lett.* **78**, 1396 (1997).

CHAPTER 13

ENERGY EFFICIENCY IN SILICON PHOTONICS

Bahram Jalali, Sasan Fathpour, and Kevin K. Tsia

Electrical Engineering Department, University of California, Los Angeles, CA, 90095-1594 USA

Silicon photonics research has been primarily motivated by its material and process compatibility with complementary metal–oxide–semiconductor (CMOS) electronics. However, full implementation of integrating optics and electronics on a common silicon platform must consider other constituents of CMOS compatibility, a critical one of which is 'heat compatibility'. A major challenge for silicon photonics is the potential aggravation of high power dissipation of very-large-scale integration (VLSI) by the photonic devices integrated on the same chip. Silicon has a fundamental problem – caused by two-photon absorption and free-carrier scattering – that renders the material lossy at the high optical intensities encountered in nanowaveguides and nonlinear optical devices. Two-photon photovoltaic effect is presented and discussed as a viable solution for achieving energy-efficient silicon photonic devices. The effect reduces the optical loss by free-carrier scattering and serendipitously harvests the optical energy lost to two-photon absorption.

1. Introduction

The prevailing vision for silicon photonics has been the introduction of photonics into the silicon CMOS manufacturing process. This vision has been motivated by the compatibility of silicon photonics with the CMOS process. To uncover the challenges that lie ahead, one is well-advised to probe the full implication of 'compatibility'. Full compatibility consists of material, process, economic and heat constituents. Material and process compatibilities are self-explanatory and are much easier to

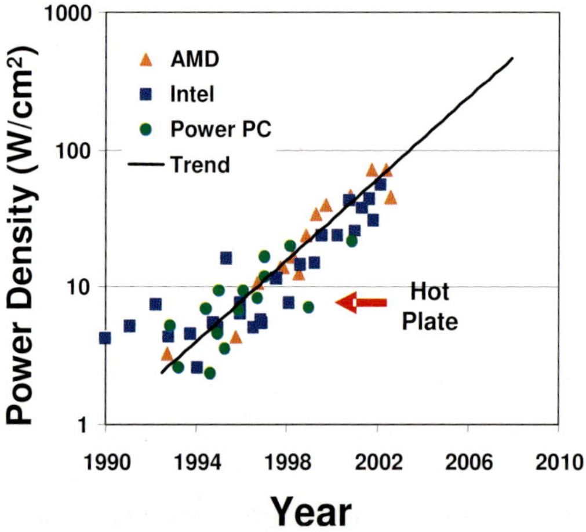

Fig. 1. The alarming increase in power density of VLSI chips. The problem of heat dissipation resulted in the recent shift of the microprocessor industry away from increasing the clock speed and in favor of multi-core processors. The figure highlights the fact that photonic devices must be able to operate on the hot VLSI substrate and that their own power dissipation must be minimal.[3]

satisfy than economic and heat compatibilities. Economic compatibility dictates highly efficient use of wafer real estate and products that have a high volume market. This constituent of compatibility is beyond the scope of this book chapter and the technological solutions offered to address it can be found elsewhere.[1,2] Heat compatibility requires that photonic devices must be able to operate on the hot very-large-scale integrated (VLSI) chip and that their own power dissipation is minimal. This chapter focuses on the heat compatibility constituent and the related energy efficiency issue faced by silicon photonics. In particular, it will be discussed how energy harvesting based on two-photon photovoltaic (TPP) effect demonstrates energy-efficient silicon photonics, particularly for nonlinear optical devices as well as high-speed nanophotonic devices.

Figure 1 shows the alarming increase in power density of VLSI chips, where today's level of 100 W/cm^2 challenges even the most advanced packaging technologies.[3] The problem of heat dissipation is so severe that it threatens to bring to halt the continued advance of the technology, as described by Moore's law.[4] This fact is highlighted by the recent momentous shift of the microprocessor industry away from increasing the clock speed and in favor of multi-core processors.[5] From a macro-scale point of view, energy consumption of millions of microchips used for electric communications in mega data centers around the world is a considerable component of global energy crisis. To highlight this issue, it should be noted that the electricity bill of these power-hungry data centers was $7 billion (corresponding to 14 gigawatts of power) in 2005 worldwide, 50% of which was consumed in information-technology equipments.[6] The observed trend over many years has been the replacement of electronics with optoelectronic communications for shorter and shorter links. The battle-line between metal and optical interconnects is currently for rack-to-rack communications and may possibly be for interchip and intrachip communications in the coming years. The unsettled question is whether optical interconnects will alleviate or aggravate the high power dissipation problem as they replace metal interconnects for such shorter links?

2. Energy Efficiency of Optical Interconnects vs. Their Metal Counterparts

There are a number of theoretical studies which attempt to answer the above vital question of energy efficiency in optical interconnects. Generally, they model the energy consumption of optical interconnects as a function of device pitch, energy per bit, power density, etc., and compare it with CMOS technology.[7-10] It is not our intention to review these works here nor do we intend to make another predication on the power limitation of optical interconnects under new assumptions. Rather, we qualitatively formulate the problem from a simple, yet fundamental, viewpoint.

The amount of information that can be sent through an optical channel increases with the bit rate for the same receiver sensitivity. On the other hand, to address the large footprint of silicon photonics devices, i.e., the aforementioned economic compatibility, smaller waveguide cross-sections or nanowaveguides are demanded. This leads to higher optical intensity for the same optical power (or the equivalent bit rate). Therefore, high optical intensity appears to be unavoidable in the

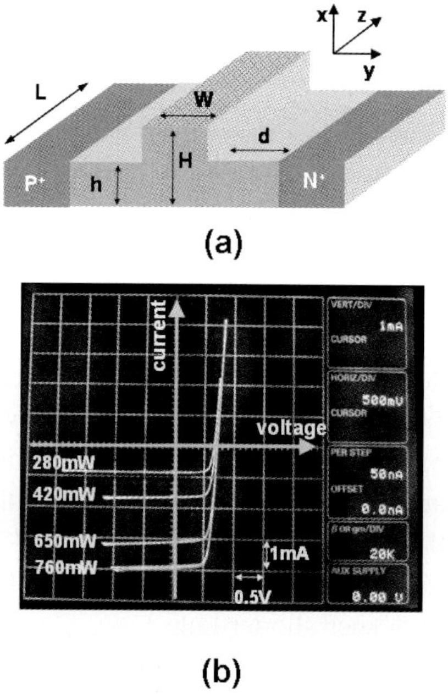

Fig. 2. (a) Schematic of a silicon waveguide with a *p-i-n* junction diode. The values of the shown geometrical dimensions are given in Table 1; (b) Current-voltage characteristics of the fabricated *p-i-n* diodes measured with a curve tracer at various coupled pump powers, showing photovoltaic effect in the fourth quadrant (Current < 0, Voltage > 0).

envisioned high-speed silicon nanophotonics schemes. Furthermore, a wide range of silicon photonic devices operate based on Raman and Kerr effects.[11-22] High optical intensity is a prerequisite for the onset of these nonlinear effects. Therefore, high optical intensity is encountered in nonlinear silicon photonics as well.

3. Energy Efficiency Crisis in Silicon Photonics

The inevitable high intensity imposes a twofold energy efficiency crisis. First, optical power is converted to heat through two-photon absorption (TPA) process. Second, free carriers generated by TPA must be actively removed from the waveguide core region, otherwise they will accumulate, cause severe free carrier absorption (FCA), and will prevent continuous-wave (CW) operation.[23,24] Carrier sweep-out is typically achieved using a reverse biased *p-n* junction.[14] This results in electrical power dissipation and adds to the optical power dissipation caused by the TPA. For example, in reference 14, about one Watt of electrical power had to be dissipated (25 V at ~ 40 mA) to achieve ~4 dB of CW optical gain and to produce ~ 8 mW of output from a Raman laser. Reducing the carrier lifetime, by introducing crystal defects, can alleviate the carrier accumulation.[25,26] However, eliminating the need for carrier sweep-out requires a lifetime that is on the order of 100 ps.[27] Presently, it is not clear whether such low lifetimes can be achieved without a pronounced increase in optical losses.[26]

We have recently proposed a new carrier sweep-out technique which not only does not require electrical power dissipation but generates electrical power too.[28,29] The technique is based on harvesting the energy of the photons lost to TPA. The core idea is that the prerequisite for carrier sweep-out from the waveguide is not negative voltage bias in a *p-n* junction but negative current. Thus, if a *p-n* junction that straddles the waveguide is biased in the fourth quadrant of its current-voltage (*I-V*) characteristics, the TPA-generated carriers are swept out by the built-in field of the junction, yet the device delivers electrical power. This novel TPP effect can be perceived as the nonlinear equivalent of the conventional photovoltaic effect of solar cells.

We have successfully applied this energy harvesting technique to three types of key silicon photonics devices, namely, Raman amplifiers,[28,29] Kerr-type wavelength converters,[30] and optical modulators.[31] Arguably, the technique can be perceived as a universal method for achieving energy-efficient silicon nonlinear devices. As discussed, high intensities can be encountered in high-speed nanophotonics, thus the same energy harvesting technique is in principle applicable to this emerging technology as well.

In the following sections, we first present a general model for energy harvesting based on TPP effect, followed by experimental results and the comparison between the two. Finally, some performance predictions and design optimization for maximum the efficiency of energy harvesting are presented.

4. Theory of Two-Photon Photovoltaic Effect

Energy harvesting in nonlinear silicon photonic devices is achieved by collecting the TPA-generated carriers by means of a *p-i-n* junction diode that is biased in the fourth quadrant (current, $I < 0$ and voltage, $V > 0$) of its *I-V* characteristics (Fig. 2b). Conceptually, TPP effect is a nonlinear equivalent of conventional photovoltaic effect used in solar cells but with certain differences. High optical intensities are required for the onset of the nonlinear effect and two photons are involved for generating one electron-hole pair, thus the collection efficiency cannot exceed 50%. Also, structural and geometrical differences with conventional photovoltaic devices demand alterations in the theoretical model that is typically used to describe solar cells.

As shown in Fig. 2a, TPP-effect devices are based on waveguides with lateral *p-i-n* junction diodes, in which the pump power is nonlinearly lost due to TPA and FCA along the z direction perpendicular to the junction x-y cross-section. Thus, unlike a solar cell which can be conveniently modeled as a one-dimensional (1-D) device,[32] the present problem requires at a minimum a 2-D approach. The nonlinear propagation of the coupled optical intensity, $I_p(z)$, is governed by:[24]

$$\frac{dI_p(z)}{dz} = -(\alpha + \alpha_{FCA}(z))I_p(z) - \beta I_p^2(z), \tag{1}$$

where β is the TPA coefficient and α is the linear absorption coefficient of the waveguide. The wavelength, λ, dependence of FCA coefficient, α_{FCA} (in cm^{-1}), is given by Soref's expression:[33]

$$\alpha_{FCA} = \Delta\alpha_e + \Delta\alpha_h = (8.5\times10^{-18}\cdot\Delta N + 6.0\times10^{-18}\cdot\Delta P)\cdot(\lambda/1.55)^2, \tag{2}$$

where ΔN and ΔP are the free electron and hole concentrations (in cm^{-3}), respectively, and are functions of $I_p(z)$ and bias voltage, V.

The carrier photogeneration rate from TPA in the waveguide core is $G = dN/dt = -(1/2E_p)dI_p/dz = \beta I_p^2/2E_p$, where E_p is the photon energy. The corresponding photocurrent per unit length is $qA_{eff}G$, where A_{eff} is the effective area of the waveguide and q is the electron charge. The total photocurrent in a waveguide with length L is expressed as

$$I_G = \frac{-q\beta A_{eff}}{2E_p} \int_0^L I_p^2(z)dz \equiv \frac{-q\beta A_{eff}}{2E_p} L_{NL} I_{p0}^2, \tag{3}$$

where I_{p0} denotes the coupled pump intensity at $z = 0$. The nonlinear photovoltaic effect is implicit in this last expression, as it is quantified in terms of the defined effective length,

$$L_{NL} \equiv \int_0^L \frac{I_p^2(z)}{I_{p0}^2} dz. \tag{4}$$

This important quantity is the nonlinear equivalent of the interaction length commonly defined in optical fibers.[34]

If the photodiode is biased in the fourth quadrant of its *I-V* characteristics (Fig. 2b), I_G can be collected with negative power dissipation, i.e., the energy of the pump lost to TPA can be harvested. In order to analyze the power generation and efficiency of this photovoltaic process, the contributions of carrier injection and recombination to the total current need to be considered. The minority carrier diffusion terms can be conveniently calculated from the celebrated Shockley equation:

$$I_D = qhL \times \left(\frac{D_n n_{p0}}{L_n} + \frac{D_p p_{n0}}{L_p} \right) (e^{qV/kT} - 1), \tag{5}$$

where h and L dimensions are depicted in Fig. 2a and the other parameters have the usual meanings.[35] Shockley equation is valid, since photogeneration in the N^+ and P^+ doped regions is negligible in the *p-i-n* diode. This contrasts the conventional solar cell theory, where photogeneration predominantly occurs in N and P doped regions and a more detailed solution of the minority carrier diffusion equations is required.[32]

Another implication of the *p-i-n* structure is that the recombination model in the intrinsic region of the present device resembles the one used for a forward biased thyristor (on state).[35] In this approximation of the Shockley-Read-Hall recombination, the rate is given by $R = \Delta N / \tau_R$, which is valid for $\Delta N = \Delta P >> n_i$. Ignoring the Auger coefficient, τ_R equals the ambipolar lifetime, $\tau_n + \tau_p$, where τ_n and τ_p are electron and hole bulk recombination lifetimes, respectively.[35] Consequently, the recombination current is given by

$$I_R = \frac{qh(W+2d)}{\tau_n + \tau_p} \int_0^L \overline{\Delta N(z)} dz \equiv \frac{qh(W+2d)LN_{eff}}{\tau_n + \tau_p}, \qquad (6)$$

where $W+2d$ is the width of the intrinsic region in Fig. 2a. The defined N_{eff} is an effective carrier density along z, similar to the discussed L_{NL}. It is implied in Eq. 6 that $\overline{\Delta N(z)}$ is an average value in the x-y plane.

Surface recombination at the Si/SiO_2 interfaces is an important effect in SOI micron- and submicron-waveguides and can indeed be a more dominant factor than bulk recombination in the effective lifetime, τ_{eff}.[36] A more elaborate analysis of TPP effect should thus consider the contribution of surface recombination to the total current. However, the values of surface recombination velocities are not easily measurable and vary considerably according to the employed fabrication technique and conditions.[37,38] Alternatively, the value of τ_{eff} can be estimated from waveguide loss measurements as well as from pump-probe experiments.[39] We have measured the lifetime in our fabricated waveguides using the latter technique, yielding a value of 15 ns. In the theoretical analysis of Eqs. 5 and 6, we assume that all the bulk and surface recombination processes are lumped into this measured value, i.e., $\tau_n = \tau_p = \tau_{eff} = 15$ ns. It is noteworthy that the effective lifetime value is smaller (~1 ns) in submicron waveguides.[40] As mentioned before, introducing lattice defects through implantation can decrease τ_{eff} too.[25,26] Effective lifetime values as low as 150 ps have been reported in submicron waveguides enhanced with argon implantation.[26]

The total photodiode current is expressed as $I_T = I_G + I_D + I_R$. The dissipated power is given by $P = V \times I_T(V)$. The bias V_m at which power generation is maximized in the fourth quadrant of power-voltage (P-V) characteristics is determined from $\partial P / \partial V = 0$. If $I_D, I_R << I_G$ at $V \approx V_m$,

$$P_m \approx -V_m \times I_G(V_m) = \frac{q\beta A_{eff}}{2E_p} V_m L_{NL} I_{p0}^2. \qquad (7)$$

The collection efficiency η_c is the number of collected carriers per number of photons consumed by TPA:

$$\eta_c = \frac{I_T}{2I_G} = \frac{1}{2} - hLE_p \times \frac{\left(\frac{D_n n_{p0}}{L_n} + \frac{D_p p_{n0}}{L_p}\right)\left(e^{qV/kT} - 1\right) + \frac{(W+2d)N_{eff}}{\tau_n + \tau_p}}{\beta A_{eff} L_{NL} I_{p0}^2}. \tag{8}$$

The factor of 2 in $I_T / 2I_G$ accounts for the fact that two absorbed photons create one electron-hole pair. Thus, η_c has a theoretical limit of 50%.

η_c is the appropriate efficiency for devices such as amplifiers and wavelength converters where energy harvesting is a useful byproduct but not the main functionality of the device. If TPP effect is intended to be used as a photovoltaic cell, power efficiency, η_P, becomes a more relevant figure of merit. First, the quantum efficiency of the device, η_q, is defined as the number of collected carriers per incident photon inside the waveguide:

$$\eta_q = \frac{E_p I_T}{qI_{p0} A_{eff}} = \frac{E_p}{I_{p0} A_{eff}} \times \left[\frac{\beta A_{eff} L_{NL}}{2E_p} I_{p0}^2 - hL \times \left(\frac{D_n n_{p0}}{L_n} + \frac{D_p p_{n0}}{L_p}\right)\left(e^{qV/kT} - 1\right) - \frac{hL(W+2d)N_{eff}}{\tau_n + \tau_p}\right]. \tag{9}$$

At the same approximate conditions that lead to Eq. 7, we obtain:

$$\eta_q(V_m) \approx \beta L_{NL} I_{p0} / 2. \tag{10}$$

It will be seen that Eqs. 7 and 10 are valid at low pump intensities. As expected, β is readily recognized as the most deterministic material property for attainable P_m and the corresponding η_c and η_q efficiencies. The dependence of η_c and η_q on *inactively* absorbed photons due to linear and free-carrier absorptions is implicit in L_{NL}. In other words, higher α and/or α_{FCA}, lead to smaller L_{NL}. If the coupling efficiency of light into the waveguide, $\eta_{coupling}$, is known, the external quantum efficiency can be written as $\eta_{ext} = \eta_{coupling} \times \eta_q$. Finally, the external

Table 1. Device Dimensions and Parameters and Material Properties

Parameter	Value	Unit
L	2.2	cm
W	1.5	μm
H	2.0	μm
d	1.9	μm
h	1.1	μm
N^+, P^+	$1\text{x}10^{19}$	cm^{-3}
E_p	0.8	eV
β	0.7	cm/GW
g_R	15	cm/GW
μ_n	1332	$cm^2/V\text{-}s$
μ_p	455	$cm^2/V\text{-}s$
τ_n, τ_p	15	ns
α	0.5	dB/cm
$\eta_{coupling}$	0.42	–

power efficiency is $\eta_P = \eta_{ext} \times qV / E_p$, which is also simply known as the "efficiency".[32] The corresponding approximate form at maximum power generation bias is

$$\eta_P \approx \frac{\eta_{coupling} q\beta}{2E_p} V_m L_{NL} I_{p0}. \tag{11}$$

Equivalently, Eq. 11 can be inferred from dividing P_m in Eq. 7 by the incident (uncoupled) optical power, $A_{eff} I_{p0} / \eta_{coupling}$.

In section 6, the quantitative predictions of the above theory is presented and compared with experimental results. A fully analytical analysis is, however, difficult to perceive because of the nonlinear nature of the problem apparent in the L_{NL} and N_{eff} quantities. Hence, these two values are evaluated using a commercial drift-diffusion simulator (ATLAS by Silvaco International), as follows. TPA was emulated by specifying a carrier generation rate at the waveguide core. The simulator provides the carrier concentrations as a function of V and I_p, which is averaged in the x-y plane of the waveguide core for being used in Eqs. 2 and 6. In order to accommodate for the nonlinear nature of the problem, a quasi-3D model was developed in which the 2-D results of ATLAS at several optical intensities were interpolated to numerically solve Eq. 1, and finally evaluate L_{NL} and N_{eff}.

A fully simulation-based analysis is also performed for comparison. The diode current per unit length, J (A/μm), is provided by ATLAS, from which the total current can be calculated from

$$I_T = \int_0^L J(I_p(z))dz.$$

It is assumed in this last expression that the ohmic loss of the contacts is negligible and that the current vector has no z-component. The latter is a fair approximation at low enough biases at which there is no voltage drop along the waveguide.

5. Energy Harvesting in Nonlinear Silicon Photonics

We have demonstrated energy harvesting in silicon Raman amplifiers and in wavelength converters based on Kerr effect.[28-30] The details of device fabrication and experiments can be found elsewhere,[28] but are briefly reviewed here for completeness. The dimensions of the fabricated devices depicted in Fig. 2a are reported in Table 1. Linear loss of $\alpha \sim 0.5$ dB/cm was measured in the waveguide using the Fabry-Perot technique.

Figure 3a presents the measured on-off Raman gain at different coupled pump powers and biasing conditions. The pump is an external-cavity tunable diode laser at 1539 nm with a linewidth of 0.1-0.2 nm amplified with an Erbium-doped fiber amplifier (EDFA). The Stokes signal is a 1673 nm distributed feedback (DFB) laser biased at an output power of ~4.5 mW. A maximum on-off gain of ~4.5 dB is obtained at a reverse bias of 15 V. A Raman gain of ~3 dB is attained when the diode is short-circuit (0 V), i.e., zero power dissipation.[41] More importantly, on-off Raman gains of more than 2 dB is measured, when the device is forward-biased at voltages ≤0.7 V. The importance of this biasing regime

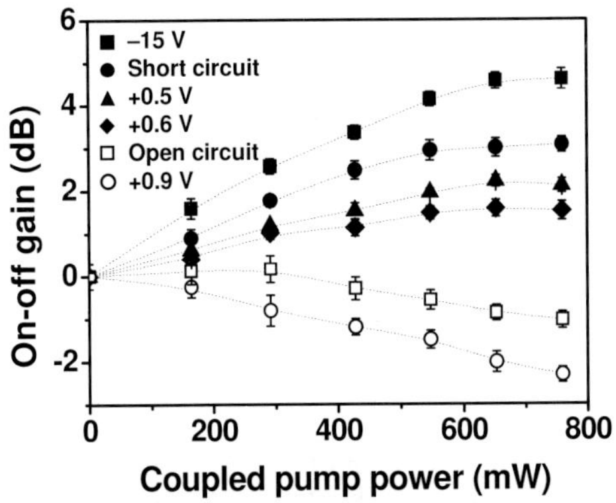

(a)

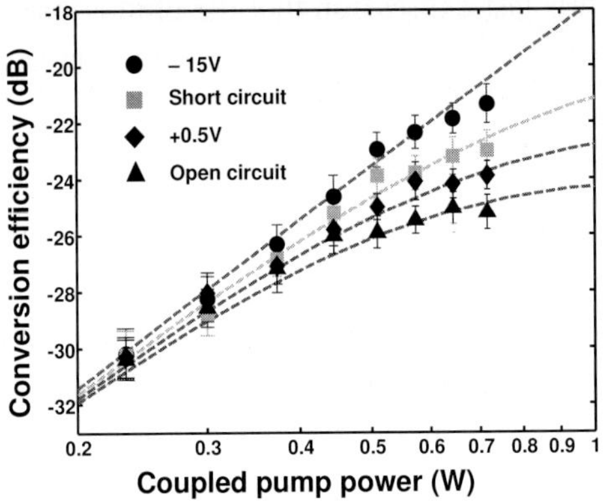

(b)

Fig. 3. (a) Measured on-off Raman gain versus coupled pump power in the devices of Fig. 1 at different biasing conditions; (b) Wavelength conversion efficiency as a function of coupled pump power under different biasing conditions. The wavelength detuning between the pump and the signal is 1 nm.

is that power dissipation is negative. For instance, ~ 2 mW of electrical power is generated at a bias of +0.5 V. In contrast, ~150 mW is dissipated at –15 V. Clearly, TPP effect offers an energy-efficient solution for nonlinear silicon photonics.

The TPA-induced photovoltaic effect is clearly evidenced in the fourth quadrant of the *I-V* characteristics in Fig. 2b, measured by a curve tracer. The attenuated pump intensity via TPA creates free carriers in the

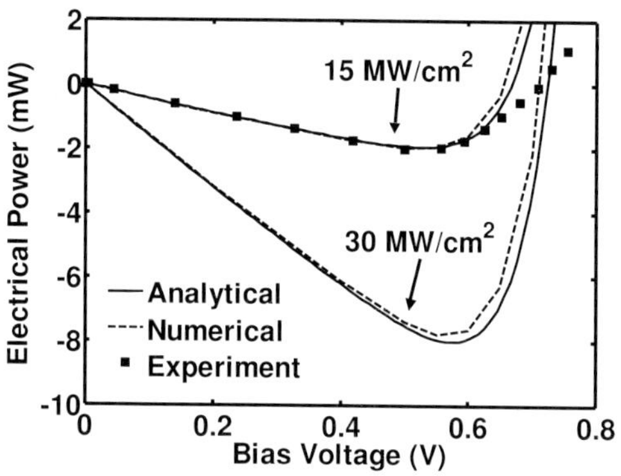

Fig. 4. Power-voltage characteristics of the diodes at two coupled pump intensities for: analytical expressions (solid line), numerical simulations (dashed line), and experiment (squares).

SOI waveguide. The collected photogenerated carriers contribute to a current component that delivers electrical power to the external circuitry. Therefore, the sweep-out of TPA generated free carriers by the built-in field of the junction can be exploited to generate electrical power.

Similar energy harvesting is demonstrated in wavelength converters, based on Kerr effect, on the same devices. The results are presented in Fig. 3b. The interesting feature is ~1.3-1.5 dB improvement in the conversion efficiency under a forward bias of +0.5 V at high pump powers as compared with the open-circuit case. The conversion efficiency can be greatly improved by engineering the waveguide dispersion and at higher optical intensities.[30] Comparison between the measured data in Figs. 3a and 3b with the corresponding theories of Raman and Kerr effects can be found elsewhere.[30,42] Finally, we have also proposed an electro-optical modulator that achieves negative static power dissipation by exploiting the TPP effect, which is not discussed further herein.[31]

6. Comparison of Theory with Experiments

The results obtained from the model developed in Section II for TPP effect are presented herein and compared with experiments. Figure 4 shows the theoretical *P-V* characteristics of the diodes in Fig. 2a with

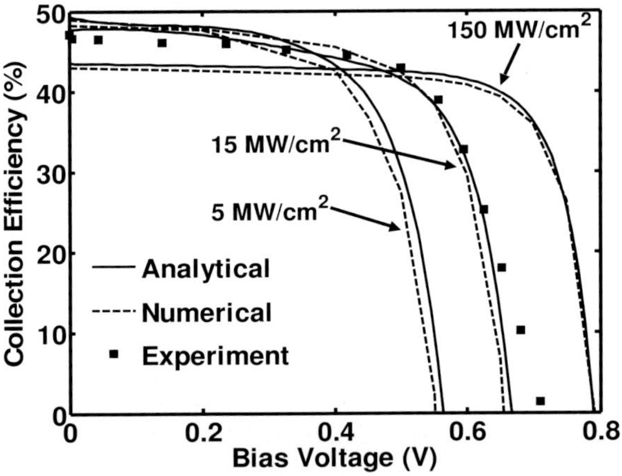

Fig. 5. Collection efficiency of the photovoltaic effect at three pump intensities for: analytical expressions (solid line), numerical simulations (dashed line), and experiment (squares).

identical dimensions to our experiments (Table 1). The material parameters used in the analysis are also summarized in the same Table. The agreement between the fully simulated results and the analytical model described in Section 4 is clear, thanks to the effective nonlinear quantities, L_{NL} and N_{eff}, borrowed for the analytical model from the simulations. The experimental *P-V* results – extracted from the measured current and voltage drop across the diode – are also presented for I_{p0} = 15 MW/cm^2. The theoretical results show excellent agreement with experiment at biases in the fourth quadrant of the *P-V* characteristics, which is the regime of interest in this work. The deviation between experiment and theory in the first quadrant is possibly due to the contact ohmic loss at high forward biases, which is ignored in the theoretical analysis.

Collection efficiency, η_c, is plotted versus voltage at different pump intensities in Fig. 5. Again, good agreement between the analytical model, the numerical simulation, and the experimental results is observed. The experimental value of η_c was extracted from the measured photocurrent and the estimated pump power inside the waveguide, obtained from waveguide loss measurements. As $V \rightarrow 0$ at lower pump intensities, η_c approaches the theoretical limit of 50% predicted by

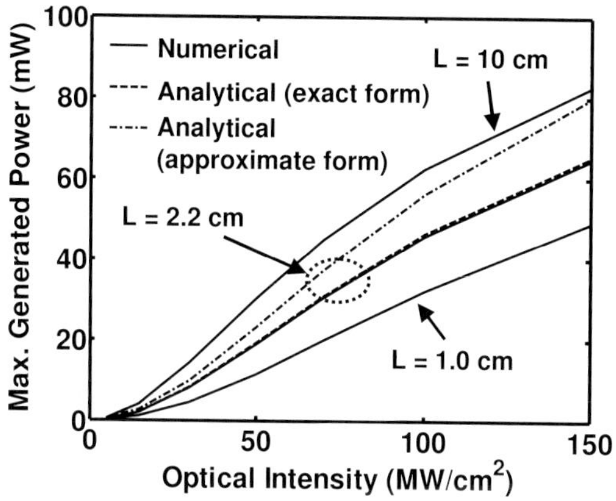

Fig. 6. Simulated maximum power generation (solid lines) versus coupled pump intensity for three different device lengths of 1.0, 2.2 and 10 cm. Also shown are the analytical values from the approximate form in Eq. 7 (dot-dashed line) and the exact expression in Eq. 12 (dashed line) for the 2.2 cm case.

Eq. 8. This is not the case for the high pump intensity of 150 MW/cm^2, at which $\eta_c \approx 43\%$ for $V \approx 0$. This decrease of η_c at $V \approx 0$ and high pump intensities is attributed to the decrease of nonlinear length, L_{NL}, as will be discussed further below. The value of η_c at maximum power generation bias, V_m, is extracted to be within 39-41%, and is nearly independent of coupled optical intensity from 5 to 150 MW/cm^2. Physically, this rather constant behavior is because the recombination current dominates the diffusive current at low biases and grows at approximately the same rate as the photogenerated current, i.e., $I_D << I_R, I_G$ and $N_{eff} \propto L_{NL} I_{p0}^2$ at $V \le V_m$ in (8).

7. Performance Predictions

Figure 6 presents the simulated maximum generated power versus coupled pump intensity for three different device lengths. The exact analytical values, obtained from

$$P_m = -V_m \times I_T(V_m), \tag{12}$$

and the approximation form in Eq. 7 are also shown (only for $L = 2.2$ cm case). The P_m values predicted by Eq. 7 deviates from the exact

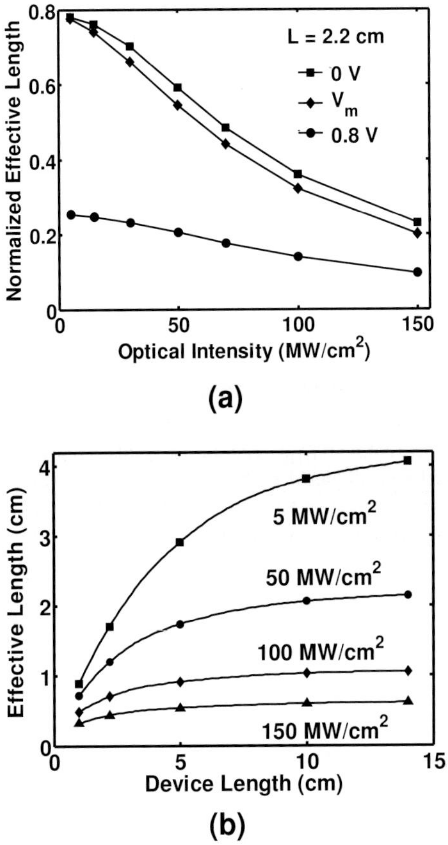

Fig. 7. (a) Normalized effective length, L_{NL}/L, versus coupled pump intensity at: short circuit or 0 V (squares), maximum power generation bias or V_m (diamonds) and a first-quadrant forward bias of 0.8 V (circles). The waveguide length is $L = 2.2$ cm and V_m varies from 0.43 to 0.62 V for the shown optical intensity range; (b) Effective length, L_{NL}, versus device length, L, at different optical intensities and at maximum power generation bias, V_m. Solid lines are fittings to the simulated data.

analytical and numerical values at higher intensities, nonetheless offers a convenient first-order estimate of the harvested power. A quadratic dependence of P_m on pump intensity may be expected from Eq. 7, which is not the case. The nonlinear decrease of L_{NL} with pump intensity is responsible for the rather linear dependence of P_m, as follows.

Figure 7a plots the dependence of the normalized effective length, L_{NL}/L, on the pump intensity at 0 V, V_m and 0.8 V for $L = 2.2$ cm. It is seen that at $V \le V_m$, the normalized length is 0.55-0.8 for intensities ≤ 50 MW/cm^2, but decreases to as low as 0.2 at 150 MW/cm^2. At a typical deep forward (first quadrant) bias of 0.8 V, current injection is so high that $L_{NL} < 0.25L$, even at very low pump intensities. This finding may be of interest in the design of variable optical attenuators and those silicon optical modulators that operate based on carrier injection in forward bias.[31,43] Moreover, a longer device does not necessarily increase the performance of the TPP effect. Figure 7b plots the effective length versus device length at different coupled pump intensities, showing a strong saturation trend, especially at high intensities. Strong saturation was also observed in the attainable harvested electrical power for $L > 10$ (not shown in Fig. 6). Physically, $L \sim 10$ cm is long enough for the absorption of virtually all the pump energy for $I_{p0} > \sim 50$ MW/cm^2.

We would now discuss the possibility of employing TPP effect for a photovoltaic cell. Targeted applications may be remote optical power delivery by fibers, free-space photonic power converters and optically powered sensors.[44,45] As mentioned before, power efficiency, η_P, is the key figure of merit for such applications.

The structure shown in Fig. 2a, with the dimensions given in Table 1, has a maximum power efficiency of ~2% attainable at I_{p0} ~ 100 MW/cm^2 and L $\ge$ 5 cm, assuming $\eta_{coupling}$ = 100% in Eq. 11. Higher efficiency can be achieved by optimizing the device structure and also by using high-reflection coating at the end facet of the waveguide to increase the average pump intensity along the waveguide. Figure 8a depicts three different structures studied for the optimization. Structure A is identical to Fig. 2a and the corresponding dimensions in Table 1. Figure 8b shows the power efficiency of structures A, B, and C at different doping levels (not all studied variations are shown). Structure C gives the maximum power efficiency of 5.5% at doping levels of $N = P \sim 1\times10^{17}$ cm^{-3}, $L = 5$, and I_{p0} ~ 100 MW/cm^2.

The above observations for micron-size waveguides are also generally valid for submicron waveguides. It is evident from Eqs. 10 and 11 that the efficiencies of TPP effect are strong functions of the optical

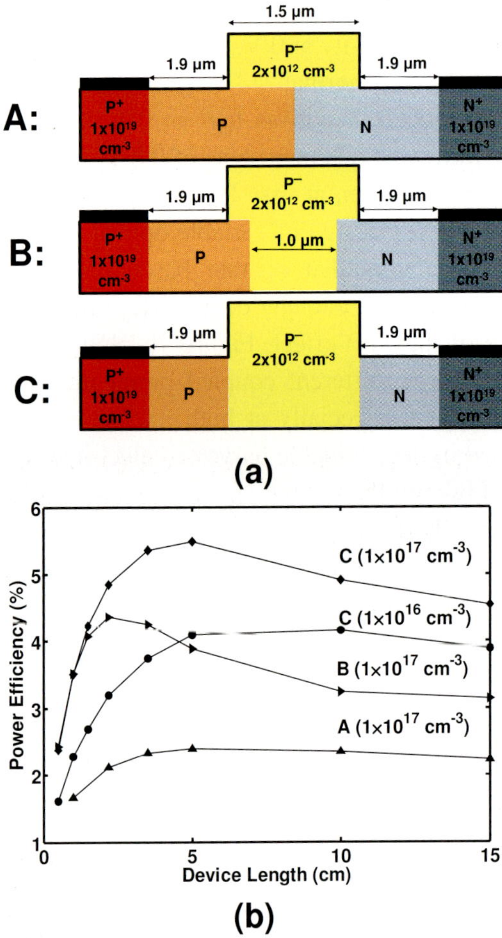

Fig. 8. (a) Schematics of the studied structures for maximizing the power efficiency of energy harvesting based on two-photon photovoltaic effect; (b) The power efficiency versus device length for different values of doping levels $N = P$, showing a maximum efficiency of 5.5% in device C and at a length of ~5 cm. In all the simulations, the input optical power intensity is 100 MW/cm^2 and it is assumed that the coupling efficiency into the waveguides is 100% and the output facet is coated with a mirror of 100% reflectivity.

intensity and the waveguide length but not the waveguide cross-sectional dimensions. This was confirmed by comparing the simulation results of micron-size waveguide given in Table 1 with submicron devices with $W = 0.5$ μm, $H = 0.6$ μm, $h = 0.25$ μm, and $d = 1.75$ μm. The results (not presented) confirm that the difference between the collection and power efficiencies of the two sets of devices is typically within a few percent of each other for the same optical intensity and device length as long as the $I_D, I_R << I_G$ at $V \approx V_m$ conditions, previously discussed in Section 4, are valid. This may not be the case in submicron waveguides if the surface recombination is very dominant (sub-nanosecond effective lifetime[26]) resulting in comparable I_R and I_G. It means that if nano-photovoltaic cells are intended, surface passivation precautions ought to be considered so that the photogenerated carriers are not recombined by the surface states.

The predicted relatively low inherent efficiency of ~ 5.5% in TPP effect is partially due to a 50% limit in the collection efficiency in Eq. 7 and partially due to the fact that two-photon scattering is generally not a very likely event as compared with interband single-photon absorption employed in solar cells. Any imaginable means that improves β can potentially enhance the power efficiency of the present approach. β is expected to be higher at shorter wavelength, e.g., 1.3 μm, as has been observed experimentally in GaAs.[46] In addition, FCA is lower at shorter wavelengths, according to Drude's model, which can increase L_{NL}. Therefore, the combination of the two effects should translate into higher limit efficiency of TPP effect, as predicted by Eqs. 10 and 11.

8. Conclusions

In summary, two-photon absorption and the resulting free-carrier scattering are omnipresent problems for energy efficiency of silicon photonics. Two-photon photovoltaic effect can be employed to harvest electrical energy from nanowaveguides and those silicon photonic devices based on nonlinear effects. A herein defined nonlinear effective length is recognized as an important parameter in the analysis and design of such devices. The electron-hole collection efficiency of the process is recognized to be ~40% and almost independent of the coupled optical intensity in the 5-150 MW/cm^2 range.

References

1. T. Indukuri, P. Koonath and B. Jalali, in *Optical Fiber Comm. Conf., OFC 2006*, Anaheim, CA, March 2006.
2. Research Highlights, *Nature* **440**, 718 (2006).
3. Courtesy of Eric Pop, Stanford University, personal communication.
4. D. J. Frank, *IBM J. Research and Development* **46**, 235 (2002).
5. International Technology Roadmap for Semiconductors (2005 Ed.): http://www.itrs.net/
6. Scott Ferguson, "Data Center Power Consumption on the Rise, Report Shows": http://www.eweek.com/article2/0,1895,2095409,00.asp
7. R. S. Tucker, *IEEE Phot. Tech. Lett.* **19**, 2036-2038 (2007).
8. P. Kapur, R. D. Kekatpure and K. C. Saraswat, *IEEE Trans. Electr. Devices* **52**, 1713 (2005).
9. J. H. Collet, F. Caignet, F. Sellaye and D. Litaize, *IEEE J. Selected Topics Quant. Electronics* **9**, 425 (2003).
10. A. V. Krishnamoorthy and D. Miller, *IEEE J. Select. Topics Quant. Electron.* **2**, 55 (1996).
11. R. Claps, D. Dimitropoulos, Y. Han and B. Jalali, *Opt. Express* **10**, 1305 (2002).
12. R. Claps, D. Dimitropoulos, V. Raghunathan, Y. Han and B. Jalali, *Opt. Express* **11**, 1731 (2003).
13. O. Boyraz and B. Jalali, *Opt. Express* **12**, 5269 (2004).
14. H. Rong, R. Jones, A. Liu, O. Cohen, D. Hak, A. Fang and M. Pannicia, *Nature* **433**, 725 (2005).
15. O. Boyraz, T. Indukuri and B. Jalali, *Opt. Express* **12**, 829 (2004).
16. I. Hsieh, X. Chen, J. I. Dadap, N. C. Panoiu, R. M. Osgood, S. J. McNab and Y. A. Vlasov, *Opt. Express* **14**, 12380 (2006).
17. L. Yin, Q. Lin and G. P. Agrawal, *Opt. Letters* **32**, 391 (2007).
18. R. Claps, V. Raghunathan, D. Dimitropoulos and B. Jalali, *Opt. Express* **11**, 2862 (2003).
19. R. L. Espinola, J. I. Dadap, R. M. Osgood, Jr., S. J. McNab and Y. A. Vlasov, *Opt. Express* **13**, 4341 (2005).
20. K. Yamada, H. Fukuda, T. Tsuchizawa, T. Watanabe, T. Shoji and S. Itabashi, *IEEE Photon. Technol. Lett.* **18**, 1046 (2006).
21. H. Rong, Y.-H. Kuo, A. Liu, M. Paniccia and O. Cohen, *Opt. Express* **14**, 1182 (2006).
22. M. A. Foster, A. C. Turner, J. E. Sharping, B. S. Schmidt, M. Lipson and A. L. Gaeta, *Nature* **441**, 960 (2006).
23. T. K. Liang and H. K. Tsang, *Appl. Phys. Lett.* **84**, 2745 (2004).
24. R. Claps, V. Raghunathan, D. Dimitropoulos and B. Jalali, *Opt. Express* **12**, 2774 (2004).
25. Y. Liu and H. K. Tsang, in *IEEE Conf. Lasers and Electro-Optics (CLEO)*, 2006, Paper CTuU6.
26. K. Yamada, H. Fukuda, T. Watanabe, T. Tsuchizawa, H. Shinojima, T. Tanabe, M. Takahashi and S. Itabashi, in *IEEE Group IV Photonics*, 2006, pp. 237–239.
27. B. Jalali, *Nature Photonics* **1**, 193 (2007).

28. S. Fathpour, K. K. Tsia and B. Jalali, *Appl. Phys. Lett.* **89**, 061109 (2006).
29. S. Fathpour, K. K. Tsia and B. Jalali, in *Optical Amplifiers and Their Applications (OAA 2006)*, 2006, paper# PD1.
30. K. K. Tsia, S. Fathpour and B. Jalali, *Opt. Express* **14**, 12327 (2006).
31. S. Fathpour and B. Jalali, *Opt. Express* **14**, 10795 (2006).
32. J. Nelson, *The Physics of Solar Cells*, Imperial College Press: London, UK, 2003.
33. R. A. Soref and B. R. Bennett, "Kramers-Kronig analysis of E-O switching in silicon," *SPIE Integr. Opt. Circuit Eng.*, SPIE vol. 704, 1986, pp. 32–37.
34. G. P. Agrawal, *Nonlinear Fiber Optics*. 3rd ed., Academic Press: New York, 2001.
35. S. M. Sze, *Physics of Semiconductor Devices*, 2nd ed., John Wiley & Sons: New York, 1981.
36. D. Dimitropoulos, R. Jhaveri, R. Claps, J. C. S. Woo and B. Jalali, *Appl. Phys. Lett.* **86**, 071115 (2005).
37. O. Palais and A. Arcari, *J. Appl. Phys.* **93**, 4686 (2003).
38. T. Kuwayama, M. Ichimura and E. Arai, *Appl. Phys. Lett.* **83**, 928 (2003).
39. R. Claps, V. Raghunathan, O. Boyraz, P. Koonath, D. Dimitropoulos and B. Jalali, *Opt. Express* **13**, 2459 (2005).
40. R. Espinola, J. Dadap, R. Osgood, Jr., S. McNab and Y. Vlasov, *Opt. Express* **12**, 3718 (2004).
41. S. Fathpour, O. Boyraz, D. Dimitropoulos and B. Jalali, in *IEEE Conf. Lasers and Electro-Optics (CLEO)*, 2006, Paper #CMK3.
42. S. Fathpour, K. M. Tsia and B. Jalali, *IEEE J.Quant. Electr.* **43**, 1211(2007).
43. J. Yu, in *IEEE Group IV Photonics Conf.*, 2004, pp. 139–141.
44. J. G. Werthen, S. Widjaja, T.-C. Wu and J. Liu, in *Proc. SPIE*, vol. 5871, 58710C-1–58710C-6, Aug. 2005.
45. J. G. Werthen, A. G. Andersson and T.-C. Wu, in *Proc. SPIE*, vol. 2872, pp. 131-138, Aug. 1996.
46. W. C. Hurlbut, K. L. Vodopyanov, P. S. Kuo, M. M. Fejer and Y.-S. Lee, in *IEEE Conf. Lasers and Electro-Optics (CLEO)*, 2006, Paper #QThM3.

CHAPTER 14

LIGHT EMITTING DEFECTS IN ION-IRRADIATED ALPHA-QUARTZ AND SILICON NANOCLUSTERS

Juhani Keinonen, Flyura Djurabekova and Kai Nordlund

Department of Physical Sciences, University of Helsinki
P.O. Box 43, FI-00014 Helsinki, Finland
E-mail: juhani.keinonen@helsinki.fi

Klaus Peter Lieb

Physikalisches Institut, Georg-August-Universität Göttingen
Friedrich-Hund-Platz 1, D-37077 Göttingen, Germany

Photoactive point defects and low-dimensional structures can be produced and tailored in α-quartz by ion implantation and irradiation. The luminescence bands are discussed in the frame work of intrinsic and ion-specific defects. The principles in the formation of damage in SiO_2 network produced by ion irradiation are discussed first. Then luminescent defects related to ion irradiation and nanocluster formation in α-quartz are described and discussed in comparison with those produced in noncrystalline silica. The microstructural properties are combined with luminescent data, and the role of photoactive defects and nanoparticles of the implants in α-quartz is discussed.

1. Introduction

Ion implantation is an important technique to dope SiO_2 with atoms and nanoparticles to obtain photoactive structures. An essential scientific and technological issue is to understand the defects responsible for the photonic properties of SiO_2.[1,2] Abundant information on photonic properties of ion-irradiated amorphous SiO_2 has been gathered during the latest decades and a large number of photoactive defect structures have

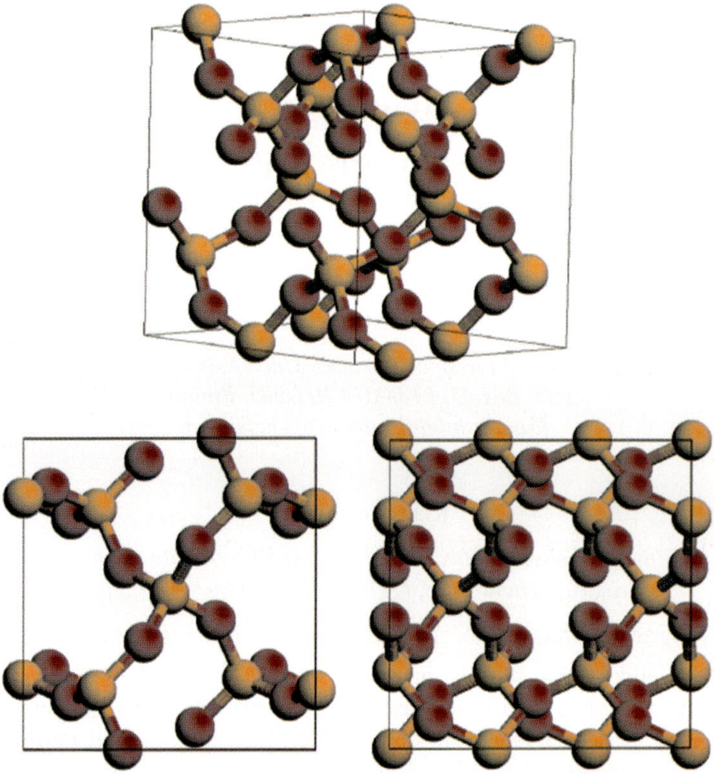

Fig. 1. Illustration of the crystalline structure of α-quartz. Shown are the unit cell (in the upper part) and the structures in the crystal directions [100] (on the left in the lower part) and [011] (on the right in the lower part). Silicon atoms are marked with the yellow color and oxygen atoms with the magenta color.

been proposed.[2–6] The situation is quite different for α-quartz, which becomes amorphous easily under the impact of low fluences of heavy ions.[7–10] Only recently, the many attempts for dynamic, laser-induced, and chemical solid-phase epitaxial growth of such amorphous layers have been successful.[11,12] The structure of crystalline α-quartz is illustrated in Figure 1.

Photoactive defects in SiO_2 are classified into intrinsic and ion-specific ones. Intrinsic defects are not related to the specific ions implanted, but to the defect structures in the SiO_2 matrix. They are produced in α-quartz and silica matrix after any ion or electron irradiation and thermal treatment.[2,6,13,14] In general, variations of ions and their fluencies and annealing conditions of the samples result in only weak variations of the luminescence arising from the intrinsic defects. The intrinsic defects consist of two groups, namely point defects and coordination ones. Vacancies and interstitials are well documented for oxygen defects.[2–6] The structures of silicon defects have been gradually elucidated.[15–17] A second kind of defects has been defined for noncrystalline silica, namely the broken bonds. The ion-specific photoactive defects in ion-irradiated and implanted SiO_2 are related to the implanted atoms or native impurity atoms and their clusters. The assignment of luminescent data to intrinsic and ion-specific defects in silica and α-quartz samples is aided by the results of modern theoretical simulations, see e.g. Refs. 18–24.

2. Ion-Irradiation Induced Damage in α-Quartz

All the ion-irradiation and implantation induced modifications of all the materials are correlated to the interactions of the energetic ions with electrons (electronic energy loss) and atoms (nuclear energy loss) of a target material. The interaction processes occur on a time scale of picoseconds and length scales as low as nanometres. The elastic collisions in the nuclear energy loss produce either highly damaged lattices following implantation or a network of damaged structures whose density depends on lattice reordering during the implantation process. In insulating materials like SiO_2, also the electronic stopping power induces radiation damage by breaking interatomic bonds.

Generally, the high deposited energy of up to tens of eV/Å, which each energetic ion deposits in the irradiated material along its track or in its collision cascade, may not only destroy the crystalline order but also induce self-organization of the microstructure, such as epitaxy, textures, and grain growth. Dynamic epitaxy, which is also called ion-beam induced epitaxial growth, is a process in which the radiation-induced damages in the crystalline matrix anneal out during the implantation process itself. Vacancy and self-interstitial diffusion in the collision cascade zone can restore the long range order of the matrix.

2.1. *Damage in the network structure*

The sensitivity of α-quartz as well as silica to the ion irradiation in producing defects is related to the network structure of SiO_2. The topology and connectivity of polyhedral network in SiO_2 can be characterized by the number of vertices in the polytope (V) and the average number of polytopes sharing each vertex (C). The numbers for SiO_2 are $\{V,C\} = \{4,2\}$. The structural freedom f of each vertex is the difference between the number of degrees of freedom d and the number of constraints h which are defined by the surrounding structure.[27–29] The structural freedom is

$$f = d - h = d - C\,\delta[1 - (\delta + 1)/2V]\,, \tag{1}$$

where δ is the dimensionality of the network-forming polytope. For SiO_2 $d = \delta = 3$ and therefore the structural freedom vanishes, i.e. $f = 0$. This is important for any transformation of an ordered network into a disordered network and *vice versa*.[30–32]

As both crystalline and amorphous SiO_2 consists of a fully connected network, it becomes most likely disordered by ion implantation and rather unlikely becomes crystalline in a thermal annealing. SiO_2 becomes disordered at a low energy (7 eV/matrix atom) deposited by the impinging ion or recoiling matrix atom, as demonstrated by comparing the energy with the values for Si (11 eV/matrix atom), SiC (26 eV/matrix atom), and Si_3N_4 (> 700 eV/matrix atom).[5,33,34]

2.2. *Phase structures in strongly damaged α-quartz*

The effects connected to the irradiation and implantation of α-quartz and silica with various types of projectiles have been studied extensively both experimentally and theoretically.[1,5,11,12,33,35–37] As a general feature in various particle irradiations, α-quartz becomes gradually disordered and evolves towards a single amorphous state *metamict*. This end product is almost independent of the irradiation process, i.e. whether neutrons, ions, electrons or photons are used for the irradiation. It is nearly indistinguishable from ordinary vitreous silica, but has a slightly higher

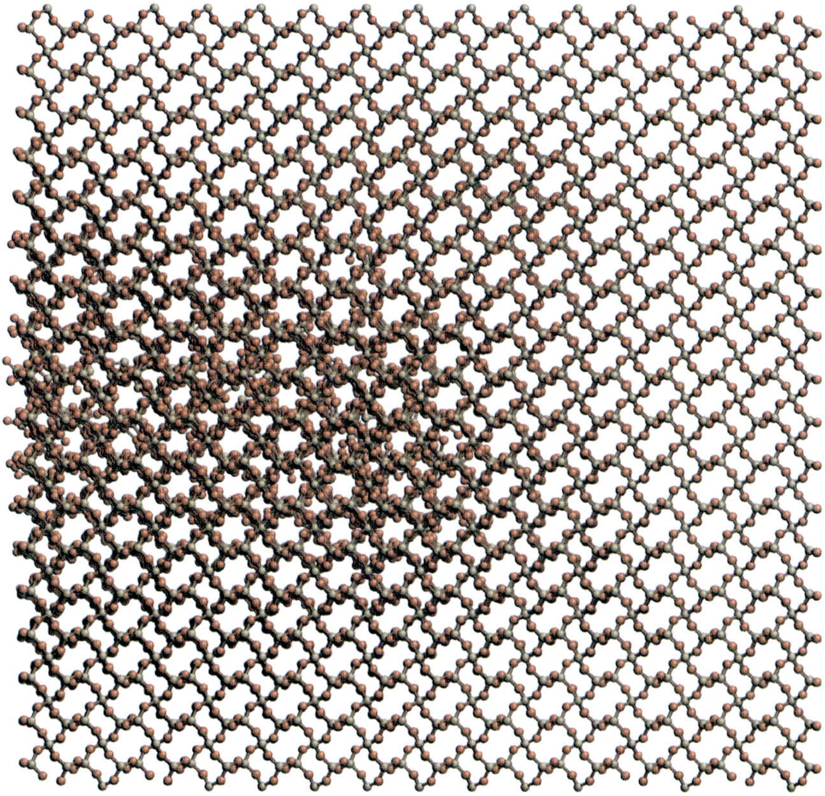

Fig. 2. Illustration of collision cascade damage in α-quartz during a Si implantation. The simulation has been done by a Molecular Dynamics technique[25,26] for one 2-keV Si atom entering α-quartz in random direction, from the left hand side in the figure. The collision cascade is shown at 0.4 ps after the entering of the ion in α-quartz.

density (3 - 4%).[37] With respect to ion energies and implanted atoms, the implanted atoms and relocated matrix atoms stay as isolated atoms and form atomic clusters or molecular-like structures. The damage regions are modified in annealing during and after the irradiation.

The lattice damage due to the deposited energy in ion implantation of α-quartz is illustrated in Figure 2. For a review of irradiation experiments on α-quartz, see Ref. 12.

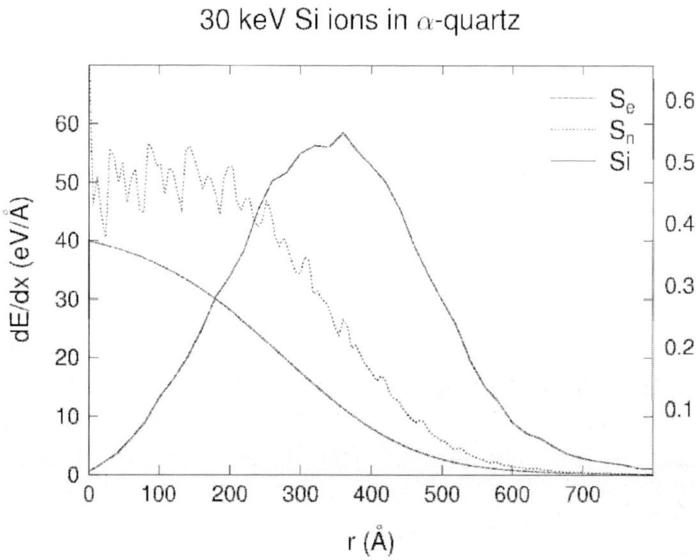

Fig. 3. Deposited energies dE/dx due to the nuclear S_n and electronic S_e stopping power and the range distribution of implanted Si atoms. The Molecular Dynamics simulations [25,26] show the deposited energies as a function of the depth for impinging 30-keV Si atoms in α-quartz in a random direction. The scale on the right hand side is an arbitrary scale for the Si-concentration distribution. For high dose implantations, the formation of nanoclusters takes place in the range region of the implanted atoms and mainly around the maximum concentration. For the nanocluster formation in implanted silica, see section 3.3.

3. Ion-Irradiation Induced Light-Emitting Defects in α-Quartz

With respect to ion energies, ion irradiation of silica introduces two types of defects. (1) A low-energy ion irradiation and implantation, governed by the nuclear stopping, remove atoms from their lattice sites in α-quartz and local SiO_2 structures in silica and introduces doped atoms. (2) A high-energy irradiation, governed by the electronic energy loss, induces the formation of transient defects, the so-called self-trapped excitons (STEs).[38] Nonradiative recombination of the STEs leads to the formation of stable point defects. The deposited energies in the electronic and nuclear stopping regions are illustrated in Figure 3.

With respect to ion energies and implanted atoms, very many types of luminescent defects are produced when the implanted atoms and relocated matrix atoms stay as dislocated isolated atoms and form atomic clusters or molecular-like structures. The defects are localized in the irradiation-induced damage, dispersed implanted atoms, atom and vacancy clusters, and at the cluster-matrix interface. The defects are modified in annealing during and after the irradiation.

The difference in the topologies of α-quartz and silica has the experimentally verified consequence that in silica irradiation-induced defects are produced much more efficiently than in α-quartz. The amorphous structure provides precursor sites for the creation of Si and O dangling bonds and new topological degrees of freedom. In ion-irradiated α-quartz, both crystalline and amorphous structures are present. The structures of the most common defects in amorphous silica are qualitatively very similar with those in α-quartz and are obtained in the far greater variety of different experimental results than those obtained in α-quartz. A review of the experiments on the luminescence of ion-irradiation induced defects in α-quartz is given in Ref. 12.

3.1. *Intrinsic point defects*

The self-trapped exciton (STE) is the most typical photoactive defect type discussed in the literature.[14,18–20] The interaction of radiation with SiO_2 causes electronic excitations that promote an electron from the valence band to the conduction band, leaving a hole in the valence band.[39] When an excited electron and a hole in the valence band remain bound together by Coulomb attraction, they are referred to as an exciton. An exciton can be trapped at a distortion in the lattice and is then called a self-trapped exciton (STE). It can be considered as a localized elementary mode of excitation with a broken symmetry of the surrounding network. The production of STEs is correlated with the presence of native matrix defects or the accumulation of irradiation-induced defects and causes an expansion of the lattice. But STEs can also be trapped at lattice distortions, which they spontaneously create themselves. They can facilitate bond breakings and lower diffusion barriers.[40] Silica glasses that support excitons under irradiation can even develop oxygen bubbles and eventually degrade to sand.[41]

Most of the experimental information on defects in silica comes from electron spin resonance studies done on the paramagnetic form of the defect produced from the intrinsic diamagnetic precursor under ionizing radiation. The high level of chemical order in SiO_2 has allowed defining very clearly the notion of vacancies and interstitials, as a missing or an extra oxygen atom on a given tetrahedron. By contrast the experimental knowledge on silicon defects is scarce. In silica the mixed ionic-covalent bonding leads to defects related to broken bonds, namely pairs formed of a threefold-coordinated silicon atom and of a singly coordinated oxygen atom.

In silica and α-quartz, the point defects can be grouped as dangling-bond-type and atomistic defects.[24] The dangling-bond-type defects include four types of defects:

(i) Paramagnetic E'-type centres. The neutral centre described by a dangling Si bond (≡Si•) is more abundant than the charged centre described by a hole trapped to the oxygen vacancy. The E' centre is by far the best-studied defect in SiO_2 because it is easily accessible to electron spin resonance.

(ii) A non-bridging oxygen-hole centre NBOHC. It contains a dangling oxygen bond (O≡Si–O•).

(iii) A two-fold coordinated Si atom centre or an ODC(II) centre (≡Si–O–Si–O–Si≡).

(iv) A peroxy radical centre POR, which contains a dangling oxygen bond (≡Si–O–O•).

Atomistic defects include three types of defects:

(i) An oxygen deficient centre in combination with a neutral oxygen vacancy ODC(I) or a silicon–silicon bond centre (≡Si–Si≡).

(ii) An interstitial oxygen or peroxy linkage POL centre (≡Si–O–O–Si≡).

(iii) An interstitial oxygen molecule O_2.

The dangling-bond-type centres are closely related to centres reported in the literature on surfaces and can be expected to be detected in amorphous silica but not in crystalline α-quartz. Atomistic defects can be expected to be roughly similar in non-crystalline silica and crystalline α-quartz.

In the following chapters, the luminescent green, blue, violet, and ultra violet (UV) bands observed in ion-irradiated α-quartz (see Ref. 12 for the original data) are discussed with respect to the intrinsic point defects (section 3.2) and ion-specific point defects (section 3.3). In sections 3.3 and 3.4, the defects related to nanoclusters in silica are compared with those obtained in α-quartz.

3.2. *Luminescence of intrinsic point defects*

Based on experimental data mainly obtained in electron and laser irradiation of silica,[14,43–51] particular defect configurations and luminescence mechanisms have been proposed for the intrinsic photonic bands.

The green band (at energy of ~2.4 eV) has been connected to oxygen-vacancy interstitial pairs (V_O;O_2)[14,50] and alternatively to irradiation-induced self-trapped excitons within the amorphous SiO_2 outgrowth at the top of the sample containing a large number of peroxy linkages (Si-O-O-Si).[43–46] The blue (~2.8 eV) and UV (~4.3 eV) bands have been observed in strongly damaged SiO_2 structures in fused quartz, SiO_2 grown on Si, and α-quartz amorphized by high-energy heavy-ion irradiation.[12] These bands have been connected to ODC(I)-type defects.[46,48] The blue band in the heavy-ion-irradiated amorphous α-quartz and in SiO_2 grown on Si is in accordance with a very strong blue band dominating the photoluminescence (PL) spectrum taken after heavy-ion irradiation of α-quartz.[52] The band is also connected to the interpretation that the radiative recombination of self-trapped excitons can produce the blue PL radiation.[53]

3.3. *Luminescence of ion-specific point defects*

Several ion-specific cathodoluminescence (CL) bands in the blue and violet regions have been reported in the literature for ion-irradiated α-quartz, namely a band connected to Ba-irradiation (~3.4 eV), three bands for Ge-irradiation (~3.0, ~3.3, and ~3.5 eV), and two bands for Na, Rb, and Cs-irradiation (~3.3 and ~3.7 eV). All the observed bands have not yet been assigned to defects.

The violet band at ~3.7 eV has been suggested[54] to arise from the presence of the defect centre $O_3{\equiv}Si{-}O{-}A$, i.e. an atomic configuration where the alkali ion (A) is connected to a dangling oxygen bond of the non-bridging oxygen-hole centre ($O_3{\equiv}Si{-}O\bullet$). The violet band at ~3.3 eV has been tentatively associated either with the positively charged alkali- (or hydrogen) compensated [A^{3+}/M^+] centre, where M^+ is H^+ or an alkali ion.[43–45] Based on the observation of a strong violet PL peak (~3.3 eV) in nanocluster experiments (see chapter 3.3.), an additional interpretation is that it is related to nanoparticles produced in implantation of group IV atoms.

3.4. *Atomistic models of embedded nanoclusters*

The implantation of group-IV atoms has been employed to produce PL nanoclusters (ncs) in silica. To understand the origin of nanocluster-related luminescence, atomistic models of Si-ncs embedded in silica have been constructed.[55] First realistic amorphous-SiO_2 (a-SiO_2) structures were constructed in two steps. At first, a Monte Carlo code[56] employing a simple Keating potential was used to construct an a-SiO_2 structure free of coordination defects. Thereafter, the resulting structure was relaxed by Molecular Dynamics method (MD) to room temperature and zero pressure using the Watanabe interatomic potential.[57] This potential has the advantage that it can describe in addition to the ideal bonding environment also coordination defects, and both pure Si as well as SiO_2 in the same functional form. This methodology was found to lead to a-SiO_2 structures with pair correlation and bond angle distribution functions in good agreement with experiments.

Si-ncs were inserted into the a-SiO_2 by cutting holes of the same radius as the desired cluster size in the middle of silica cells. In order to avoid the formation of unreasonably short bonds at the initial stages of the MD run, the Si-nc was compressed evenly over the volume by about 3 Å before the insertion. Thereafter a series of annealing runs combined with pressure monitoring was used to obtain a low-energy interface structure. A statistical variation was provided by random shifts of the

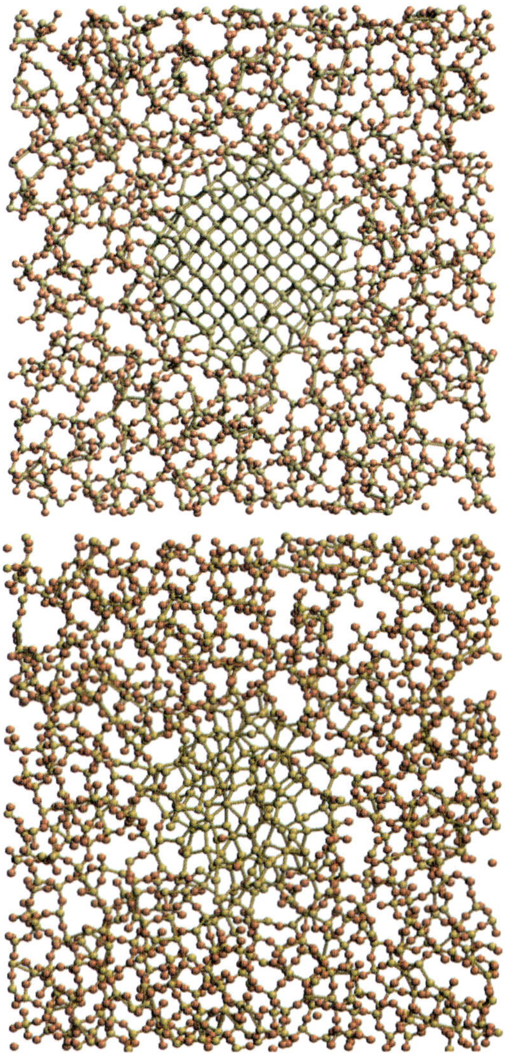

Fig. 4. Molecular Dynamics simulation[65] of the damage in an irradiation of Si-nc embedded in silica. The upper part shows an unirradiated nanocluster and the lower part a nanocluster irradiated with 1-keV Si-atoms to a dose corresponding to a deposited energy of 5 eV/(Si atom in nc), at which dose the amorphization takes place.

insertion position of the cluster. The resulting Si-ncs are illustrated in the upper part of Figure 4.

Table 1. Fraction of undercoordinated and overcoordinated atoms as well as suboxide Si atoms relative to the total number of atoms in the interface region for Si nanoclusters in a-SiO_2 created with a molecular dynamics approach. Also the average interface layer thickness is given. The values are averages over 10 different cells.

$d_{Si\text{-}nc}$ (nm)	Undercoord. (%)	Overcoord. (%)	Suboxides (%)	Thickness (Å)
2.4	8.4 ± 0.7	1.6 ± 0.2	10.2 ± 0.5	9.9 ± 0.8
3.6	10.5 ± 0.45	2.3 ± 0.2	11.1 ± 0.7	6.9 ± 0.2

The final structure was analyzed with respect to coordination defects and suboxide bonds (Si atoms with 4 bonds both to Si and O atoms). The thickness of the interface region was determined from the distance distribution (relative to the centre of the nanocluster) of coordination defects.

Some of the coordination defect analysis results for two different sizes of ncs are given in Table 1, after annealing at 1400 K. The interface layer thicknesses of 7-9 Å, depending on nanocluster size, are in excellent agreement with an experimental value of 8 Å (Ref. 58) for clusters in the same size range. The data in the table show that the total amount of coordination defects never exceeds 15% of the entire number of interface atoms. The number of overcoordinated defects is much smaller than that of the undercoordinated ones, while the fraction of suboxide bonds is similar to that of undercoordinated defects. Comparison of the results of the two sizes shows that the fraction of coordination defects or suboxide atoms is not strongly dependent on the cluster size. The interface layer thickness is clearly reduced going from a 2.4 nm to a 3.6 nm cluster since the larger cluster induces less strain on the matrix due to the larger radius of curvature.

3.5. *Luminescence of ion-specific point defects associated with nanoclusters*

Experimental investigations of the ncs with diameters of 1 – 10 nm consisting Si, Ge, and Sn in SiO_2 structures have shown that the emission mechanism is an electron transition located at the interface between a nc and SiO_2 matrix.[2,4] Violet and blue PLs observed in SiO_x layers have been reported to be due to radiative centres identified as neutral oxygen vacancies (NOVs or ODC(I) centres).[59–62] An optical transition of electrons in the centres has been interpreted to be the origin of the luminescence. The transition from the ground (singlet) to an excited (triplet) state is well known.[63] The NOV structure in the frame work of the oxygen-deficient Ge centre (GODCs) (≡Ge–Si≡) has been obtained by implanting Ge ions into SiO_2.[64] The GODCs and Sn-implanted ncs (≡Sn–Si≡) in SiO_2 emit the violet PL radiation (~3.3 eV),[65] while Si-ion implantation leads to the formation of Si NOVs (≡Si–Si≡) connected to the emission of the blue PL (~2.8 eV).[53] The violet (~3.3eV) line from SiO_x nanoparticles has been associated to Si–Si vibrations.[66] The violet band observed in the solid-phase epitaxial growth of ion-irradiated α-quartz has also been interpreted to be connected to the formation of the NOV structure ≡Si–Si≡.[54] The PL intensity of the single-crystalline Ge and Sn nanoparticles, observed in a high-resolution transmission electron microscopy, has been assigned to NOVs around the nanoparticles.[65,66] The formation of the NOV structures between group-IV atoms and indicated by the intensive violet (~3.3 eV) light, provides a strongly damaged SiO_2 matrix. The violet band appears to arise from either an intrinsic NOV structure (in the case of (≡Si–Si≡)), or an ion-specific one (in the cases of (≡Si–Ge≡) and (≡Si–Sn≡)).

3.6. *Quantum confinement and interface defects*

Light emission of Si nanoclusters (Si-ncs) is one of the prime examples of nanoscale phenomena related to quantum confinement.[67] The Si-nc band gap depends on the cluster size (1 to 10 nm), which influences the optical properties. After the first report on optical gain of Si-nc it was found that the amplified light pulses can be very short, which is of great importance for technological applications.[68] The preceding chapters have shown that the defects related to the luminescence emission in Si-ncs[66] are also related to defects obtained in strongly damaged ion-irradiated α-quartz. Now, the roles of the quantum confinement and interface defects are compared to what is known for luminescence emission in Si-ncs.[66]

The light emission from a quantum confinement is characterized by a band centered at about 500 nm (2.5 eV) and band in the wavelength range 600 - 900 nm (2.1 – 1.4 eV). The first band has been related to a defect structure. The second broad band is connected to the presence of the Si-nc, but the exact origin of the emission is not clear. The structure of the Si-nc characterized by three regions, namely the central region made of amorphous or crystalline silicon, the interface region made of substoichiometric and stressed silica and the embedding amorphous dielectric. Resonant photoluminescence experiments[68] indicate crystalline silicon phonon assisted recombinations. These data are interpreted in the frame work of a pure quantum confinement process. Other experiments have shown that the emission is strongly dependent on the exposure to ambient oxygen, which seems to support a key role played by silicon-oxygen bond in the emission processes. The most reasonable conclusion is that both mechanisms are co-present in Si-nc.

Figure 4 illustrates how the ion irradiation produces damage in the Si-nc and how the Si-nc dissolves in the silica. This demonstrates that the Si-nc – silica interface is in a dynamic change during an ion irradiation and that the end product is a complex defect structure.

4. Summary

Intrinsic and ion-specific defects in ion-irradiated and implanted α-quartz produce luminescent bands in the same wave length regions. The bands have been observed mostly in the blue and violet regions. Ion and electron irradiations produce intrinsic-defect bands which are related to low damage in electron irradiation and to the one corresponding to the amorphized silica matrix produced in the ion irradiation.

Although quantum confinement effects play a crucial role in emission from nanoclusters, the role of the surface and its defect structure cannot be discarded. To understand not only the quantum confinement-related microscopy of mechanisms but also the defects on the atomistic scale in the nanoclusters, at its surface, and at the nanocluster/dielectric interface require experiments in the whole wavelength range of 300 – 1000 nm (4.1 eV – 1.2 eV). Such systematic experiments have not been reported in the literature.

The demand to know atomic scale structure of Si-SiO_2 interface has grown as the size of electronic devices has shrunk into nanoscale dimension. This has led to theoretical studies on interface models with local structural properties related to electronic states. These studies will be of use to study photonic properties of Si-ncs embedded in amorphous and crystalline silica, as also indicated by Figure 4.

Acknowledgments

We are grateful for the financial support of the Academy of Finland in the FinnNano Program "Optical and Surface Properties of Nanoparticles".

References

1. P. Mazzoldi and G. W. Arnold, Eds., *Ion Beam Modifications of Insulators* (Elsevier, Amsterdam, 1987).
2. R. A. B. Devine, J.-P. Duraud and E. Dooryhee, Eds., *Structure and Imperfections in Amorphous and Crystalline Silicon Dioxide* (John Wiley & Sons, New York, 2000).
3. R. A. B. Devine, Ed., *The Physics and Technology of Amorphous* SiO_2 (Plenum Press, New York, 1988).
4. P. D. Townsend, P. J. Chandler and L. Zhang, *Optical Effects of Ion Implantation* (Cambridge University Press, Cambridge, 1994).
5. E. Lell, N. J. Kreidl and J. R. Hensler, *Radiation Effects in Quartz, Silica and Glasses* (Pergamon, New York, 1996).
6. L. Skuja, M. Hirano, H. Hosono and K. Kajihara, *Phys. Stat. Sol. (c)* **2**, 15 (2005).
7. L. M. Wang, S. X. Wang, W. L. Gong, R. C. Ewing and W. J. Weber, *Mat. Sci. Eng. A* **253**, 106 (1998).
8. K. Trachenko, M. Pruneda, E. Artacho and M. T. Dove, *Phys. Rev.* **B71**, 174207 (2005).
9. K. Trachenko, I. Todorov, M. T. Dove, E. Artacho and B. Smith, *Phys. Rev.* **B73**, 184104 (2006).
10. F. Harbsmeier and W. Bolse, *J. Appl. Phys.* **83**, 4049 (1998).
11. K. P. Lieb, in *Encyclopedia on Nanoscience and Nanotechnology*, Ed. H. S. Nalwa (American Scientific Publishers, 2004) vol. 3, p. 233.
12. K. P. Lieb and J. Keinonen, *Contemp. Phys.* **47**, 305 (2006).
13. E. Bridges, G. Davies, J. Robertson and A. M. Stoneham, *J. Phys.: Cond. Matter* **2**, 2875 (1990).
14. M. A. Stevens-Kalceff and M. R. Philips, *Phys. Rev.* **B52**, 3122 (1995).
15. G. Roma and Y. Limoge, *Phys. Rev.* **B70**, 174101 (2004).
16. N. Richard *et al.*, *J. Non-Cryst. Solids* **351**, 1825 (2005).

17. L. Martin-Samos, Y. Limoge, J. P. Crocombette, G. Roma, N. Richard, E. Anglada and E. Artacho, *Phys. Rev.* **B71**, 014116 (2005).
18. R. M. Van Ginhoven, H. Jónsson, K. A. Peterson, M. Dupuis and L. R. Corrales, *J. Chem. Phys.* **118**, 6582 (2003).
19. R. M. Van Ginhoven, H. Jónsson and L. R. Corrales, *Phys. Rev.* **B71**, 024208 (2005).
20. S. Ismail-Beigi and S. G. Louie, *Phys. Rev. Lett.* **95**, 156401 (2005).
21. S. Mukhopadhyay, P. V. Sushko, A. M. Stoneham and A. L. Shluger, *Phys. Rev.* **B70**, 195203 (2004).
22. S. Mukhopadhyay *et al.*, *Phys. Rev.* **B71**, 235204 (2005).
23. A. M. Stoneham, L. Gavartin and A. L. Shluger, *J. Phys.: Condens. Matter* **17**, S2027 (2005).
24. P. V. Sushko, S. Mukhopadhyay, A. S. Mysovsky, V. B. Sulimov, A. Taga and A. L. Shluger, *J. Phys.: Condens. Matter* **17**, S2115 (2005).
25. K. Nordlund, computer code parkas (2006).
26. M. P. Allen and D. J. Tildesley, in *Computer Simulation of Liquids* (Oxford University Press, Oxford, 1989)
27. A. R. Cooper, *Phys. Chem. Glasses* **19**, 60 (1978).
28. P. K. Gupta and A. R. Cooper, *J. Non-Cryst. Solids* **123**, 14 (1990).
29. P. K. Gupta, *J. Non-Cryst. Solids* **192**, 79 (1995).
30. L. W. Hobbs, *Nucl. Instr. Meth.* **B91**, 30 (1994).
31. L. W. Hobbs, *J. Non-Cryst. Solids* **192/193**, 79 (1995).
32. L. W. Hobbs, A. N. Sreeram, C. E. Jesurum and B. A. Berger, *Nucl. Instr. Meth.* **B116**, 18 (1996).
33. E. Kasper, I. Eisele and H. C. Evan, Eds., *Mat. Sci. Eng.* **B89** (2002).
34. J. P. Rino, I. Ebbsjö, R. K. Kalia *et al.*, *Phys. Rev.* **B47**, 3053 (1993).
35. L. Douillard and J. P. Duraud, *J. Phys. III France* **6**, 1677 (1996).
36. L. Douillard and J. P. Duraud, *Nucl. Instr. Meth.* **B107**, 212 (1996).
37. F. Piao, W. G. Oldham and E. E. Haller, *J. Non-Cryst. Solids* **276**, 61 (2000).
38. K. S. Song and R. T. Williams, in *Self-Trapped Excitons* (Springer, Berlin, Heidelberg, New York, 1992).
39. A. J. Fisher, W. Hayes and A. M. Stoneham, *J. Phys. Condens. Matter* **2**, 6707 (1990).
40. J. Song, L. R. Corrales, G. Kresse and H. Jónsson, *Phys. Rev.* **B64**, 134102 (2001).
41. W. J. Weber, R. C. Ewing, C. R. A. Catlow, *et al.*, *J. Mater. Res.* **13**, 1434 (1998).
42. R. Tohmon, Y. Shimogaichi, H. Mizuno, Y. Ohki, K. Nagasawa and Y. Hama, *Phys. Rev. Lett.* **62**, 1388 (1989).
43. L. N. Skuja and W. Entzian, *phys. stat. sol. (a)* **96**, 191 (1986).
44. L. Skuja, *J. Non-Cryst. Solids* **239**, 16 (1998).
45. L. Skuja, B. Güttler, D. Schiel and A. R. Silin, *Phys. Rev.* **B58**, 4296 (1998).
46. M. A. Stevens-Kalceff, *Phys. Rev.* **B57**, 5674 (1998).
47. M. A. Stevens-Kalceff, *Phys. Rev. Lett.* **88**, 3137 (2000).

48. H.-J. Fitting, T. Barfels, A. N. Trukhin and B. Schmidt, *J. Non-Cryst. Solids* **279**, 51 (2001).
49. H.-J. Fitting, T. Barfels, A. N. Trukhin, B. Schmidt, A. Gulans and A. von Czarnowski, *J. Non-Cryst. Solids* **303**, 218 (2002).
50. M. Yoshikawa, K. Matsuda, Y. Yamaguchi, T. Matsunobe, Y. Nagasawa, H. Fujino and T. Yamane, *J. Appl. Phys.* **92**, 7153 (2002).
51. M. A. Stevens-Kalceff and J. Wong, J., *J. Appl. Phys.* **97**, 113519 (2005).
52. J. M. Costantini, F. Brisard, G. Biotteau, E. Balanzat and B. Gervais, *J. Appl. Phys.* **88**, 1339 (2000).
53. G. G. Siu, X. L. Wu, Y. Gu and X. M. Bao, *Appl. Phys. Lett.* **74**, 1812 (1999).
54. J. Keinonen, S. Gąsiorek, P. K. Sahoo, S. Dhar and K. P. Lieb, *Appl. Phys. Lett.* **88**, 261102 (2006).
55. F. Djurabekova, M. Backman and K. Nordlund, *Nucl. Instr. Meth. Phys. Res. B* (2007), accepted for publication.
56. S. von Alfthan, A. Kuronen and K. Kaski, *Phys. Rev.* **B68**, 073203 (2003).
57. T. Watanabe, D. Yamasaki, K. K. Tatsumura and I. Ohdomari, *Appl. Surf. Sci.* **234**, 207 (2004).
58. L. Khriachtchev, S. Novikov and O. Kilpelä, *J. Appl. Phys.* **87**, 7805 (2000).
59. K. Awazu and H. Kawazoe, *J. Appl. Phys.* **68**, 3584 (1990).
60. L. Skuja, A. R. Silin and A. G. Boganov, *J. Non-Cryst. Solids* **63**, 431 (1984).
61. S. Munekuni, T. Yamanaka, Y. Shimogaichi, R. Tohmon, Y. Ohki, K. Nagasawa and Y. Hama, *J. Appl. Phys.* **68**, 1212 (1990).
62. H. S. Bae, *et al., J. Appl. Phys.* **91**, 4078 (2002).
63. N. Kuzuu, Y. Matsumoto and M. Murahara, *Phys. Rev.* **B48**, 6952 (1983).
64. J. M. J. Lopes, F. C. Zawislak, M. Behar, P. F. P. Fichtner, L. Rebohle and W. Skorupa, *J. Appl. Phys.* **94**, 6059 (2003).
65. J. M. J. Lopes, F. C. Zawislak, P. F. P. Fichtner, F. C. Lovey and A. M. Condó, *Appl. Phys. Lett.* **86**, 023101 (2005).
66. L. Pavesi, L. Dal Negro, C. Mazzoleni, G. Franzò and F. Priolo, *Nature* **408**, 440 (2000).
67. A. G. Cullis, L. T. Canham and P. D. J. Calcott, *J. Appl. Phys.* **82**, 909 (1997).
68. D. Kovalev, H. Heckler, G. Polisski and F. Koch, *phys. stat. sol. (b)* **215**, 871 (1999).

CHAPTER 15

AUGER PROCESSES IN SILICON NANOCRYSTALS ASSEMBLIES

Dmitri Kovalev

Department of Physics, University of Bath,
Bath BA2 7AY, United Kingdom
E-mail: D.Kovalev@bath.ac.uk

Minoru Fujii

Department of Electrical and Electronic Engineering,
Graduate School of Engineering, Kobe University,
Rokkodai, Nada, Kobe 657-8501, Japan
E-mail: fujii@eedept.kobe-u.ac.jp

Many properties of materials often undergo significant changes in a certain size range. For example, different forms of silicon nanocrystal assemblies have relatively high photoluminescence quantum yield due to morphological and quantum size effects. This chapter describes the main decay channel for photogenerated electron-hole pairs confined in silicon nanocrystals at high levels of optical excitation. We will argue that nonradiative Auger decay processes inherently limit quantum yield of photoluminescence of this system due to efficient competition with the radiation recombination. We further discuss implications of doping of silicon nanocrystals for their light emitting properties.

1. Introduction

Recombination of electrons and holes (*e-h* pairs) in semiconductors is a process by which both carriers annihilate each other: the electrons fall in one or multiple steps into the empty state which is associated with the hole. Both carriers eventually disappear in the process. This leads to one simple classification of the recombination processes: in the case of radiative recombination this energy is emitted in the form of a photon, in the case of nonradiative recombination via nonradiative centers it is transferred to energy of phonons or local vibration modes. These processes compete directly with each other and the ratio of their probabilities defines the quantum yield (QY) of the photo- or electroluminescence. Therefore in semiconductor optics the reduction of the number of lattice or surface nonradiative defects is a critical issue. However, there is an additional omnipresent intrinsic recombination process: nonradiative Auger recombination of *e-h* pairs. Auger recombination is a process in which an electron and a hole recombine in a band-to-band transition, but now the resulting energy is given off to another electron or hole. The involvement of a third particle affects the recombination rate so that Auger recombination has to be treated differently from a band-to-band recombination.[1,2]

2. Auger Recombination Processes

2.1. *Auger recombination in bulk semiconductors*

Band-to-band recombination depends on the density of available electrons and holes. Since both carrier types need to be available in the recombination process, the rate is expected to be proportional to the product of n and p. Auger recombination involves three particles: an electron and a hole which recombine in a band-to-band transition and give off the resulting energy to another electron or hole. We demonstrate this effect in Fig. 1. The expression for the net recombination rate is therefore similar to that of band-to-band recombination but includes the density of electrons or holes which receive the released energy from the electron-hole annihilation. Therefore, its rate scales as n^2p or np^2 and starting from certain concentration of generated carriers nonradiative Auger recombination will dominate recombination statistics.[1,2]

The Auger process is a predominant nonradiative process in semiconductors at high carrier or doping concentrations. In semiconductors Coulomb interaction between electrons and holes

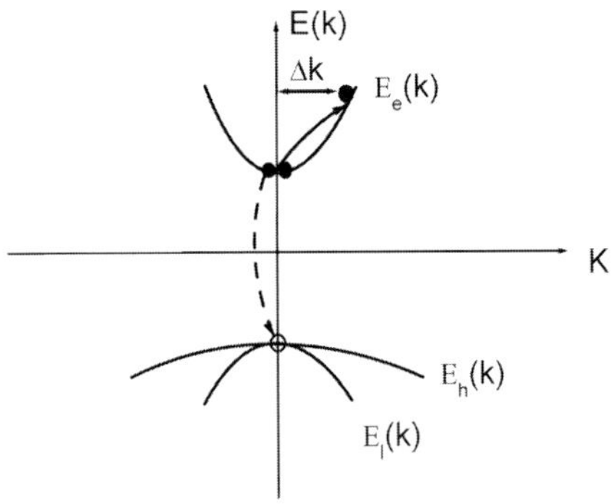

Fig. 1. Sketch of the Auger recombination event. Energy of an annihilated *e-h* pair is transferred to the remaining electron. Gain in crystal quasimomentum Δk is indicated by the horizontal arrow. Similar process is possible for two holes and an electron.

(responsible for Auger processes) is much smaller than in atoms due to dielectric screening and is weaker than their interaction with the electromagnetic field (responsible for the light emission). For direct bandgap semiconductors due to very fast radiative recombination it becomes only important at carrier concentrations above 10^{19} cm^{-3}. However already special care is taken to suppress this effect in semiconductor quantum-well lasers at high injection levels.[3–5]

Bulk Si, due to its indirect band-gap structure and nonpolar lattice is a spectacularly poor light emitter. Since the radiative time of the indirect transitions is extremely long and the transport of excitons is efficient, the main decay channel for free excitons or *e-h* pairs at low temperatures and low excitation levels is their capture by doping impurities forming bound exciton states or by nonradiative defects. This results in a very low QY of the light emission. Even at liquid He temperatures it is of the order of 10^{-4}-10^{-6}.

While reduction of photoluminescence (PL) QY due to carriers capture to nonradiative centers is an obvious effect, the effect of carries capture to shallow neutral impurities requires further explanation. Bound exciton is a localized state and at first glance its PL QY should be equal to 100%. Nevertheless, its value has been found to be extremely low as well. Schmid[6] found that there is a clear correlation between measured lifetimes of excitons bound to the neutral donors Li, P, and As and neutral acceptors B, Al, Ga, and In and the binding energy of the impurity. Depending on the impurity binding energy and, therefore, on spatial localization of carriers measured bound exciton lifetimes range from nanoseconds (for the case of strong localization) to microseconds for smaller spatial localization. The spatial confinement of three carriers in the vicinity of a charged impurity center leads to a very high effective *e-h* concentration of the order of ~10^{18}-10^{19} cm^{-3}. Therefore, the quantum efficiency for bound exciton transitions is determined by the ratio of the indirect optical transition probability to the nonradiative Auger process probability and is extremely small.[6,7]

Nonradiative Auger process is also inherent limitation for the light emission from intrinsic bulk Si. In 1968, Blinov *et al.* [8] have used the free carrier absorption and photoluminescence as measures of the carrier concentration in optically excited silicon. The decay of light emission was not studied in detail but the recombination was observed to be very rapid during the excitation, i.e. at high carrier concentration. Their results showed that a third-order (Auger) recombination process was active at high carrier concentrations. Free carrier absorption measurements in bulk Si gave results which also indicated a third-order process. Afterwards all details affecting the efficiency of Auger process: doping concentration, type of doping, excitation power dependence, temperature influence etc. have been subjected to in-depth scrutiny.[9–13] It has been demonstrated that for intrinsic bulk Si when surface recombination centers are well-passivated Auger processes start to play important role at carriers concentration as low as 10^{15} cm^{-3}.[14,15] It is interesting to note that Auger processes not only limit the QY of bulk Si PL but also the efficiency of Si-based solar cells. Green [16] and Tiedje *et al.* [17] showed that Auger recombination limits peak energy conversion efficiency to ~ 30%.

In a bulk semiconductor the efficiency of Auger processes is restricted by the conservation of the total momentum. At low temperatures to compensate the final large crystal quasimomentum of the excited hot electron (Δk in Fig. 1) or hole, a phonon or phonons have to be emitted or absorbed to conserve crystal quasimomentum. This scenario is very similar to that for indirect optical transitions. However, Si is a nonpolar material and, therefore, electron-phonon interaction in Si is weak and the probability of phonon emission/absorption is small. At elevated temperatures, however, there is a finite probability to find a required combination of initial crystal quasimomenta of carriers distributed within the band according to Maxwell distribution and this can fulfill the momentum conservation law. This probability depends exponentially on the temperature and is usually regarded as the kinematic temperature threshold:[2] in bulk semiconductors the probability of Auger effect rises with temperature.

2.2. *Auger recombination in low-dimensional semiconductors*

Auger processes play an important role in light emitting properties of all low-dimensional semiconductor structures. According to the uncertainty principle localization of carriers in geometrical space would imply immediately their delocalization in the momentum space. Thus, the momentum conservation requirements for Auger processes are partially lifted[2] and the kinematic temperature threshold inherent for bulk semiconductors is reduced. Nonthreshold Auger recombination can have an important influence on the optoelectronic characteristics of quantum well and quantum wire lasers. Momentum conservation is violated at heterosurfaces in nanosized quantum well structures. This causes the non-threshold Auger recombination to occur and at low temperatures it becomes the leading recombination channel.[2] Finally, for highest level of carrier localization, realized in semiconductor nanocrystals/quantum dots, again due to the violation of momentum conservation, this threshold can be almost absent.

It is important to note that at low excitation levels when no more than one *e-h* pair is present in a nanocrystal, the Auger rate is zero in contrast to bulk semiconductors, where the Auger rate $W_{Auger}=An^2p+Bp^2n$ is not zero at any carrier concentration. In that sense, at low temperature and weak excitation power, the absence of carrier transport in nanocrystal assemblies makes the Auger process less efficient compared with a bulk semiconductor. On the other hand, at intense excitation, when a double occupation of a nanocrystal can be achieved, the local concentration of 2 *e-h* pairs in nanocrystals due to the geometrical confinement is very high: of the order of 10^{18}-10^{20} cm^{-3}. This estimate combined with the size dependence of quasi-momentum conservation law gives the overall size dependence of the Auger rate: $W_{Auger} \sim a^{-7}$, where a is the radius of the nanocrystal.[2] Even in direct band-gap quantum dots with radii of the order of a few nanometers this rate is expected to be larger than that of radiative transitions [18] and to a large extent determines the recombination statistics.

Fluorescence intermittency, also called blinking, is a widely observed property of single fluorophores, from organic dye molecules to colloidal semiconductor nanocrystals and polymer nanoparticles. Under continuous optical excitation, the light emission from individual nanocrystals switches to "on" and "off" states, with "on" or "off" events lasting from microseconds to minutes. This behavior has been widely studied in spherical CdSe nanocrystals, both experimentally [19] and theoretically.[20] Later a similar behavior has been observed in other nanocrystal systems. The random and uncontrollable nature of blinking is a major obstacle to optoelectronic applications of nanocrystals such as lasers and single-photon sources. Consequently, understanding of this phenomenon may facilitate many single-NC applications. Although the mechanism of blinking is not fully understood, it is believed that nanocrystals become "dark" when one of the charge carriers is either trapped at the surface of the nanocrystals or excited at the level where it can escape into the environment, leaving a hole localized in the nanocrystal core.

The most prominent feature of Auger recombination is the energy gain of the carriers. If two *e-h* pairs occupy simultaneously the same nanocrystals, one pair will recombine nonradiatively via the Auger process. One of the carriers left gains the energy of the recombined pair and if this energy is large enough it can escape from the crystallite to the nanocrystal surrounding or be trapped at the surface of a nanocrystal. The unpaired charge carrier in the core can then effectively quench PL by means of Auger recombination. In this picture, proposed by the authors of Ref. 20, the nanocrystal remains "dark" until the trapped electron returns to the interior of the nanocrystals. This model is also consistent with experiments performed in core-shell quantum dot structures where wide bandgap shell prevents ionization of nanocrystals.

All the effects described above are based on the competition between radiative and nonradiative decay channels. In Si nanocrystals, the radiative decay rate is tremendously increased with respect to bulk Si. However, even for nanosilicon radiative decay of excitons is still very slow and can not compete with fast Auger processes. As we shall see this gives rise to a large variety of optical phenomena: PL saturation, spectral hole burning, autoionization of Si nanocrystals and strong dependence of the emission QY for doped nanocrystal systems. It also inherently limits the luminosity of different systems containing Si nanocrystals.

3. Silicon Nanocrystals Assemblies: Main Observations

3.1. *Morphological properties of Si nanocrystals assemblies*

To prepare nanostructured materials two simple approaches can be used. Nanostructures can be prepared from atomic or molecular precursors in gas or liquid phase. Another approach relies on reducing the dimensions of bulk materials. This can be done using standard methods: photolithography, electron beam lithography, chemical or electrochemical etching, *etc*. Historically, nanostructured Si was first produced by Arthur Uhlir at Bell Labs in the 1950's.[21] Despite the observation of PL from porous silicon (PSi) at cryogenic temperatures in 1984,[22] wide attention to optical properties of PSi and other nanosilicon-containing systems was drawn only after reports on visible light emission at room temperature by Canham[23] and on blue shift of the absorption by Lehmann and Gösele.[24]

An ideal functional device should combine electronic and optoelectronic components in the same chip. However, bulk Si, being a main material for the semiconductor industry, due to its indirect band-gap electronic structure, is a very inefficient light emitter. The reduction of size has been considered as the most promising approach to improve the efficiency of light emission from silicon-based structures. The key idea is that size reduction should result, due to quantum size effects, in a tunable energy of excitons and a partial breaking of the indirect band-gap nature of bulk Si.[25,26]

Si nanostructures can be produced according to different technological procedures. We demonstrate a few of them in Fig. 2.

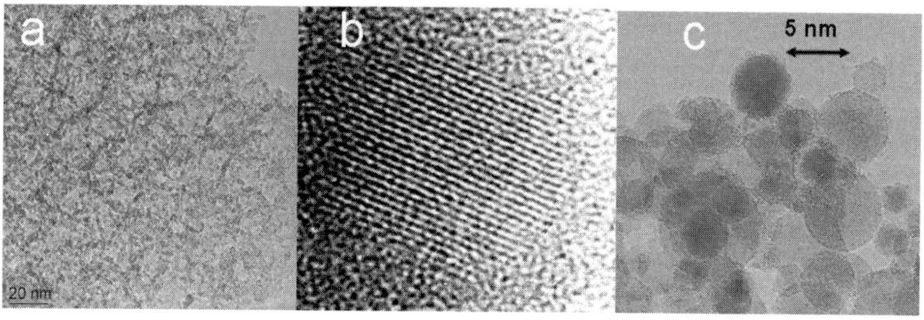

Fig. 2. HRTEM images of different types of Si nanocrystals assemblies. (a) Porous Si powder. (b) Si nanocrystals in SiO_2 matrix prepared according to the procedure described in Ref. 31. (c) Si nanospheres synthesized from the gas phase.

Structural investigations have confirmed that they all consist of Si nanocrystals of different size (typically a few nm) and shape that retain the diamond lattice structure of bulk Si. The most widely discussed system is PSi[23–27] prepared via anodization of bulk Si wafers in HF-based solutions. Other examples include Si nanocrystal assemblies prepared via ion implantation in a SiO_2 matrix,[28] by reactive Si deposition onto quartz,[29] by plasma enhanced chemical vapor deposition [30] and by magnetron sputtering[31] (for further details see comprehensive review [32]). Recently, a laser pyrolysis technique has been introduced to achieve relatively narrow size distribution of Si nanocrystals.[33] Probably, the simplest preparation procedure of nanosilicon proposed so far is stain etching of bulk crystalline or polycrystalline Si wafers and films.[34]

The interest in luminescence properties of Si nanocrystal assemblies was caused not only by the demonstration in 1990 that PSi can emit visible PL very efficiently at room temperature[23] but also by the confidence that Si-based efficient light-emitting device operating in the visible range can be realized. Since this time, much progress has been made. All features of the structural and electronic properties of the material have been subjected to in-depth scrutiny.[25–27] Optical properties of other nanosilicon-based systems were also studied in great details and they were found to be almost identical to those of PSi: similar tunable spectral range of the PL, exciton lifetimes and their spectral dispersion, similar temperature dependence of exciton lifetimes, *etc.*

3.2. *Optical properties of Si nanocrystals*

Bulk Si is very inefficient light emitter due to very small oscillator strength of optical transitions and efficient capture of excitons or *e-h* pairs to nonradiative centers or doping impurities. In a nanocrystal, the situation changes. Firstly, spatial confinement according to the prediction of quantum confinement model has to shift both absorbing and luminescing states of nanocrystals to higher energies due to a rising of the minimum allowed kinetic energy of confined charge carriers. Secondly, according to the uncertainty principle, the geometrical confinement leads to a delocalization of carriers in the crystal quasi-momentum space thus allowing zero phonon optical transitions.[25,35,36] Thirdly, due to the better overlap of electron and hole envelope wave functions one can expect a strong enhancement of the oscillator strength of optical transitions [25,26] and of the *e-h* exchange interaction inducing a splitting of the exciton levels.[25,37] Finally, since photoexcited carriers are strongly geometrically localized in nanocrystals they were created, recombination has the geminate character. Therefore the recombination statistics for Si nanocrystals assemblies is quite different from that used for bulk crystals. Measured PL lifetimes can be considered to a large extent to be radiative. This can be understood by regarding Si nanocrystal assemblies as a granular-like materials consisting of luminescing (internal quantum yield is equal to 1) and dark nanocrystals. Crystallites of the first type belong to those which do not contain nonradiative centers while the second ones have at least one surface nonradiative defect. Under this assumption the observed PL decay time is the time of radiative recombination.[38,39]

Clear evidence that emitting states of different types of Si nanocrystals assemblies are driven to higher energies by the confinement is an observation that energy of PL maxima correlates very well with Si nanocrystal size distribution.[26,32] The PL, depending on the mean size of Si nanocrystals, can be continuously tuned with small increments over a very wide spectral range from the Si band-gap to almost the green region. Despite PL measurements performed on individual Si nanocrystals have confirmed their atomic-like emission,[40] PL from assembles of Si nanocrystals is very broad due to residual nanocrystals size and shape

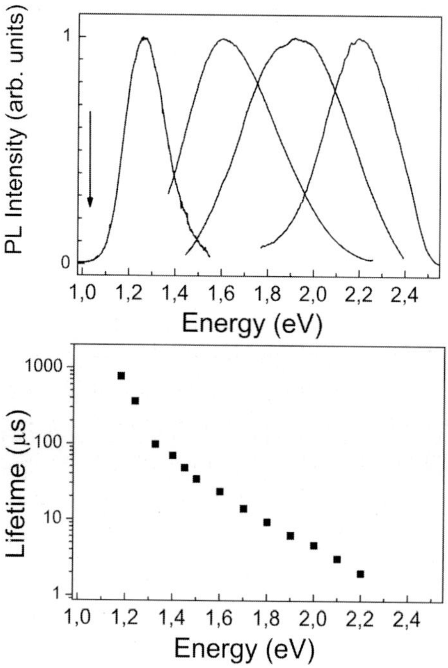

Fig. 3. Upper part: PL spectra of PSi samples having different nanocrystal size distributions. Lower part: spectral dispersion of singlet exciton lifetime. T=200 K. $E_{ex.}$=3.67 eV.

distribution.[25,26] We demonstrate in Fig. 3 the spectral range of the emission from different PSi samples (larger energy implies smaller nanocrystals sizes).

In bulk silicon, the radiative lifetime of electron-hole excitations is extremely long and can not be measured directly since the nonradiative processes dominate the recombination statistics. The contribution of nonradiative processes results in an extremely low optical emission quantum yield. Already first measurements of the temporal evolution of the PL emitted by PSi demonstrated that the exciton lifetime, depending on the temperature, is in the microsecond – millisecond time domain.[41] For the comparison, in low-dimensional direct band gap semiconductors radiative exciton lifetimes are in the nanosecond time domain.[42] The geometrical confinement of the exciton results not only in a blue shift of the optical transitions but also in a modification of the oscillator strengths of no-phonon and phonon-assisted processes. In smaller nanocrystals (having larger confinement energies) both zero-phonon transition (due to an efficient breakdown of crystal quasi-momentum conservation rule) and phonon-assisted transition probabilities (due to a better overlap of *e-h* wave functions) should be significantly increased. This can be seen experimentally as a shortening of the exciton radiative lifetime. We illustrate this effect in the lower part of Fig. 3 which demonstrates the spectral dependence of the PL decay time measured at 200 K when the PL quantum yield has a maximum value. Lifetimes measured for other systems containing Si nanocrystals have very similar values. In this temperature range the PL decay time to a large extent can be considered as radiative.[38,39] The PL decay time varies from almost a millisecond in the vicinity of the bulk Si band gap down to a microsecond for the green spectral range. The PL decay times measured for other systems containing Si nanocrystals have very similar values. A very strong spectral dispersion of the exciton recombination time is a direct consequence of quantum confinement effects. In smaller nanocrystals the emission energy and the oscillator strength of radiative transitions are larger in accordance with predictions of quantum confinement theory.

The study of the temporal PL behavior is very important since it allows the prediction of possible optical behavior under certain excitation conditions. Long lifetime of excitons confined in Si nanocrystals, in the range of 10^{-3} s - 10^{-6} s, implies that under CW excitation there is a large probability that a second photon can be absorbed before the exciton recombination and two *e-h* pairs can occupy Si nanocrystals simultaneously. A simple estimate shows that already at ~ 1 W/cm^2 of CW optical excitation in average each crystallite is already occupied by at least one *e-h* pair. Since Auger recombination time is in nanosecond time domain[43] this process will take over radiative recombination. This leads to a variety of optical phenomena. In the next chapter we will discuss fundamental consequences of this simple fact.

4. Auger Processes in Si Nanocrystals

4.1. *Nonlinear optical phenomena governed by Auger processes*

Already first experiments performed on PSi exhibited PL saturation at high excitation power. Afterwards, voltage quenching of the PL[44] and the voltage tunable electroluminescence[45] phenomena have been considered as additional evidence for the importance of the Auger effect. Nonradiative Auger lifetimes are orders of magnitude shorter than the radiative lifetime.[43] Therefore a simple assumption that under optical excitation nanocrystals can exist only in three possible states: unoccupied, containing one *e-h* pair and 2 *e-h* pairs gives the PL power law:[46]

$$I_{PL}\,(\hbar\omega_{\text{det.}}, I_{ex.}) = \frac{N\ (\hbar\omega_{\text{det.}})\cdot\sigma(\hbar\omega_{\text{det.}},\hbar\omega_{ex.})\cdot I_{ex.}}{1+\tau_R(\hbar\omega_{\text{det.}})\cdot I_{ex.}} \qquad (1)$$

Here N is a number of luminescing nanocrystals, τ_R is the radiative exciton lifetime, $\sigma(\hbar\omega_{\text{det.}},\hbar\omega_{ex.})$ is nanocrystal absorption cross-section dependent on excitation and emission energy and $I_{ex.}$ is exciting light intensity. The higher occupation numbers are neglected since the rate of the Auger process significantly exceeds the optical generation rate.

In Fig. 4 we show some examples of the measured PL intensities at different excitation and detection energies as a function of the excitation intensity. At low excitation levels the PL intensity is a linear function of the exciting light power while at higher intensities the PL saturation regime occurs. The excitation levels, at which a transition to the PL saturation is observed, are extremely sensitive to the excitation energy. The solid lines represent fits according to the steady state solution of equations with only one adjustable parameter $\sigma(\hbar\omega_{\text{det.}},\hbar\omega_{ex.})$. We found that all experimental curves can be fitted very precisely at all excitation and detection energies used. In general, the absorption cross-section $\sigma(\hbar\omega_{\text{det.}},\hbar\omega_{ex.})$ is a function of the detection and excitation energies. In indirect semiconductors the density of electronic states varies very rapidly with the increase of the energy. This is why PL saturation occurs at very small excitation intensities when short

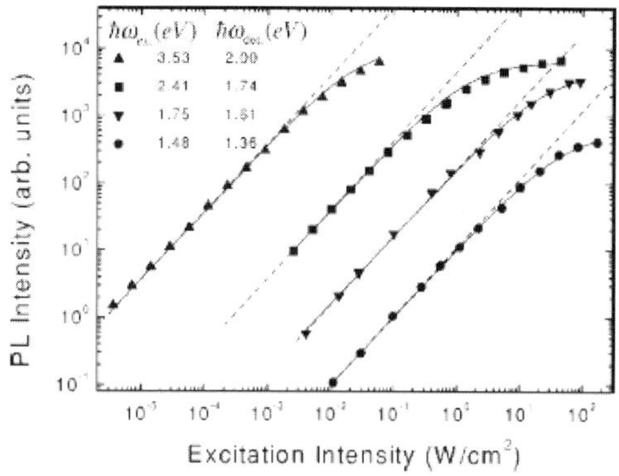

Fig. 4. The dependencies of the PL intensity versus excitation intensity. Excitation and detection energies are indicated. Solid lines are fits according to written formula. Dashed lines are linear power laws. The data sets are arbitrarily scaled for clarity. Reproduced with permission from Ref. 46, D. Kovalev et al., *Phys. Rev. B* **61**, 4485 (2000). Copyright (2000) by the American Physical Society http://link.aps.org/abstract/PRB/v61/p4485.

wavelength light is used for PL excitation. PL saturation implies that the QY of the PL is excitation intensity-dependent and it is obviously a disadvantage for technical applications: the luminosity of a device based on silicon nanocrystals is inherently limited to one photon per crystallite per exciton lifetime, which is still in a microsecond range even at room temperature.

During Auger process in doubly occupied nanocrystals one of *e-h* pairs annihilates but remaining second pair can participate in the recombination process. Therefore if pulsed excitation is used, PL decay time should be constant at any level of optical excitation. This arises from the fact that all Auger events take place only during the illumination period while after switching off the exciting light *e-h* pairs remaining in Si nanocrystals recombine with the constant radiative decay rate. This is not the case when PL saturation experiments are performed in a pump-probe technique. We have used an intense CW laser to achieve a permanent average occupation of nanocrystals by *e-h* pairs while monitoring the response to a weak probe pulse or modulated CW excitation source.[47] Lifetimes, measured as a function of the pump beam intensity, do not change at low pump beam intensity but decrease in the PL saturation regime (see Fig. 5b). The dependence of the measured

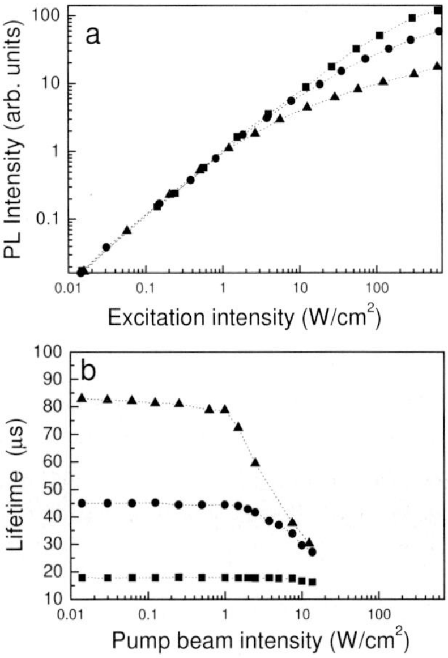

Fig. 5. (a) PL intensity versus CW excitation intensity for three different detection energies. $E_{ex.}$=2.54 eV, T=300 K. (b) PL lifetimes versus CW pump intensity for the same detection energies. The PL is excited by a 300 ps pulsed N_2 laser at an energy of 3.67 eV. T=300 K. Detection energies: triangles – 1.55 eV, circles – 1.77 eV, squares – 2.07 eV. Reproduced with permission from Ref. 26, D. Kovalev et al, *Phys. Stat. Sol.* (b) **215**, 871 (1999). Copyright Wiley-VCH Verlag GmbH & Co. KGaA.

lifetimes on the CW laser intensity has a form which is very similar to that for PL saturation regime:

$$\tau_{meas} = \frac{\tau_{rad.}(\hbar\omega_{\text{det.}})}{1+\tau_{rad.}(\hbar\omega_{\text{det.}})\cdot I_{pump.}\cdot\sigma(\hbar\omega_{\text{det.}},\hbar\omega_{pump})} \qquad (2)$$

where $I_{pump.}$ is the intensity of the pump beam.

There is an obvious correlation between the PL saturation (Fig. 5a) and measured exciton lifetimes in the presence of the additional CW illumination (see Fig. 5b). In the PL saturation regime when the CW generation rate exceeds the recombination rate, the generation of *e-h* pairs by the CW pump source in a nanocrystal already containing another pair created by the probe pulse introduces an effective nonradiative Auger decay channel. One of the *e-h* pairs is lost for the nonradiative recombination, whereas the remaining one further contributes to the CW PL background. This results in a progressive shortening of measured exciton lifetimes with increase of the pump beam intensity since the rate of Auger processes is proportional to the generation rate of additional *e-h* pairs excited by the pump beam. Again, the effect is stronger for nanocrystals having larger sizes and, therefore, larger absorption cross-sections and longer radiative lifetimes.

As it has been mentioned before, one of the most important features of Auger process is the energy gain. The energy of the annihilated *e-h* pair is transferred to remaining electron or hole. If gained energy is large enough the remaining carriers can escape from the nanocrystals to the nanocrystal network or the surrounding matrix. As a result, the nanocrystal is charged, and any further photoexcitation would imply three carriers localized in a small volume. Again due to very high effective concentration of carriers all further radiative recombination will be suppressed due to Auger processes. At low temperatures the nanocrystal remains ionized for long time, resulting in a persistent degradation of the PL (photodarkening effect).[19,20,48] However, at elevated temperatures, due to thermally activated transport, an ejected carrier can return to the interior of the nanocrystals and make them neutral, i.e. luminescing again.

This effect is observed experimentally in persistent PL hole burning experiments. After measuring the usual PL spectrum using a weak probe laser beam, the sample is exposed to high intensity laser beam which energy falls inside the PL band. The intense pump beam is used to create a long term photodegradation of the luminescence via an Auger ionization of Si nanocrystals. Since this light can be absorbed only by nanocrystals having their bandgap energies below the energy of the pump beam, only their emission is affected. After each light soaking step (with varied exposure time), the strong pump beam is switched off and the luminescence is measured again using the same weak probe beam.

The corresponding PL spectra are shown in Fig. 6a. A clear decrease in PL efficiency is recorded for emission energies below that of the light soaking laser beam, while almost no influence is detected above it. The amplitude of the spectral hole depends on the light soaking time. The spectral shape remains nearly constant, whereas the amplitude is growing

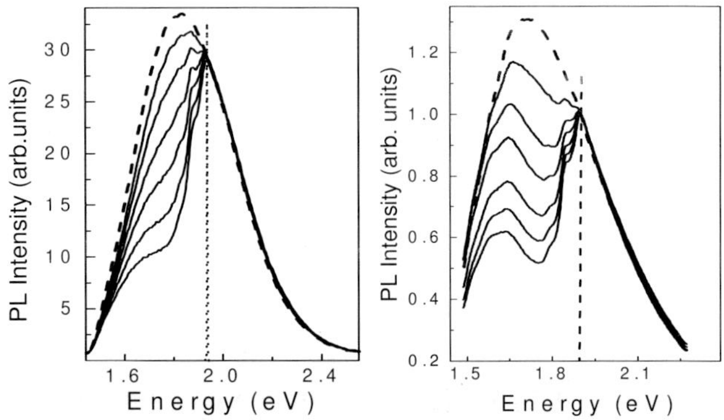

Fig. 6. (a) Persistent hole burning spectra of porous Si luminescence for different resonant light soaking time (solid lines, the burning energy is indicated by the vertical dashed line). The illumination exposure times are 0, 3, 10, 40, 100, 220 and 400 minutes, respectively. (b) Thermal healing of the spectral hole. Each spectrum has been measured at 4.2 K after progressive heating with 50 K steps. Dashed lines show the initial spectrum and the final one measured after the heating to 300 K. Reproduced with permission from Ref. 26, D. Kovalev *et al*, *Phys. Stat. Sol.* (b) **215**, 871 (1999). Copyright Wiley-VCH Verlag GmbH & Co. KGaA.

with an increase of the exposure time to the resonant light. Two phonon replicas known from the resonant PL measurements are observed.[49] The degradation becomes less efficient at low detection energies: the energy gained by the carrier is equal to E_{gap} and at small E_{gap} the energy may not be enough to overcome the potential barrier. At low temperatures, the spectrum of the "hole" and its magnitude do not change in a time scale of hours.

The simplest way to neutralize the nanocrystals is a thermal activation of the carrier transport. Electrons or holes ejected from the Si nanocrystals during the Auger ionization process are either captured by surface traps or transferred to neighboring nanocrystals. When the sample is heated the trapped carriers can be thermally reexcited into the nanocrystals. This process restores the initial optical properties of the sample, since the nanocrystals become neutral, and leads to an efficient hole healing effect. In Fig. 6b we show how the amplitude of the spectral hole depends on the thermocycling process. All the spectra have been measured at liquid He temperature after sweeping up the ambient temperature with 50 K increments. Increase of the temperature of the

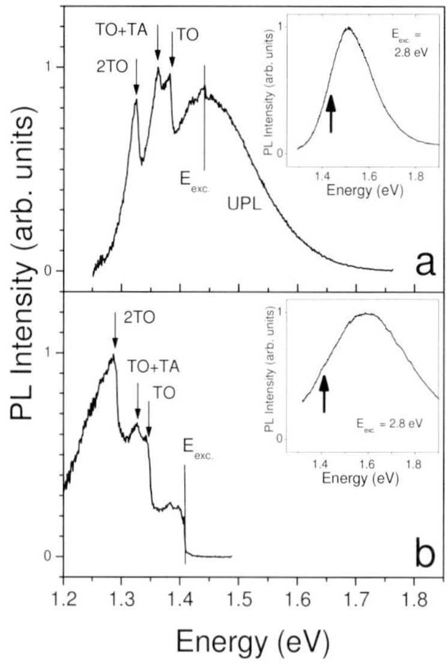

Fig. 7. Resonantly excited PL spectra of H-terminated porous Si (a) and Si nanocrystals having SiO_2 shell (b). The position of the exciting light within the distribution of luminescing Si nanocrystals is almost the same (shown by arrows). Insets: PL from these samples under high energy excitation. $E_{ex.}$ = 2.8 eV. T= 1.5 K. Reproduced with permission from Ref. 26, D. Kovalev et al, *Phys. Stat. Sol.* (b) **215**, 871 (1999). Copyright Wiley-VCH Verlag GmbH & Co. KGaA.

sample to 300 K washes out the hole burnt in the PL spectrum completely.

As it has been mentioned before one of the main features of Auger process is the transfer of the energy equal to nanocrystals bandgap to one of the remaining photogenerated carriers. If this energy is large enough this carrier can be injected to the neighboring nanocrystal having larger bandgap. Remaining nanocrystals can further participate in Auger process and remaining carrier can also be injected in the same nanocrystal. As a result emission *above* the excitation energy can take place. We demonstrate this effect in Fig. 7a. It shows a typical low-temperature PL spectrum of hydrogen-passivated PSi samples excited on the low-energy side of the emission band (shown in the inset of Fig. 7a). The Stokes PL consists of a set of peaks assigned to transitions involving no-phonon and momentum-conserving phonon transitions (indicated in the figure by arrows) in the absorption-emission cycle.[49] However, the spectrum also shows an intense upconverted emission which grows progressively relative to the Stokes PL with decreasing excitation energy. The efficiency of the upconversion is very large: at the lowest excitation energy used, the intensity of the UPL reaches the intensity of the Stokes PL.

We can verify our conjecture about possible energy gain by carriers by performing identical experiments using nanocrystals isolated by high potential barriers. Figure 7b shows the resonant PL spectrum for a system of Si nanocrystals surrounded by a thick SiO_2 shell. These nanocrystals have been produced by Si ion implantation in a SiO_2 layer, which ensures the presence of such a shell. Although all relevant experimental parameters like the excitation intensity, PL quantum yield, shape of nonresonant PL spectra and the position of the excitation energy within the size distribution are almost identical, no up-converted PL can be detected. At a macroscopic Si-SiO_2 interface the energy spacing between the top of the Si valence band and the bottom of the SiO_2 conduction band is equal to 4.3 eV.[50] The maximum energy for Auger-electrons is twice the band gap energy of the crystallite in which they are created and is smaller than 4.3 eV. In addition, the spacing between the crystallites is supposed to be larger than in microporous layers (containing interconnected nanocrystals). These two factors make the Auger carrier transfer impossible.

Up to now, we have considered optical phenomena mediated by Auger processes under optical excitation, when more than two *e-h* pairs can be generated simultaneously in Si nanocrystals. The Auger recombination can be considered as an inherent and an undesirable phenomenon which limits potential performance of Si nanocrystal-based optoelectronics. However, the authors of Ref. 51 exploited the potential of Auger processes to modulate the PL output from nanosilicon-based devices electrically. Indeed, while operation frequency of any device based on ordinary optical or electrical injection schemes is of the order of inverse exciton lifetime (~ 10 kHz), the electrical injection of excess charge carrier will result in a fast Auger damping of the PL. Therefore, if the fast electrical injection can be realized, the modulation frequency can be as fast as inverse Auger recombination time, which is in the GHz range. In Ref. 51 the reduction in the PL decay time under an applied gate bias has been observed and attributed to the energy exchange between nanocrystals. The authors suggested that electrically charged nanocrystals are rendered "dark" by Auger recombination and do not contribute to the recorded PL. The remaining ensemble of neutral nanocrystals emits light with a reduced lifetime due to the exchange of excitons with the "dark" charged nanocrystals. Unfortunately estimates that such an exchange occurs give a characteristic lifetime of about 10 μs at the extremes of applied gate bias what is still comparable with the exciton lifetime.

4.2. *Influence of dopant atoms on the emission properties of Si nanocrystals*

The introduction of shallow impurities in a semiconductor significantly modifies its optical and electrical transport properties. A precise control of an impurity profile is a key issue to achieve desirable functions in almost all kinds of semiconductor devices. While the electronic structure of shallow impurities in bulk semiconductors has been studied for a long time and is fairly well understood, in semiconductor nanostructures it is still a subject of intensive research.[52,53] The understanding of the electronic structure of shallow impurities in Si nanostructures is very important from the point of view of fundamental physics and is indispensable for realization of Si nanostructure-based electronic and opto-electronic devices. Unfortunately, contrary to many theoretical works,[53–58] experimental work on shallow impurities in Si nanocrystals make a poor progress because of difficulties in preparing impurity-doped Si nanocrystals in a controlled manner. The main difficulty arises from the fluctuation of impurity number per a nanocrystal in a nanocrystal assembly. For Si nanocrystals as small as a few nanometers in diameter, the expression of the doping level in the form of "impurity concentration" is not suitable and it should be expressed as "impurity numbers" because it changes digitally. For example, doping of one impurity atom to a nanocrystal having 3 nm in diameter (1.4×10^{-20} cm^3, about 700 atoms) corresponds to an impurity concentration of 7.0×10^{19} atoms/cm^3. At this doping level, bulk Si is degenerate and exhibits metallic behavior. Therefore, in nanometer-sized nanocrystals, addition or subtraction of a single impurity atom changes the electronic structure and the resultant optical and electrical transport properties drastically. Because of statistical fluctuation of the number of impurities in a nanocrystal assembly, all experimentally observed phenomena are broadened and smeared.

PL spectroscopy is one of the powerful tools to study the electronic structure of impurity-doped Si nanocrystals. Although the spectra are usually significantly broadened due to several kinds of inhomogeneities, one can still see the effect of shallow impurity doping on the spectra. In this section, we will show how PL properties of Si nanocrystals are modified by the impurity-related Auger processes in Si nanocrystals. We also discuss a possible route to avoid the Auger process.

4.2.1. *Preparation of impurity doped Si nanocrystals and evidence of impurity doping*

Usually, to grow Si nanocrystals, a thermal treatment at rather high temperatures is required. During the treatment, it is very plausible that impurity atoms are pushed out of nanocrystals by a so-called self-purification effect. In order to avoid this effect, impurities should be doped both into nanocrystals and surrounding matrices. This kind of Si nanocrystal samples can be prepared by annealing a mixture of Si, SiO_2 and oxides of impurity atoms in an inert gas atmosphere. In our case, the mixture is prepared by a co-sputtering method.[59] In the case of P (B) doped Si nanocrystals Si, SiO_2 and P_2O_5 (B_2O_3) are simultaneously sputter-deposited and they are annealed in N_2 gas atmosphere at 1100-1250°C. During the annealing, Si nanocrystals are grown in phosphosilicate (PSG) [borosilicate (BSG)] thin films. The P (B) concentration in PSG (BSG) thin films is at maximum several percents. Similarly, if Si, SiO_2, B_2O_3 and P_2O_5 are sputter-deposited simultaneously, B and P co-doped Si nanocrystals are grown in borophosphosilicate glass (BPSG) matrices. Note that impurity concentration in nanocrystals is different from that of matrices because the segregation coefficient, i.e., the ratio of equilibrium concentration of impurity atoms in Si and SiO_2, is not 1; it depends strongly on the kind of impurities.

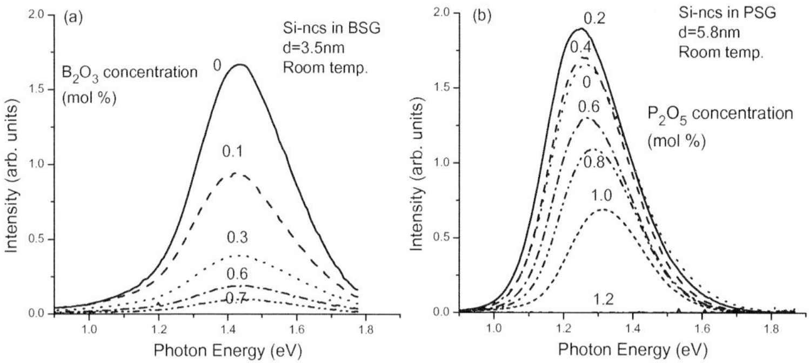

Fig. 8. PL spectra of (a) B doped and (b) P doped Si nanocrystals at room temperature. B and P concentrations are changed. (a) Reproduced with permission from Ref. 64, A. Mimura et al., Solid State Commun. 109, 561 (1999). Copyright Elsevier Science Ltd. (b) Reproduced with permission from Ref. 61, M. Fujii et al., Phys. Rev. Lett., 89, 206805 (2002). Copyright (2002) by the American Physical Society. http://link.aps.org/abstract/PRL/v89/e206805.

Doping of Si nanocrystals by shallow impurities can be confirmed by various methods. For P doped Si nanocrystals, electron spin resonance provides clear evidence of the doping. P doped Si nanocrystals show very broad band with the *g*-value of 1.998 due to conduction electrons in nanocrystals.[59,60] When P concentration is "relatively" low, in average about one per a nanocrystal, a hyper-fine structure due to the interaction of a donor electron and a P nucleus can be seen.[61] Infrared absorption due to intra-conduction and intra-valence band transitions is also observed for *n*- and *p*-type shallow impurity-doped Si nanocrystals, respectively.[59, 62]

4.2.2. *Luminescence from p- or n-doped Si nanocrystals*

As it can be expected from the discussion in previous sections, the introduction of extra carriers by impurity doping makes the three-body Auger process possible even if Si nanocrystals are excited very weakly. In other words, PL quantum efficiency of impurity doped Si nanocrystals is considered to be much lower than that of "pure" Si nanocrystals.[43] Since we measure ensembles of many Si nanocrystals with different impurity concentrations (numbers), with increasing average impurity concentration in a nanocrystal assembly, the PL intensity is expected to decrease continuously. This effect is really observed both in *p*- and *n*-type doped Si nanocrystals. Figure 8 shows the examples. In the case of B doping (Fig. 8a), the PL intensity decreases monotonously by increasing B concentration. The same trend can be seen when P is used for doping (Fig. 8b). In both cases, the decrease of the PL intensity is accompanied by the shortening of the lifetime.

If we closely look at Fig. 8b, when the P concentration is relatively low, the PL intensity is increased compared to that of undoped Si nanocrystals. This effect can be seen more clearly when the size of Si nanocrystals is smaller.[59,63] The increase of the PL intensity indicates that nonradiative recombination centers are terminated by P doping. In fact, defect-related infrared emission band observed at low temperatures is significantly quenched by P doping. Furthermore, defect-related ESR signal almost disappears.[63] Two scenarios are considered as the mechanism of the quenching of non-radiative recombination centers. The first one is that the number of dangling bonds on the surface of Si nanocrystals is decreased because of higher flexibility of bond angles and lengths in PSG than SiO_2. The second scenario is that electrons supplied by P doping inactivate the nonradiative recombination centers. The definite evidence of these mechanisms is not yet obtained but the second one seems to be plausible because this effect is observed only when *n*-type impurities are doped. It should be noted here that, the impurity concentration dependence of the PL intensity depends strongly on the size and the concentration of Si nanocrystals as well as an annealing parameters such as temperature, duration, atmosphere, cooling speed, etc.[64]

4.2.3. *Luminescence from p- and n-type impurities co-doped Si nanocrystals*

Because of the very efficient Auger process, shallow impurity-doped Si nanocrystals are very bad candidates for light-emitting applications. However, the additional localization of carriers by impurities is a possible approach to enhance oscillator strength of excitons in Si nanocrystals. The Auger process can be avoided if isoelectronic impurities are used. Unfortunately, Si does not have proper isoelectronic

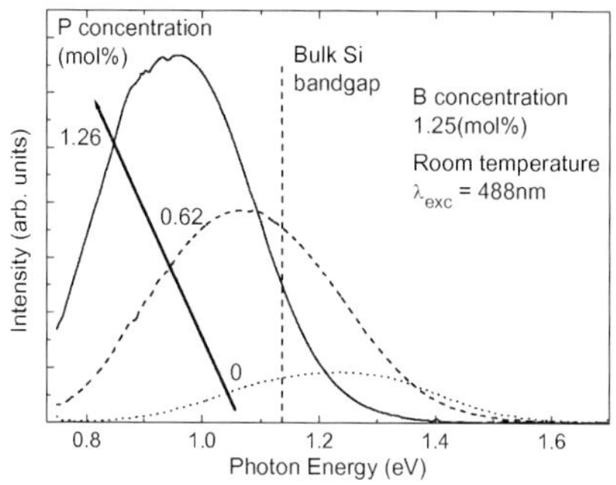

Fig. 9. PL spectra of B doped (dotted curve) and B and P doped (dashed and solid curves) Si nanocrystals at room temperature. B concentration is fixed, while P concentration is changed. Reproduced with permission from Ref. 66, M. Fujii *et al.*, J. Appl. Phys. **94**, 1990 (2003). Copyright American Institute of Physics.

impurities that can strongly localize excitons and enhance the PL intensity. An alternative approach is to use simultaneously *p*- and *n*-type impurities for doping of Si nanocrystals. The co-doping may result in further localization of excitons without being afraid of the Auger process. The major technological problem is how to dope exactly the same number of B and P atoms into Si nanocrystals. If the number is not exactly equal, excitons will again recombine nonradiatively due to the Auger process. Although a precise control of the number of *p*- and *n*-type impurities in a Si nanocrystal is almost impossible, recent calculations demonstrated preferential formation of nanocrystals with equal number of *p*- and *n*-type impurities. The group of Ossicini[57,65] demonstrated by first-principles calculations that the formation energy of Si nanocrystals is drastically decreased when pairs of B and P are doped into Si nanocrystals; formation energy of a pair of B and P doped Si nanocrystals ($Si_{147}H_{100}$ clusters) is about 1 eV smaller than that of B doped ones and about 0.7 eV smaller than that of P doped ones. This indicates that if Si nanocrystals are grown by the above procedure, nanocrystals with equal number of B and P are preferentially grown.

Figure 9 shows PL spectra of B and P doped Si nanocrystals.[66,67] B concentration in a sample is almost fixed while P concentration is changed. Without P doping, i.e., when only B is doped, the PL intensity is very weak due to the efficient Auger process. By doping P

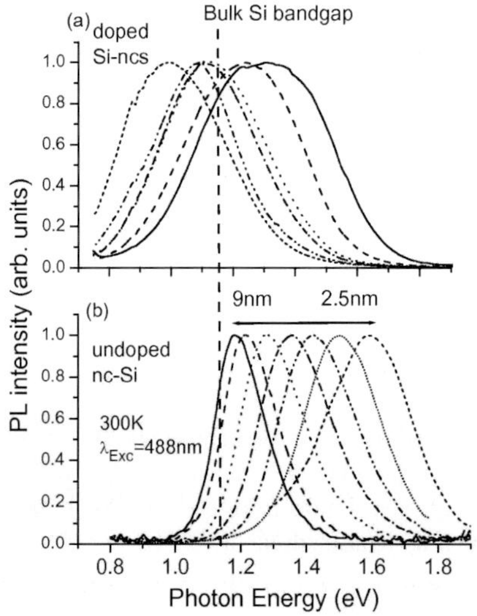

Fig. 10. Normalized PL spectra of (a) p- and n-type impurities co-doped and (b) undoped Si nanocrystals at room temperature. The lowest possible PL energy of undoped Si nanocrystals is the bulk Si band gap, while that of co-doped Si nanocrystals is extended to 0.9 eV.

simultaneously, the intensity recovers significantly and the peak shifts to lower energy. It should be stressed here that the luminescence energy is far below the band gap energy of bulk Si crystals. The low energy shift is considered to be due to the optical transition between donor and acceptor levels, and the recovery of the PL intensity is due to compensation of carriers. Therefore, simultaneous B and P doping of Si nanocrystals seem to be realized by the present process and they have rather high PL quantum efficiency. Note that the PL spectra are measured at room temperature and the temperature quenching from liquid He temperature to room temperature is less than 50%. This small temperature quenching of the PL is completely different from that of donor-acceptor pair emission from heavily doped bulk Si crystals.[68] Probably, large ionization energy of impurities and restriction of carrier transport within nanocrystals result in the very small dependence of the PL on temperature.

An efficient PL from *p*- and *n*-type impurities co-doped Si nanocrystals indicates that impurity control is another parameter to tailor optical properties of Si nanocrystals. As can be seen in Fig. 10b, the tunable range of PL energy of undoped Si nanocrystals is limited above the bulk Si band gap because the PL shift is caused only by the quantum size effects. The range can be extended to 0.9 eV for impurity co-doped Si nanocrystals (Fig. 10a) without significant reduction of the PL intensity. This wide tunability controllability of the PL energy may open up new application of Si nanocrystals, especially in the filed of optical telecommunication.

5. Conclusions

Auger processes seem to be an inherent and undesirable limitation for the luminosity of light emitting devices based on Si nanocrystals. The key restriction is based on uncertainty in crystalline quasimomentum: while geometrical confinement increases the probability of radiative recombination due to enhancement of zero-phonon processes, it also simultaneously increases the probability of Auger processes. Due to an enormous difference in their rates radiative processes can not compete with Auger nonradiative recombination. Consequently, this process strongly affects basic optical properties of Si nanocrystals. Furthermore, to realize electrically pumped efficient light emitting device doping of Si nanocrystals is required. As it has been demonstrated, fundamental issues of light emission from Si nanocrystals assemblies relevant to the doping-mediated Auger processes should be taken into account.

References

1. P. T. Landsberg, *Recombination in Semiconductors* (University Press, Cambridge, 1991).
2. V. A. Kharchenko and M. Rosen, *J. Lumin.* **70**, 158 (1996).
3. G. P. Agraval and N. K. Dutta, Long Wavelength Semiconductor Lasers (Van Nostrand-Reinhold Company, New York, 1986).
4. G. Fuchs, C. Schiedel, A. Hangleiter, V. Härle and F. Scholz, *Appl. Phys. Lett.* **62**, 396 (1993).
5. S. Seki, H. Oohasi, H. Sugiura, T. Hirono and K. Yokoyama, *Appl. Phys. Lett.* **67**, 1054 (1995).
6. W. Schmid, *Phys. Stat. Sol.* **84**, 529 (1977).

7. J. Dziewior and W. Schmid, *Appl. Phys. Lett.* **31**, 346 (1977).
8. L. M. Blinov, E. A. Bobrova, V. S. Vavilov and G. N. Galkin, Fiz. Tverdogo Tela **9**, 3221 (1967); *Soviet Phys. Solid State* **9**, 2537 (1968).
9. N. G. Nilsson, *Physica Scripta*. **8**, 165-176, (1973).
10. K. G. Svantesson and N. G. Nilsson, *J. Phys. C: Solid State Phys.* **12**, (1979).
11. J. Schmidt, M. Kerr and P. P. Altermatt, *J. Appl. Phys* **88**, 1494 (2000).
12. A. Hangleiter and R. Hacker, *Phys. Rev. Lett.* **65**, 215 (1990).
13. R. Hacker and A. Hangleiter, *J. Appl. Phys.* **75**, 7570 (1994).
14. E. Yablonovitch and T. Gmitter, *Appl. Phys. Lett.* **49**, 587 (1986).
15. P. T. Landsberg, *Appl. Phys. Lett.* **50**, 745 (1987).
16. M. A. Green, *IEEE Transactions on Electron Devices* **ED-31**, 671 (1984).
17. T. Tiedje, E. Yablonovitch, G. D. Cody and B. G. Brooks BG., *IEEE Transactions on Electron Devices* **ED-31**, 711 (1984).
18. D. I. Chepic, Al. L. Efros, A. I. Ekimov, M. G. Ivanov, V. A. Kharchenko, I. A. Kudriavtsev and T. V. Yaseva, *J. Lumin.* **47**, 113 (1990).
19. M. Nirmal, B. O. Dabbousi, M. G. Bawendi, J. J. Macklin, J. K. Trautman, T. D. Harris and L. Brus, *Nature* **383**, 802 (1996).
20. Al. Efros and M. Rosen, *Phys. Rev. Lett.* **78**, 1110 (1997).
21. A. Uhlir, *Bell Syst. Tech.* **35**, 333 (1956).
22. C. Pickering, M. I. J. Beale, D. J. Pearson and R. G. Greef, *Physica* **C17**, 6535 (1984).
23. L. T. Canham, *Appl. Phys. Lett.* **57**, 1046 (1990).
24. V. Lehmann and U. Goesele, *Appl. Phys. Lett.* **58**, 856 (1991).
25. A. G. Cullis, L. T. Canham and P. D. J. Calcott, *J. Appl. Phys.* **82**, 909 (1997).
26. D. Kovalev, H. Heckler, G. Polisski and F. Koch, *Phys. Stat. Sol. (a)* **215**, 871 (1999).
27. O. Bisi, S. Ossicini and L. Pavesi, *Surf. Sci. Rep.* **38**, 1 (2000).
28. M. L. Brongersma, P. G. Kik, A. Polman, K. S. Min and H. A. Atwater, *Appl. Phys. Lett.* **76**, 351 (2000).
29. L. Khriachtchev, M. Räsänen, S. Novikov and J. Sinkkonen, *Appl. Phys. Lett.* **79**, 1249 (2001).
30. L. Del Negro, M. Cazzanelli, L. Pavesi, S. Ossicini, D. Pacifici, G. Franzo and F. Priolo, *Appl. Phys. Lett.* **82**, 4636 (2003).
31. M. Fujii, M. Yoshida, Y. Kanzawa, S. Hayashi and K. Yamamoto, *Appl. Phys. Lett.* **71**, 1198 (1997); J. Ruan, P. M. Fauchet, L. Del Negro, M. Cazzanelli and L. Pavesi, *Appl. Phys. Lett.* **83**, 5479 (2003).
32. J. Heitmann, F. Müller, M. Zacharias and U. Gösele, *Adv. Mater.* **17**, 795 (2005).
33. F. Huisken, G. Ledoux, O. Guillois and C. Reynaud, *Adv. Mater.* **14**, 1861 (2002).
34. A. J. Steckl, J. Xu and H. C. Mogul, *Appl. Phys. Lett.* **62**, 2111 (1992); G. Anaple, R. Burrows, Y. Wu and P. Boolchand, *J. Appl. Phys.* **78**, 4273 (1995).
35. M. S. Hybertsen, *Phys. Rev. Lett.* **72**, 1514 (1994); M. S. Hybertsen in "*Porous Silicon Science and Technology*" Ed. by J.-C.Vial and J. Derrien, Springer (1995)].

36. D. Kovalev, M. Ben-Chorin, J. Diener, B. Averboukh, G. Polisski and F. Koch, *Phys. Rev. Lett.* **79**, 119 (1997).
37. K. J. Nash, P. D. J. Calcott, L. T. Canham and R. J. Needs, *Phys. Rev.* **B51**, 17698 (1995).
38. J. Diener, D. Kovalev, G. Polisski, H. Heckler and F. Koch, *Phys. Stat. Sol. (b)* **214**, R13 (1999).
39. R. J. Walters, J. Kalkman, A. Polman, H. A. Atwater and M. J. A. de Dood, *Phys. Rev.* **B73**, 132302-1 (2006).
40. I. Sychugov, R. Juhasz, J. Valenta and J. Linnros, *Phys. Rev. Lett.* **94**, 087405 (2005); I. Sychugov, R. Juhasz, J. Valenta, A. Zhang, P. Pirouz and J. Linnros, *Appl. Surf. Science* **252**, 5249 (2006).
41. J. C. Vial, A. Bsiesy, F. Gaspard, R. Herino, M. Ligeon, F. Muller, R. Romestain and R. M. Macfarlane, *Phys. Rev.* **B45**, 14171 (1992).
42. Spectroscopy of Isolated and Assembled Semiconductor Nanocrystals, ed. by L. E. Brus, Al. L. Efros, T. Itoh, *J. Lumin.* **70**, (1996).
43. C. Delerue, M. Lanoo, G. Allan, E. Martin, I. Mihalcescu, J. C. Vial, R. Romestain, F. Muller and A. Bsiesy, *Phys. Rev. Lett.* **75**, 2228 (1995).
44. A. Bsiesy, F. Muller, I. Mihalcescu, M. Ligeon, F. Gaspard, R. Herino, R. Romestain and J. C. Vial, in *Light Emission from Silicon*, Ed. By J. C. Vial, L. T. Canham and W. Lang (Elsevier-North-Hollanf, Amsterdam, 1993), p. 29.
45. A. Bsiesy, F. Muller, M. Ligeon, F. Gaspard, R. Herino, R. Romestain and J. C. Vial, *Phys. Rev. Lett.* **71**, 637 (1993).
46. D. Kovalev, J. Diener, H. Heckler, G. Polisski, N. Kuenzner and F. Koch, *Phys. Rev.* **B61**, 4485 (2000).
47. Al. L. Efros, M. Rosen, B. Averboukh, D. Kovalev, M. Ben-Chorin and F. Koch, *Phys. Rev.* **B56**, 3875 (1997).
48. V. Grivickas, J. Linnros and J. A. Tellefsen, *Thin Solid Films* **25**5, 208 (1995).
49. P. D. J. Calcott, K. J. Nash, L. T. Canham, M. J. Kane and D. Brumhead, *J. Phys. Condens. Matter* **5**, L91 (1993); D. Kovalev, H. Heckler, M. Ben-Chorin, G. Polisski, M. Schwartzkopff and F. Koch, *Phys. Rev. Lett.* **81**, 2803 (1998).
50. R. Williams, *Phys. Rev.* **A140**, 569 (1965).
51. R. J. Walters, P. G. Kik, J. D. Casperson, H. A. Atwater, R. Lindstedt, M. Giorgi and G. Bourianoff, *Appl. Phys. Lett.* **85**, 2622 (2004).
52. P. O. Holtz and Q. X. Zhao, *Impurities Confined in Quantum Structures* (Springer-Verlag, Berlin Heidelberg, 2004).
53. S. Ossicini, F. Iori, E. Degoli, E. Luppi, R. Magri, R. Poli, G. Cantele, F. Trani and D. Ninno, *IEEE J. sel. top. quantum electron.*, **12**, 1585 (2006).
54. G. Allan, C. Delerue, M. Lannoo and E. Martin, *Phys. Rev.* **B52**, 11982 (1995)
55. Z. Shou, R. A. Friesner and L. Brus, *J. Am. Chem. Soc.* **125**, 15599 (2003).
56. D. V. Melnikov and J. R. Chelikowsky, *Phys. Rev. Lett.* **92**, 046802 (2004).
57. G. Cantele, E. Degoli, E. Luppi, R. Magri, D. Ninno, G. Iadonisi and S. Ossicini, *Phys. Rev.* **B72**, 113303 (2005).

58. Q. Xu, J. Luo, S. Li, J. Xia, J. Li and S. Wei, *Phys. Rev.* **B75**, 235304 (2007).
59. A. Mimura, M. Fujii, S. Hayashi, D. Kovalev and F. Koch, *Phys. Rev.* **B62**, 12625 (2000).
60. K. Sumida, K. Ninomiya, M. Fujii, K. Fujio, S. Hayashi, M. Kodama and H. Ohta, *J. Appl. Phys.* **101**, 033504 (2007).
61. M. Fujii, A. Mimura, S. Hayashi, Y. Yamamoto and K. Murakami, *Phys. Rev. Lett.* **89**, 206805 (2002).
62. M. Fujii, Y. Yamaguchi, Y. Takase, K. Ninomiya and S. Hayashi, *Appl. Phys. Lett.* **87**, 211919 (2005).
63. M. Fujii, A. Mimura, S. Hayashi, K. Yamamoto, C. Urakawa and H. Ohta, *J. Appl. Phys.* **87**, 1855 (2000).
64. A. Mimura, M. Fujii, S. Hayashi and K. Yamamoto, *Solid State Commun.* **109**, 561 (1999).
65. S. Ossicini, E. Degoli, F. Iori, E. Luppi, R. Magri, G. Cantele, F. Trani and D. Ninno, *Appl. Phys. Lett.* **87**, 173120 (2005).
66. M. Fujii, K. Toshikiyo, Y. Takase, Y. Yamaguchi and S. Hayashi, *J. Appl. Phys.* **94**, 1990 (2003).
67. M. Fujii, Y. Yamaguchi, Y. Takase, K. Ninomiya and S. Hayashi, *Appl. Phys. Lett.* **85**, 1158 (2004).
68. M. Levy, P. Y. Yu, Y. Zhang and M. P. Sarachik, *Phys. Rev.* **B49**, 1677 (1994).

CHAPTER 16

BIOLOGICAL APPLICATIONS OF SILICON NANOSTRUCTURES

Sharon M. Weiss

Department of Electrical Engineering and Computer Science
Vanderbilt University, Nashville, TN, USA
E-mail: sharon.weiss@vanderbilt.edu

The importance of silicon photonic devices extends beyond passive structures for light guiding and light emission. Nanostructured silicon devices have emerged as gas, chemical, and biological sensors, as well as drug delivery agents, molecular separation media, and interfaces for tissue engineering. While early silicon-based sensing devices were based on bulk silicon properties, more recent devices take advantage of the unique properties of nanostructured silicon-based devices. This chapter highlights several advances in the area of biological applications of silicon nanostructures.

1. Introduction

The integration of silicon nanophotonic structures with chemical and biological species for the detection, delivery, growth, and separation of these chemical and biological molecules has paralleled the development of nanoscale silicon optical components. Nanostructured silicon holds several advantages for advanced sensing applications, including the capability for high sensitivity detection, a low analyte volume requirement, reduced size, the potential for high density sensor arrays, compatibility with existing CMOS technology, and, in some cases, a significantly larger surface area than planar substrates. Section 2 describes the various silicon structures used for biological applications. Section 3 addresses gas, chemical, and biological optical sensing applications of nanostructured silicon, which represent the largest body of work in the area of silicon nanophotonics for biological applications. Section 4 highlights work in the emerging areas of drug delivery, molecular separation, and tissue engineering.

While the thrust of this chapter is on silicon nanophotonics for biological applications, there is a plethora of methods other than optical techniques used for transducing signals from silicon sensors, with the most popular being electrical and mechanical detection schemes. For example, detection of chemical and biological species using field effect transistors (FETs) has evolved over the past few decades.[1-5] Most recently, nanoscale FETs based on silicon nanowires and nanotubes opened the possibility for small footprint, high density sensor arrays.[5] Other electronic sensing techniques that utilize a change in conductance of silicon nanowires[6] or porous silicon structures[7] have also been demonstrated. Micro- and nanocantilevers are another technology that has emerged from standard silicon microelectronics fabrication methods to become highly sensitive devices capable of detecting the presence of a variety of species including solvents, DNA, proteins, and cells.[8-15]

2. Silicon Nanostructures

Several different approaches have been followed to achieve biofunctionality on a silicon-compatible platform. Figure 1 illustrates three types of silicon nanomaterials and nanostructures that have been utilized for biological applications: porous silicon, ring resonators, and slot waveguides. The fabrication techniques for each of these materials and structures are briefly described in the following subsections.

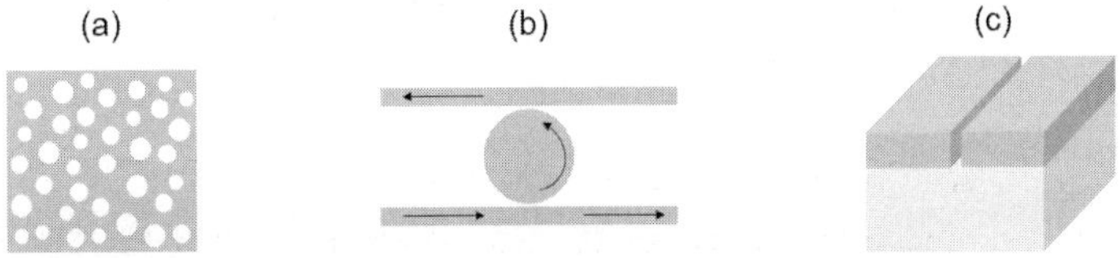

Fig. 1. Schematic of silicon nanostructures: (a) porous silicon, (b) ring resonator, and (c) slot waveguide.

2.1. *Porous silicon*

Porous silicon is a material that consists of nanometer to micron-sized diameter pores within a crystalline silicon matrix. The pore diameter, in addition to the porosity and thickness of the porous layer, can be tuned depending on the formation conditions. The most common method of fabrication is by electrochemical etching in an electrolyte containing hydrofluoric acid. While first reported in 1956,[16] strong interest in the potential usefulness of porous silicon did not begin until several years later, after the discovery of luminescence from the nanostructured material.[17] Several review articles and books have been published that comprehensively describe porous silicon formation mechanisms, porous silicon morphology, and control of porous silicon physical properties.[18-24]

For biological applications, one of the greatest attributes of porous silicon is its enormous internal surface area, which can range up to a few hundred square meters per cubic centimeter. Since many more biological molecules can attach to porous silicon compared to planar materials, its structure provides a significant advantage for biosensing. Moreover, the pores can be used as hosts and delivery units for drugs, scaffolds for bone growth, and filters for molecular separations.

When the pore diameter is much smaller than the wavelength of light, porous silicon is treated as an effective medium for which the refractive index is a weighted average of the refractive indices of silicon and air.[25-27] The refractive index of porous silicon can be tuned by simply adjusting the porosity. A porosity range of 30% to 90% porosity gives an approximate refractive index range of 3.30 to 1.15, depending on the wavelength of light and morphology of the porous silicon. Multilayer thin film structures, such as Bragg mirrors,[28] rugate filters,[29] microcavities,[30] and waveguides[31] can be formed by simply varying the porosity appropriately during electrochemical etching. Since the etching preferentially proceeds at the pore tips where the electric field is concentrated and carriers are present, sharp interfaces between porous silicon layers of different porosity are possible.[25] Previously formed porous silicon layers are largely depleted of charge carriers, which minimizes continued etching of those layers as the formation of subsequent layers proceeds. Figure 2 schematically illustrates the etching conditions necessary to form a two layer porous silicon waveguide. Light is trapped by total internal reflection in the low porosity (high refractive index) layer 1. The thickness of layer 1 determines the number of modes present in the waveguide. High porosity (low refractive index) layer 2 must be thick enough to prevent light leakage into the substrate.

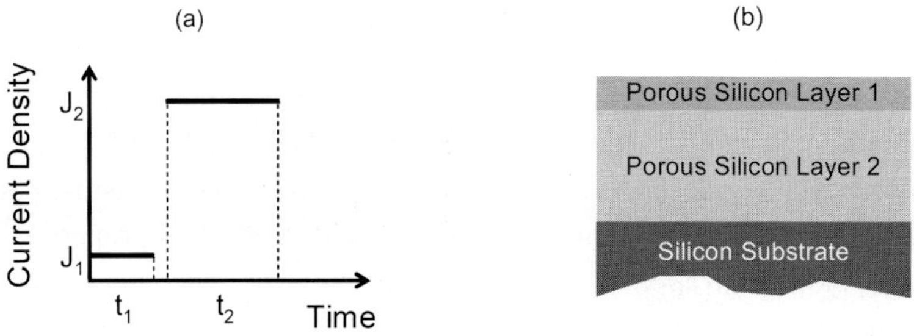

Fig. 2. (a) Electrochemical etching conditions to produce a (b) two-layer porous silicon waveguide. Application of the lower current density J_1 for a time period t_1 forms the low porosity (high refractive index) porous silicon layer 1. Application of the higher current density J_2 for a time period t_2 forms the high porosity (low refractive index) porous silicon layer 2. Light is guided in porous silicon layer 1.

Figure 3 shows schematics and associated optical spectra for four different porous silicon nanostructures that have been employed as biosensors. The simplest is a single porous silicon layer (Fig. 3a). The reflectance spectrum is characterized by thin film Fabry-Perot fringes (Fig. 3b). When gases, chemicals, or biomolecules are infiltrated into the pores, the refractive index of the porous silicon changes (i.e., the effective medium consists of silicon, air, and gas/chemical/biological molecules). This refractive index change is detected as a shift of the fringes; the fringe spacing is directly related to the optical thickness of the porous silicon thin film. Alternating layers of low and high porosity porous silicon ($\lambda/4$ optical thickness) lead to a Bragg mirror structure shown in Fig. 3c. When chemical or biological material is present inside the porous silicon layers, the wavelength shift of the high reflectance stop-band (Fig. 3d) can be monitored to confirm the presence of chemical or biological material. A variation of the Bragg mirror is the rugate filter for which the porosity varies sinusoidally and the side lobes in the optical spectrum are markedly suppressed.[32, 33] It is easier to detect small shifts of sharper features in an optical spectrum. Hence, the porous silicon microcavity, shown in Fig. 3e, can be a very sensitive biosensor. Light is trapped in the central defect region of the microcavity for a

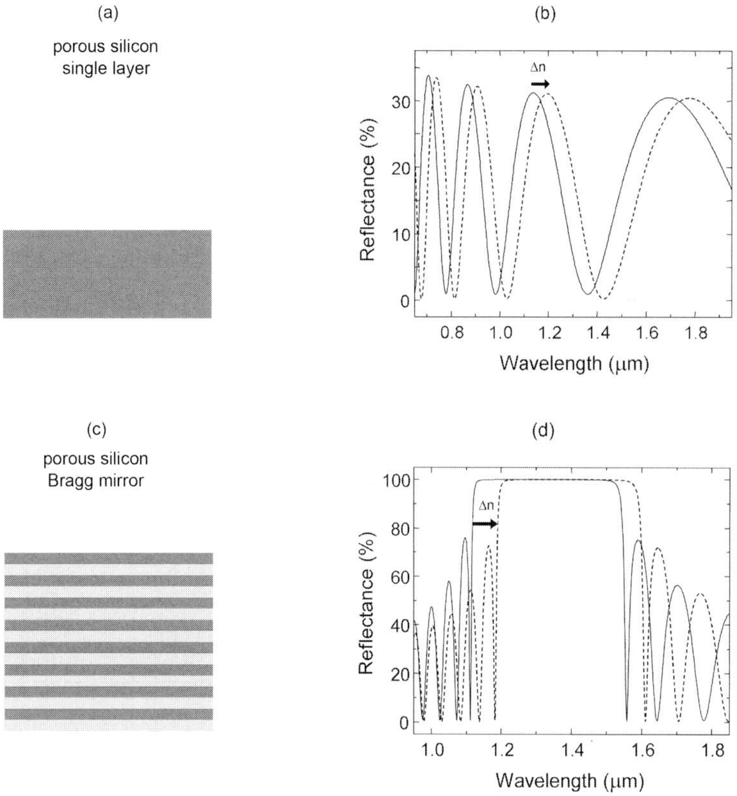

Fig. 3. Schematics and associated optical spectra of porous silicon (a, b) single layer; (c, d) Bragg mirror; (e, f) microcavity; and (g, h) waveguide. For the schematics, the darker layers represent low porosity porous silicon layers and the brighter layers represent high porosity porous silicon layers. For the optical spectra, the solid lines correspond to measurements of porous silicon structures with air in the pores and the dotted lines correspond to porous silicon structures with chemical or biological molecules in the pores. The spectral shifts are easiest to detect when resonant features are present in the spectrum. In each case, the shift is due to the increase in effective refractive index from the presence of gases, chemicals, or biological molecules in the pores.

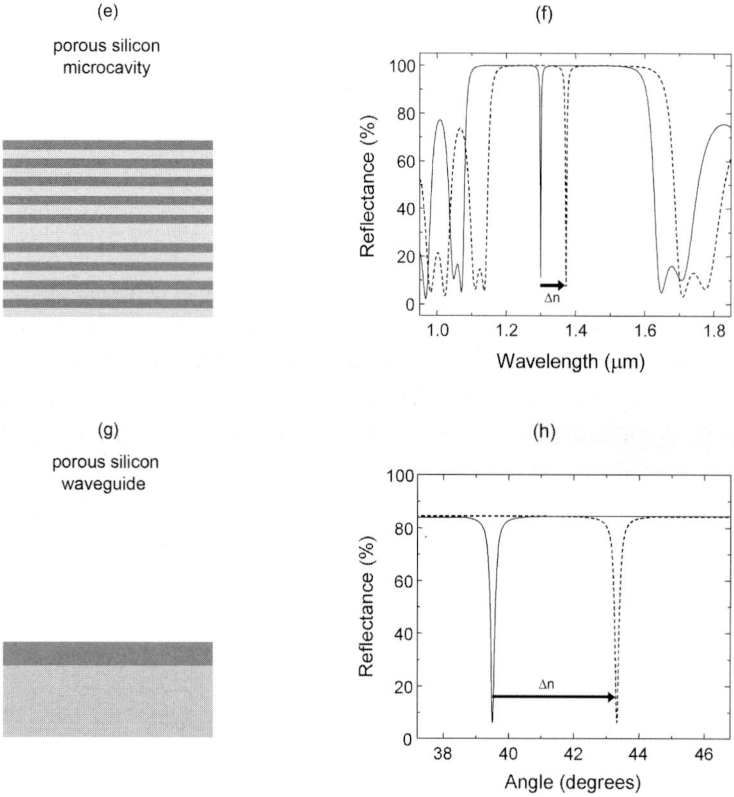

Fig. 3. *Continued.*

duration that is proportional to the cavity quality (Q) factor*, producing a resonance in the optical spectrum (Fig. 3f). The porous silicon waveguide structure, shown in Fig. 3g, also possesses a sharp resonance in the optical spectrum (Fig. 3h). The resonance corresponds to the condition for which light is coupled into the waveguide, typically via a prism or grating on top of the waveguide.

* $Q = \lambda/\Delta\lambda$, where λ is the resonance wavelength and $\Delta\lambda$ is the full-width-at-half-maximum of the resonance.

2.2. *Ring resonators*

Using standard silicon processing techniques, ring resonators, such as those shown in Fig. 4, have been developed for chip-scale silicon photonic components including optical filters, switches, and lasers.[34-37] Variations of the silicon ring resonator include lateral and vertical coupled rings, disks, toroids, and cascaded resonators. The high Q-factors and small mode volumes of these structures lead to strong electric field enhancement in the resonators. While the overall dimensions of the resonators are on the order of microns, the gap between the resonators and adjacent bus waveguides is typically on the nanometer scale. Light in the bus waveguide can be evanescently coupled through the gap into a whispering gallery mode of the resonator if the proper phase matching conditions exist.[38] Light is trapped within the ring by total internal reflection. Figure 5 shows a typical transmission spectrum of a microring resonator. Over the past few years, the utility of microring and mircrodisk resonators has expanded to biological applications. Resonance shifts, intensity changes, and changes to the quality factor have been used to detect the presence of chemical and biological molecules attached to microring and microdisk resonators, as detailed in section 3.2.

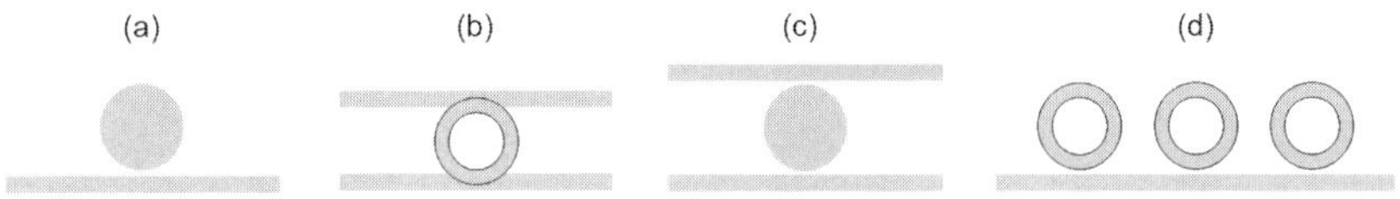

Fig. 4. Schematics of various microring and microdisk resonators: (a) laterally coupled microdisk with single bus waveguide, (b) vertically coupled microring with input and output waveguides, (c) laterally coupled microdisk with input and output waveguides, and (d) laterally coupled cascaded microrings with single bus waveguide. Any combination of ring/disk, one/two waveguides, and lateral/vertical coupling is possible.

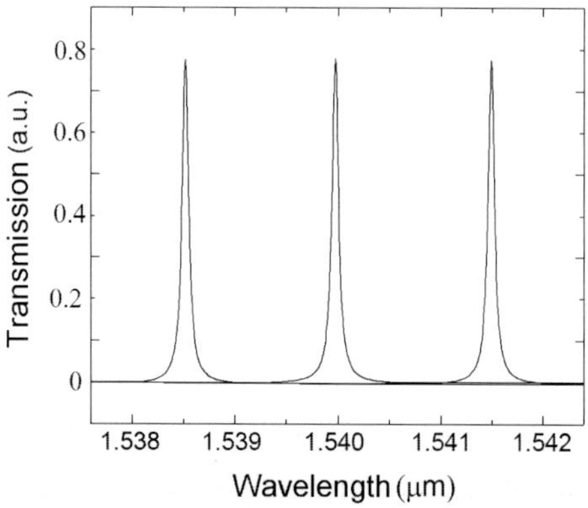

Fig. 5. Illustration of a typical optical spectrum of a microring resonator. The periodic peaks correspond to resonant modes of the cavity and depend on the optical path length of light inside the cavity.

2.3. *Slot waveguides*

In contrast to traditional waveguides in which light is guided by total internal reflection in a high index region surrounded by lower index materials, slot waveguides allow light to be guided in a nanometer-scale slot of low index material sandwiched between two strips of high index material.[39] Due to the large discontinuity of the electric field at high index contrast interfaces, light can be concentrated in the low index slot. A schematic of a slot waveguide and electric field distribution is shown in Fig. 6. The slot waveguide mode is theoretically lossless and the intensity of light propagating in the slot is an order of magnitude higher than that which can be achieved in conventional waveguides. A silicon slot waveguide in the form of a ring resonator has been demonstrated as a light emitting diode for which erbium incorporated into the silicon dioxide slot region emits light near 1.54 μm.[40] The large field confinement in the slot is also advantageous for increasing the sensitivity of optical sensors as small changes in refractive index of the slot due to the presence of gases, chemicals, or biomolecules can be more easily detected. Recently, multiple slot waveguides and multiple slot ring resonator devices were shown to have even stronger optical confinement in the low index slots than single slot structures.[41]

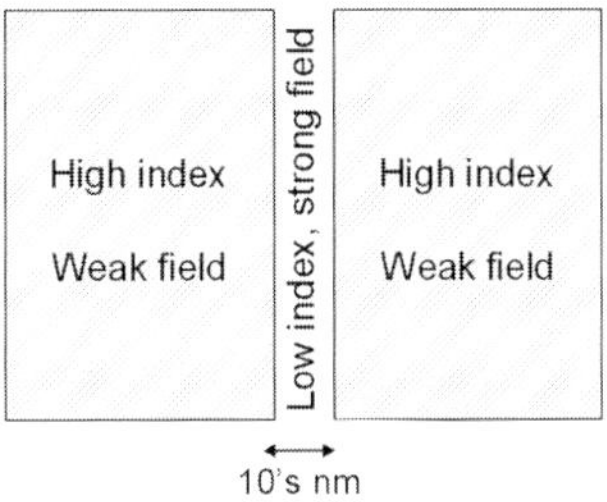

Fig. 6. Top view schematic of slot waveguide. Due to the field discontinuity and the high index contrast interface, light is confined and enhanced in the slot region.

3. Sensing Applications: Detection of Gases, Chemicals, DNA, Viruses, Proteins, and Cells

The most widespread biological application of silicon nanophotonic structures is for sensing. Well-established silicon processing techniques have allowed the fabrication of compact, high-Q factor devices that can be used to detect trace amounts of gases, chemicals, and biological material. The key to achieving high sensitivity is to increase the interaction between target species and locally enhanced electric fields. Porous silicon structures utilize an extremely large surface area to significantly increase the number of molecules in a given volume of the sensing medium compared to sensors based on planar substrates. Ring resonators possess very large optical path lengths without requiring a large footprint since light circulates in the resonator many times where it can interact multiple times with molecular species on the surface of the ring. Slot waveguides utilize nanosized regions of electric field enhancement and confinement to achieve low detection limits. For the detection of biomolecules, the size of the molecule detected is limited in the cases of nanoporous silicon and slot waveguide sensors. To achieve maximum sensitivity of detection, the biomolecules must bind inside the nanoscale pores and slots. Due to electrostatic forces, the pore and slot openings need to be much larger than the biomolecule size.[42] Ring resonators do not face this limitation as the biomolecules bind to a planar surface and are detected via the evanescent field escaping from the ring.

3.1. *Porous silicon structures for optical sensing applications*

Porous silicon-based photonic structures have been the most prolific sensors among silicon nanophotonic devices. Many of the initial demonstrations of porous silicon sensors were for the detection of gases and vapors, such as nitric oxide, TNT vapors, or vapor from organic solvents, using a single layer of porous silicon.[43-63] Porous silicon Bragg mirrors, rugate filters, and microcavities have also been used for the detection of gaseous species.[64-76] Due to its enormous surface area, porous silicon can absorb a large number of molecules, which affects the porous silicon dielectric constant, PL intensity, capacitance and conductivity. When organic vapor is introduced into the pores, air is displaced and capillary condensation partly fills the pores with organic liquid. Porous silicon gas sensors tend to be reversible[43] with substantially faster baseline recovery times (<30 seconds) than many other competing sensor platforms.[51] Optical detection methods of gaseous species using porous silicon sensors include measuring a shift in the reflectance spectra due to modification of the effective refractive index of the films (e.g., Fig. 3b), monitoring the PL intensity, which is quenched upon condensation of molecules in the pores, and using ellipsometry to determine effective refractive index change. The sensitivity of porous silicon gas and vapor sensors, which depends on the ambient conditions, is typically in the ppm range; however, detection limits as low as 1 ppb for TNT vapors,[50] 70 ppb for nitrogen dioxide[44] and 250 ppb for ethanol vapors[58] have been reported.

Porous silicon structures have also been used to determine the presence of liquids, such as organic solvents and pesticides, due to PL quenching, a bulk refractive index change that shifts the reflectance spectrum, or modification of waveguide scattering losses.[63,77-87] An interesting set of experiments was performed by Anderson *et al.* to compare the sensitivities of a porous silicon single layer, Bragg mirror, and microcavity to bulk refractive index changes.[88] Sucrose solutions of various concentrations were introduced into the pores in order to change the refractive index of the liquid in the pores from 1.3330 to 1.3344. While it was predicted that the microcavity would be as much as twice as sensitive to small refractive index changes, the experimental readout from the microcavity structure was unstable; the minimum detectable refractive index change, given for both the single layer and Bragg mirror, was 7×10^{-5}. DeLouise et al. experimentally demonstrated sensitivity on the order of 10^{-4} for porous silicon microcavities exposed to a variety of solvents.[89]

Unlike the detection of gaseous or liquid substances, the detection of biological species allows for highly selective optical sensing. The selectivity of the sensors is due to chemical functionalization of the pores that allows only a specific target species to bind to probe molecules immobilized in the pores. A major breakthrough in porous silicon sensing came in 1997 when Sailor and co-workers demonstrated the detection of 16-base pair DNA (detection limit ~ 9 fg/mm^2) and proteins (detection limit ~ 10 fM for Streptavidin binding to biotin) in porous silicon single layers.[90] It is interesting to note that instead of observing a Fabry-Perot fringe red shift due to the expected increase in refractive index upon biomolecule binding in the pores, a blue shift was observed at pM DNA concentration. This effect has been attributed to oxidation and corrosion of porous silicon resulting from interaction with negatively charged DNA, which reduces the porous silicon effective optical thickness.[91] Fringe red shifts are observed at higher biomolecule concentrations when the effective refractive index change due to biomolecule attachment is larger than the effective optical thickness decrease due to porous silicon oxidation and corrosion.

Several additional demonstrations of porous silicon biomolecular sensing have been performed, based on PL and reflectance changes, with single layer structures,[72,92-103] as well as rugate filters,[104] microcavities,[89,105-111] and waveguides.[112-114] De Stefano *et al.*[100] and Di Francia *et al.*[98] detected the presence of ~10 μM DNA (25-base pair and 16-base pair, respectively) in porous silicon single layer structures based on a shift of Fabry-Perot fringes and a change in PL intensity, respectively. Chan *et al.* demonstrated the recognition and binding of 1μM concentration of 24-base DNA to its complement, and ~200 fM concentration of bacteriophage lambda viral DNA to a 30-nucleotide complementary DNA sequence immobilized in a porous silicon microcavity.[105,106] The detection method for this experiment was monitoring a red shift of the cavity modes in the PL spectrum upon DNA binding. The reflectance properties of porous silicon microcavities have been monitored by Ouyang *et al.*[108] and DeLousie *et al.*[89] to detect 1 μM streptavidin attached to biotin probes and ~ 5 μM glutathione-S-transferase (GST), a 50 kDa enzyme, covalently attached on a glutaraldehyde-treated surface, respectively. Most recently, porous silicon waveguide sensors, of the type shown in Fig. 7, have been demonstrated for 50 μM DNA detection (24-base pair), with an anticipated detection limit in the nM range.[112-114] A prism is used to couple light into the waveguide and the intensity of reflected light as a function of the coupling angle is measured (Fig. 3h). Only light at a particular incident angle can be coupled into the waveguide.[112] DNA hybridization inside the porous silicon waveguide changes the refractive index of the waveguide and, hence, changes the angle at which light couples into the waveguide. Porous silicon waveguides combine the advantages of electric field localization, to produce a sharp resonance in the optical spectrum for low detection limits, and a thin active sensing layer, to achieve fast sensor response times with small analyte volumes.

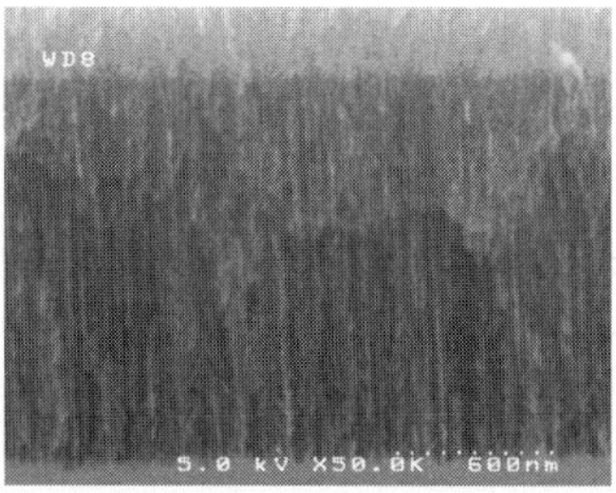

Fig. 7. SEM image of porous silicon waveguide consisting of a 310 nm, 56% porosity layer (n ~ 2.08) and a 1330 nm, 84% porosity layer (n ~ 1.41). Light is guided in the high refractive index, top porous silicon layer. Hence, the waveguide is most sensitive to detecting DNA hybridization in the thin top layer. (Image courtesy of G. Rong and S. M. Weiss, Vanderbilt University)

3.2. *Ring resonator sensor applications*

In 2001, theoretical papers by Blair and Chen,[115] and Boyd and Heebner,[116] introduced the idea of biosensing using ring resonator cavities, such as those illustrated in Fig. 4. Blair and Chen proposed the use of cylindrical microdisk cavities to achieve an order of magnitude increase in the sensitivity of biomolecule detection over traditional planar waveguides. Additionally, the required sample volume for the microdisks is several orders of magnitude smaller. In their design, Blair and Chen use fluorescently labeled biomolecules that are captured by a biospecific monolayer on top of the microdisk. The presence of the biomolecule is detected by monitoring the fluorescence signal from the label. Long photon lifetimes of high-Q whispering gallery modes increases the probability of the fluorophores absorbing photons, and consequently enhances the fluorescence emission for easier detection.

Boyd and Heebner theoretically demonstrated that microdisk resonators could be designed to detect as few as 100 dopamine molecules. This small detection limit can be achieved with a high finesse[†] cavity that allows light to interact many times with each biomolecule as the light recirculates within the disk. The detection scheme, in this case, is label-free (i.e., no fluorescent tag is required on the biomolecule to be detected). Biomolecules attached to the resonator surface absorb light. As a result, the transmission characteristics of light in the waveguide coupled to the disk resonator change. By simple extension, it is possible to design an array of microdisk resonators to detect trace amounts of several different biomolecules.

Over the past few years, there have been experimental realizations of ring resonator sensors. Ksendzov and Lin demonstrated sub-nM detection of avidin using a racetrack-shaped ring resonator made of Si_xN_y on SiO_2 functionalized with biotin.[117] The dimensions of the resonator, however, fail to qualify this device as a nanophotonic component. In order to minimize losses due to lateral coupling, the separation between the ring and waveguides was 1.6 μm and the radius of curvature of the ring was 2 mm. Silica resonators on silicon wafers have allowed demonstrations of nanophotonic biosensing. Armani and Vahala were able to distinguish the chemically similar species D_2O and H_2O at a level of parts per million per volume using planar arrays of ultra-high-Q

[†] The finesse of a cavity, determined by its losses, is defined as its free spectral range (spacing of resonant modes) divided by the full-width-at-half-maximum of its resonances.

($Q > 10^7$) silica microtoroids.[118] Light was evanescently coupled to the microtoriods via a tapered fiber. Since the absorption of H_2O is larger than that of D_2O and changes in absorption modify the intrinsic quality factor of the microtoroid cavities, the presence of trace amounts of D_2O in H_2O can be detected. Most recently, Armani, Vahala, and co-workers demonstrated label-free, single-molecule detection of interleukin-2 with their ultra-high-Q microtoroids ($Q \sim 10^8$).[119] It is proposed that light circulating in the resonator with a high intensity due to the ultra-high cavity-Q locally heats molecules attached to the toroid, which causes a red shift of the resonant wavelength through the thermo-optic effect. Yalcin *et al.* used vertically coupled microring resonators ($Q \sim 10^4$) made of Hydex (glass-based material compatible with standard silicon processing techniques) to detect avidin-biotin binding at μM concentrations based on a shift of the resonant mode wavelength.[120] For detection of liquids, their system is capable of detecting refractive index changes on the order of 10^{-5}. Very recently, Ramachandran *et al.* used similar vertically coupled ring resonators made of Hydex to detect 500 nM of 25-base pair DNA and 10^5 CFU/ml of *E. coli* O157:H7 bacterial cells.[121] It was further shown, for different concentrations of IgG protein, that target binding on an antibody-modified microring was dependent on the mass-transport rate.

3.3. *Slot waveguide sensor applications*

The use of slot waveguide structures for chemical and biological sensing applications is an emerging field. In 2006, Bernini *et al.* performed a theoretical analysis of the fluorescence excitation and collection efficiencies for evanescent wave-based planar waveguide sensors and slot waveguide sensors employing fluorescent labels for biomolecule detection.[122] It was found that, under ideal conditions, the excitation efficiency of fluorophores in slot waveguides was up to 40% greater than the efficiency in slab waveguides due to the stronger field confinement in the slot. The slot waveguide also showed improved collection efficiency of fluorescence from a thin layer of fluorophores attached to the surface of the waveguide. In 2007, a detailed theoretical study of silicon-on-insulator slot waveguides and slot rib‡ waveguides, including fabrication tolerances, was reported for compact chemical and biological sensing applications.[123] Through a series of full-vectorial 2D FEM calculations, Dell'Olio and Passaro show that the slot waveguides can achieve improved sensitivity compared to silicon-on-insulator wire waveguides[124] and other nanometer guiding structures.

An experimental demonstration of an integrated chemical sensor based on a slot-waveguide microring resonator was reported recently by Barrios *et al.*[125] Using standard silicon processing techniques, a silicon nitride slot-waveguide microring resonator was fabricated on a silicon-on-insulator wafer. A silicon dioxide top cladding layer with a window opened for the ring resonator was also deposited. The slot width in the microring was 200 nm. By dropping liquids with varying refractive indices onto the slot-waveguide microring and monitoring the transmission spectra, a detection limit on the order of 10^{-4} refractive index units was shown. Simulations suggested that improved sensor performance could be demonstrated by achieving a more complete filling of the slot.

4. Drug Delivery, Molecular Separation, and Tissue Engineering

Drug delivery, molecular separation, and bone scaffolding represent emerging biological application areas for nanoscale silicon. Nanostructured silicon holds great promise for targeted drug delivery and monitoring.[126-138] Canham and co-workers showed that highly porous silicon is biocompatible and can dissolve in body fluids through conversion to non-toxic silicic acid.[126,128] Thus, drug molecules loaded into nanopores are released over time as the porous silicon dissolves. The rate of dissolution can be tuned by adjusting the porous silicon porosity and pore size. One company, pSiVida, is actively pursuing the development of porous silicon, termed BioSilicon™, for drug delivery applications. Using standard silicon processing techniques, implantable silicon biocapsules containing a high density of uniform nanoscale pores

‡ Slot rib waveguides incorporate a thin silicon planar layer placed below the slot structure.

have also been fabricated for the sustained release of drugs.[131,132,139] For example, using 9 μm thick membranes with nanopore densities below 10^7 cm^{-2}, glucose and insulin were able to diffuse through the nanopores while the protein IgG was blocked, illustrating the controlled release of small molecules while serum molecules are prevented from entering the biocapsule.[139]

Micromachined nanopores can additionally be used for molecular separations and are well-suited for low sample volume, lab-on-a-chip applications. Elwenspoek and co-workers theorized and demonstrated the fabrication of arrays of 25 nm pores in a 10 nm thick silicon nitride film suspended on a silicon matrix, showing very high gas permeabilities through the ultrathin structure.[140] Unfortunately, fabrication limitations prevented the testing of this structure in fluid environments. More recently, ultrathin (5-25 nm), on-chip porous nanocrystalline silicon membranes, of the type shown in Fig. 8, have been developed using standard silicon microelectronic processing techniques.[141] Because the transport rate across a barrier is proportional to the barrier thickness,[140] these ultrathin porous nanocrystalline silicon membranes (~1000 times thinner than commercial polymeric membranes) provide highly efficient removal of cells, cellular debris, and viruses. By tuning the pore diameter between 5 and 100 nm, separation of moderate sized proteins or nucleic acids based on molecular size and/or charge is possible. The porous nanocrystalline silicon membranes have separated two common blood proteins of different molecular weight, IgG and BSA (bovine serum albumin), that are too similar in molecular weight to be efficiently separated with existing membrane-based techniques.[141]

Over the past several years, in addition to potential drug delivery and molecular separation applications, nanostructured silicon has emerged as a viable candidate for tissue engineering.[142] Existing porous biodegradable polymeric systems and porous ceramic systems suffer from poor control over pore size distribution and, in some cases, mechanical deficiencies. Hydroxyapatite (HA) nucleation, a precursor for bone growth, has been demonstrated on porous silicon, suggesting bioactivity for the material.[143,144] By adjusting its surface chemistry and morphology, porous silicon can be appropriately tailored to interface with living cells.[145-149] A recent study found that porous silicon with pore diameters of 1 μm provided the most favorable balance between biodegradability and stability for primary osteoblast growth and mineralization.[150]

Fig. 8. TEM image of a 15-nm thick porous nanocrystalline silicon membrane for molecular separations. Pores are the white objects and the black structures are silicon nanocrystals with an orientation that diffracts the electron beam. The membrane is pure silicon. This morphology is formed during a thermal annealing step that crystallizes an amorphous silicon film. (Image courtesy of C. C. Striemer, University of Rochester).

5. Conclusions and Outlook

In this chapter, the utility of nanostructured silicon was examined for applications beyond light guiding and emission for optical communications and photonic circuits. Over the past decade, nanostructured silicon has been shown to have several advantages for biological applications. Due to the extensive fabrication capabilities for silicon-based devices, it is possible to fabricate highly efficient silicon-based optical sensors for the detection of gases, chemicals, and biomolecules. Sensing applications of porous silicon structures, silicon ring resonators, and silicon slot waveguides were discussed. The key to the enhanced sensitivity of these structures is an increased interaction between the electric field and the molecules to be detected. Two factors that may contribute to this interaction are an increased surface area that accommodates a higher density of molecules and a locally enhanced electric field intensity. Porous silicon structures have the distinct advantage of a large surface area, which allows for the attachment of more biomolecules than is possible on planar surfaces. Resonant porous silicon microcavities and waveguides provide the dual advantage of a large surface area and a locally enhanced electric field. With their high finesse cavities that allow light to circulate in the resonator and interact with molecular species many times, ring resonators enable highly sensitive, compact sensor devices. The Q-factors of ring resonators are well in excess of what can be achieved with porous silicon resonant structures. Slot waveguides also have been demonstrated with enhanced field intensities in nanoscale regions. As the experimental conditions are optimized to achieve complete slot filling, slot waveguides will be very competitive for chemical detection on an integrated silicon platform.

Beyond sensing applications, nanostructured silicon structures have also emerged as viable candidates for drug delivery, molecular separations, and tissue engineering. These silicon structures offer significant morphological, chemical, and integration capabilities beyond their organic counterparts. In general, the application of silicon technology to biological challenges is an area of increasing research attention. It is likely that the next decade will bring continued breakthroughs and the emergence of more biotechnology companies developing nanostructured silicon for biological applications.

Acknowledgments

This work was supported in part by the National Science Foundation (ECCS-0722143). Contributions from Guoguang Rong and Christopher Striemer are gratefully acknowledged.

References

1. P. Bergveld, *Sens. Act.* **B88**, 1 (2003).
2. D. S. Kim, J. E. Park, J. K. Shin, P. K. Kim, G. Lim and S. Shoji, *Sens. Act.* **B117**, 488 (2006).
3. A. K. Wanekaya, W. Chen, N. V. Myung and A. Mulchandani, *Electroanal.* **18**, 533 (2006).
4. P. Estrela and P. Migliorato, *J. Mater. Chem.* **17**, 219 (2007).
5. Y. Cui, Q. Q. Wei, H. K. Park and C. M. Lieber, *Science* **293**, 1289 (2001).
6. Z. Li, Y. Chen, X. Li, T. I. Kamins, K. Nauka and R. S. Williams, *Nano Lett.* **4**, 245 (2004).

7. M. Archer, M. Christophersen and P. M. Fauchet, *Biomed. Microdevices* **6**, 203 (2004).
8. B. Ilic, D. Czaplewski, H. G. Craighead, P. Neuzil, C. Campagnolo and C. Batt, *Appl. Phys. Lett.* **77**, 450 (2000).
9. M. K. Baller, H. P. Lang, J. Fritz, C. Gerber, J. K. Gimzewski, U. Drechsler, H. Rothuizen, M. Despont, P. Vettiger, F. M. Battiston, J. P. Ramseyer, P. Fornaro, E. Meyer and H. J. Guntherodt, *Ultramicrosc.* **82**, 1 (2000).
10. J. Fritz, M. K. Baller, H. P. Lang, H. Rothuizen, P. Vettiger, E. Meyer, H. J. Guntherodt, C. Gerber and J. K. Gimzewski, *Science* **288**, 316 (2000).
11. R. Raiteri, M. Grattarola, H. J. Butt and P. Skladal, *Sens. Act.* **B79**, 115 (2001).
12. G. H. Wu, H. F. Ji, K. Hansen, T. Thundat, R. Datar, R. Cote, M. F. Hagan, A. K. Chakraborty and A. Majumdar, *Proc. Nat. Acad. Sci. U.S.A.* **98**, 1560 (2001).
13. F. M. Battiston, J. P. Ramseyer, H. P. Lang, M. K. Baller, C. Gerber, J. K. Gimzewski, E. Meyer and H. J. Guntherodt, *Sens. Act.* **B77**, 122 (2001).
14. C. A. Savran, S. M. Knudsen, A. D. Ellington and S. R. Manalis, *Anal. Chem.* **76**, 3194 (2004).
15. P. S. Waggoner and H. G. Craighead, *Lab Chip* **7**, 1238 (2007).
16. A. Uhlir, *Bell Syst. Tech. J.* **35**, 333 (1956).
17. L. T. Canham, *Appl. Phys. Lett.* **57**, 1046 (1990).
18. L. Canham, Ed., *Properties of Porous Silicon* (INSPEC, London, UK, 1997).
19. V. Lehmann, Electrochemistry of Silicon: Instrumentation, Science, Materials and Applications (Wiley-VCH, Weinheim, Germany, 2002).
20. G. Amato, C. Delerue and H.-J. v. Bardeleben, Eds., *Structural and Optical Properties of Porous Silicon Nanostructures* (Gordon and Breach Science Publishers, Amsterdam, 1997).
21. R. L. Smith and S. D. Collins, *J. Appl. Phys.* **71**, R1 (1992).
22. D. J. Lockwood, *Solid State Comm.* **92**, 101 (1994).
23. A. G. Cullis, L. T. Canham and P. D. J. Calcott, *J. Appl. Phys.* **82**, 909 (1997).
24. M. P. Stewart and J. M. Buriak, *Adv. Mater.* **12**, 859 (2000).
25. W. Theiss, *Surf. Sci. Rep.* **29**, 95 (1997).
26. J. E. Lugo, J. A. delRio and J. Taguena-Martinez, *J. Appl. Phys.* **81**, 1923 (1997).
27. J. J. Saarinen, S. M. Weiss, P. M. Fauchet and J. E. Sipe, *J. App. Phys.*, in press.
28. G. Vincent, *Appl. Phys. Lett.* **64**, 2367 (1994).
29. M. G. Berger, R. Arens-Fischer, M. Thonissen, M. Kruger, S. Billat, H. Luth, S. Hilbrich, W. Theiss and P. Grosse, *Thin Solid Films* **297**, 237 (1997).
30. L. Pavesi, *Rivista Del Nuovo Cimento* **20**, 1 (1997).
31. A. Loni, L. T. Canham, M. G. Berger, R. ArensFischer, H. Munder, H. Luth, H. F. Arrand and T. M. Benson, *Thin Solid Films* **276**, 143 (1996).
32. B. G. Bovard, *Appl. Opt.* **32**, 5427 (1993).
33. E. Lorenzo, C. J. Oton, N. E. Capuj, M. Ghulinyan, D. Navarro-Urrios, Z. Gaburro and L. Pavesi, *Appl. Opt.* **44**, 5415 (2005).
34. M. Soltani, S. Yegnanarayanan and A. Adibi, *Opt. Express* **15**, 4694 (2007).

35. H. S. Rong, Y. H. Kuo, S. B. Xu, A. S. Liu, R. Jones and M. Paniccia, *Opt. Express* **14**, 6705 (2006).
36. B. E. Little, J. S. Foresi, G. Steinmeyer, E. R. Thoen, S. T. Chu, H. A. Haus, E. P. Ippen, L. C. Kimerling and W. Greene, *IEEE Photon. Techonl. Lett.* **10**, 549 (1998).
37. D. X. Dai and S. L. He, *Opt. Comm.* **279**, 89 (2007).
38. A. Yariv, *Electron. Lett.* **36**, 321 (2000).
39. V. R. Almeida, Q. F. Xu, C. A. Barrios and M. Lipson, *Opt. Lett.* **29**, 1209 (2004).
40. C. A. Barrios and M. Lipson, *Opt. Express* **13**, 10092 (2005).
41. R. Sun, P. Dong, N.-n. Feng, C.-y. Hong, J. Michel, M. Lipson and L. Kimerling, *Opt. Express* **15**, 17967 (2007).
42. G. Rong and S. M. Weiss, *Proc. of SPIE* **6769**, 676909 (2007).
43. J. M. Lauerhaas, G. M. Credo, J. L. Heinrich and M. J. Sailor, *J. Am. Chem. Soc.* **114**, 1911 (1992).
44. J. Harper and M. J. Sailor, *Anal. Chem.* **68**, 3713 (1996).
45. R. B. Bjorklund, S. Zangooie and H. Arwin, *Appl. Phys. Lett.* **69**, 3001 (1996).
46. M. T. Kelly, J. K. M. Chun and A. B. Bocarsly, *Nature* **382**, 214 (1996).
47. S. Zangooie, R. Bjorklund and H. Arwin, *Sens. Act.* **B43**, 168 (1997).
48. S. E. Letant and M. J. Sailor, *Adv. Mater.* **12**, 355 (2000).
49. J. Gao, T. Gao and M. J. Sailor, *Appl. Phys. Lett.* **77**, 901 (2000).
50. S. Content, W. C. Trogler and M. J. Sailor, *Chem. Eur. J.* **6**, 2205 (2000).
51. S. E. Letant, S. Content, T. T. Tan, F. Zenhausern and M. J. Sailor, *Sens. Act.* **B69**, 193 (2000).
52. H. Sohn, S. Letant, M. J. Sailor and W. C. Trogler, *J. Am. Chem. Soc.* **122**, 5399 (2000).
53. G. Di Francia, V. La Ferrara, L. Quercia and G. Faglia, *J. Porous Mater.* **7**, 287 (2000).
54. L. Quercia, F. Cerullo, V. La Ferrara, G. Di Francia, C. Baratto and G. Faglia, *Phys. Stat. Sol.* **A182**, 473 (2000).
55. J. Dian, T. Holec, I. Jelinek, J. Jindrich, J. Valenta and I. Pelant, *Phys. Stat. Sol.* **A182**, 485 (2000).
56. E. Gross, D. Kovalev, N. Kunzner, V. Y. Timoshenko, J. Diener and F. Koch, *J. Appl. Phys.* **90**, 3529 (2001).
57. S. E. Letant and M. J. Sailor, *Adv. Mater.* **13**, 335 (2001).
58. J. Gao, T. Gao, Y. Y. Li and M. J. Sailor, *Langmuir* **18**, 2229 (2002).
59. G. L. Wang and H. Arwin, *Sens. Act.* **B85**, 95 (2002).
60. B. H. O, R. Liu, Y. Y. Li, M. J. Sailor and Y. Fainman, *IEEE Photon. Techonl. Lett.* **15**, 834 (2003).
61. M. Rocchia, E. Garrone, F. Geobaldo, L. Boarino and M. J. Sailor, *Phys. Stat. Sol.* **A197**, 365 (2003).
62. H. H. Lin, T. Gao, J. Fantini and M. J. Sailor, *Langmuir* **20**, 5104 (2004).

63. T. Chvojka, V. Vrkoslav, I. Jelinek, J. Jindrich, M. Lorenc and J. Dian, *Sens. Act.* **B100**, 246 (2004).
64. P. A. Snow, E. K. Squire, P. S. J. Russell and L. T. Canham, *J. Appl. Phys.* **86**, 1781 (1999).
65. S. Zangooie, R. Jansson and H. Arwin, *J. Appl. Phys.* **86**, 850 (1999).
66. P. Allcock and P. A. Snow, *J. Appl. Phys.* **90**, 5052 (2001).
67. Z. Gaburro, N. Daldosso, L. Pavesi, G. Faglia, C. Baratto and G. Sberveglieri, *Appl. Phys. Lett.* **78**, 3744 (2001).
68. T. A. Schmedake, F. Cunin, J. R. Link and M. J. Sailor, *Adv. Mater.* **14**, 1270 (2002).
69. L. De Stefano, L. Moretti, I. Rendina and A. M. Rossi, *Sensors and Actuators a-Physical* **104**, 179 (2003).
70. L. De Stefano, I. Rendina, L. Moretti and A. M. Rossi, *Mat. Sci. Eng.* **B100**, 271 (2003).
71. L. De Stefano, N. Rendina, L. Moretti, S. Tundo and A. M. Rossi, *Appl. Opt.* **43**, 167 (2004).
72. L. De Stefano, L. Moretti, A. Lamberti, O. Longo, M. Rocchia, A. M. Rossi, P. Arcari and I. Rendina, *IEEE Trans. Nanotechnol.* **3**, 49 (2004).
73. L. De Stefano, L. Moretti, I. Rendina and A. M. Rossi, *Phys. Stat. Sol.* **A201**, 1011 (2004).
74. M. S. Salem, M. J. Sailor, F. A. Harraz, T. Sakka and Y. H. Ogata, *J. Appl. Phys.* **100** (2006).
75. I. A. Levitsky, W. B. Euler, N. Tokranova and A. Rose, *Appl. Phys. Lett.* **90** (2007).
76. M. S. Salem, M. J. Sailor, T. Sakka and Y. H. Ogata, *J. Appl. Phys.* **101** (2007).
77. H. F. Arrand, T. M. Benson, A. Loni, R. Arens-Fischer, M. G. Krueger, M. Thoenissen, H. Lueth, S. Kershaw and N. N. Vorozov, *J. Lumin.* **80**, 119 (1998).
78. H. F. Arrand, T. M. Benson, A. Loni, R. Arens-Fischer, M. Kruger, M. Thonissen, H. Luth and S. Kershaw, *IEEE Photon. Techonl. Lett.* **10**, 1467 (1998).
79. S. Fellah, F. Ozanam, N. Gabouze and J. N. Chazalviel, *Phys. Stat. Sol.* **A182**, 367 (2000).
80. V. Mulloni, Z. Gaburro and L. Pavesi, *Phys. Stat. Sol.* **A182**, 479 (2000).
81. V. Mulloni and L. Pavesi, *Appl. Phys. Lett.* **76**, 2523 (2000).
82. H. K. Min, H. S. Yang and S. M. Cho, *Sens. Act.* **B67**, 199 (2000).
83. C. Pacholski, M. Sartor, M. J. Sailor, F. Cunin and G. M. Miskelly, *J. Am. Chem. Soc.* **127**, 11636 (2005).
84. L. Rotiroti, L. De Stefano, N. Rendina, L. Moretti, A. M. Rossi and A. Piccolo, *Biosens. Bioelectron.* **20**, 2136 (2005).
85. L. De Stefano, L. Moretti, I. Rendina and L. Rotiroti, *Sens. Act.* **B111**, 522 (2005).
86. M. Rocchia, A. M. Rossi and G. Zeppa, *Sens. Act.* **B123**, 89 (2007).

87. J. Chapron, S. A. Alekseev, V. Lysenko, V. N. Zaitsev and D. Barbier, *Sens. Act.* **B120**, 706 (2007).
88. M. A. Anderson, A. Tinsley-Bown, P. Allcock, E. A. Perkins, P. Snow, M. Hollings, R. G. Smith, C. Reeves, D. J. Squirrell, S. Nicklin and T. I. Cox, *Phys. Stat. Sol.* **A197**, 528 (2003).
89. L. A. DeLouise, P. M. Kou and B. L. Miller, *Anal. Chem.* **77**, 3222 (2005).
90. V. S. Y. Lin, K. Motesharei, K. P. S. Dancil, M. J. Sailor and M. R. Ghadiri, *Science* **278**, 840 (1997).
91. C. Steinem, A. Janshoff, V. S. Y. Lin, N. H. Volcker and M. R. Ghadiri, *Tetrahedron* **60**, 11259 (2004).
92. D. van Noort, S. Welin-Klintstrom, H. Arwin, S. Zangooie, I. Lundstrom and C. F. Mandenius, *Biosens. Bioelectron.* **13**, 439 (1998).
93. S. Zangooie, R. Bjorklund and H. Arwin, *Thin Solid Films* **313**, 825 (1998).
94. V. M. Starodub, L. L. Fedorenko, A. P. Sisetskiy and N. F. Starodub, *Sens. Act.* **B58**, 409 (1999).
95. K. P. S. Dancil, D. P. Greiner and M. J. Sailor, *J. Am. Chem. Soc.* **121**, 7925 (1999).
96. H. Arwin, M. Gavutis, J. Gustafsson, M. Schultzberg, S. Zangooie and P. Tengvall, *Phys. Stat. Sol.* **A182**, 515 (2000).
97. A. M. Tinsley-Bown, L. T. Canham, M. Hollings, M. H. Anderson, C. L. Reeves, T. I. Cox, S. Nicklin, D. J. Squirrell, E. Perkins, A. Hutchinson, M. J. Sailor and A. Wun, *Phys. Stat. Sol.* **A182**, 547 (2000).
98. G. Di Francia, V. La Ferrara, S. Manzo and S. Chiavarini, *Biosens. Bioelectron.* **21**, 661 (2005).
99. L. De Stefano, L. Rotiroti, I. Rendina, L. Moretti, V. Scognamiglio, M. Rossi and S. D'Auria, *Biosens. Bioelectron.* **21**, 1664 (2006).
100. L. De Stefano, P. Arcari, A. Lamberti, C. Sanges, L. Rotiroti, I. Rea and I. Rendina, *Sensors* **7**, 214 (2007).
101. A. M. Rossi, L. Wang, V. Reipa and T. E. Murphy, *Biosens. Bioelectron.* **23**, 741 (2007).
102. M. P. Schwartz, C. Yu, S. D. Alvarez, B. Migliori, D. Godin, L. Chao and M. J. Sailor, *Phys. Stat. Sol.* **A204**, 1444 (2007).
103. M. P. Schwartz, S. D. Alvarez and M. J. Sailor, *Anal. Chem.* **79**, 327 (2007).
104. F. Cunin, T. A. Schmedake, J. R. Link, Y. Y. Li, J. Koh, S. N. Bhatia and M. J. Sailor, *Nature Mater.* **1**, 39 (2002).
105. S. Chan, P. M. Fauchet, Y. Li, L. J. Rothberg and B. L. Miller, *Phys. Stat. Sol.* **A182**, 541 (2000).
106. S. Chan, Y. Li, L. J. Rothberg, B. L. Miller and P. M. Fauchet, *Mat. Sci. Eng.* **C15**, 277 (2001).
107. S. Chan, S. R. Horner, P. M. Fauchet and B. L. Miller, *J. Am. Chem. Soc.* **123**, 11797 (2001).

108. H. Ouyang, M. Christophersen, R. Viard, B. L. Miller and P. M. Fauchet, *Adv. Funct. Mater.* **15**, 1851 (2005).
109. L. De Stefano, I. Rea, I. Rendina, L. Rotiroti, M. Rossi and S. D'Auria, *Phys. Stat. Sol.* **A203**, 886 (2006).
110. M. Lee and P. M. Fauchet, *Opt. Express* **15**, 4530 (2007).
111. H. Ouyang, L. A. DeLouise, B. L. Miller and P. M. Fauchet, *Anal. Chem.* **79**, 1502 (2007).
112. J. J. Saarinen, S. M. Weiss, P. M. Fauchet and J. E. Sipe, *Opt. Express* **13**, 3754 (2005).
113. G. Rong and S. M. Weiss, *Proc. of SPIE* **6477**, 647717 (2007).
114. G. Rong and S. M. Weiss, *Biosens. Bioelectron.* **23**, 1572 (2008).
115. S. Blair and Y. Chen, *Appl. Opt.* **40**, 570 (2001).
116. R. W. Boyd and J. E. Heebner, *Appl. Opt.* **40**, 5742 (2001).
117. A. Ksendzov and Y. Lin, *Opt. Lett.* **30**, 3344 (2005).
118. A. M. Armani and K. J. Vahala, *Opt. Lett.* **31**, 1896 (2006).
119. A. M. Armani, R. P. Kulkarni, S. E. Fraser, R. C. Flagan and K. J. Vahala, *Science* **317**, 783 (2007).
120. A. Yalcin, K. C. Popat, J. C. Aldridge, T. A. Desai, J. Hryniewicz, N. Chbouki, B. E. Little, O. King, V. Van, S. Chu, D. Gill, M. Anthes-Washburn and M. S. Unlu, *IEEE J. Sel. Top. Quantum Electron.* **12**, 148 (2006).
121. A. Ramachandran, S. Wang, J. Clarke, S. J. Ja, D. Goad, L. Wald, E. M. Flood, E. Knobbe, J. V. Hryniewicz, S. T. Chu, D. Gill, W. Chen, O. King and B. E. Little, *Biosens. Bioelectron.* **23**, 939 (2008).
122. R. Bernini, N. Cennamo, A. Minardo and L. Zeni, *IEEE Sens. J.* **6**, 1218 (2006).
123. F. Dell'Olio and V. M. N. Passaro, *Opt. Express* **15**, 4977 (2007).
124. A. Densmore, D. X. Xu, P. Waldron, S. Janz, P. Cheben, J. Lapointe, A. Delage, B. Lamontagne, J. H. Schmid and E. Post, *IEEE Photon. Techonl. Lett.* **18**, 2520 (2006).
125. C. A. Barrios, K. B. Gylfason, B. Sanchez, A. Griol, H. Sohlstrom, M. Holgado and R. Casquel, *Opt. Lett.* **32**, 3080 (2007).
126. L. T. Canham, *Adv. Mater.* **7**, 1033 (1995).
127. J. T. Santini, M. J. Cima and R. Langer, *Nature* **397**, 335 (1999).
128. L. T. Canham, M. P. Stewart, J. M. Buriak, C. L. Reeves, M. Anderson, E. K. Squire, P. Allcock and P. A. Snow, *Phys. Stat. Sol.* **A182**, 521 (2000).
129. J. L. Coffer, J. L. Montchamp, J. B. Aimone and R. P. Weis, *Phys. Stat. Sol.* **A197**, 336 (2003).
130. Y. Y. Li, F. Cunin, J. R. Link, T. Gao, R. E. Betts, S. H. Reiver, V. Chin, S. N. Bhatia and M. J. Sailor, *Science* **299**, 2045 (2003).
131. S. L. Tao and T. A. Desai, *Adv. Drug Delivery Rev.* **55**, 315 (2003).
132. S. E. Letant, T. W. van Buuren and L. J. Terminello, *Nano Lett.* **4**, 1705 (2004).
133. E. J. Anglin, M. P. Schwartz, V. P. Ng, L. A. Perelman and M. J. Sailor, *Langmuir* **20**, 11264 (2004).

134. J. Salonen, L. Laitinen, A. M. Kaukonen, J. Tuura, M. Bjorkqvist, T. Heikkila, K. Vaha-Heikkila, J. Hirvonen and V. P. Lehto, *J. Controlled Release* **108**, 362 (2005).
135. L. Vaccari, D. Canton, N. Zaffaroni, R. Villa, M. Tormen and E. di Fabrizio, *Microelectron. Eng.* **83**, 1598 (2006).
136. I. Batra, J. L. Coffer and L. T. Canham, *Biomed. Microdevices* **8**, 93 (2006).
137. T. Limnell, J. Riikonen, J. Salonen, A. M. Kaukonen, L. Laitinen, J. Hirvonen and V. P. Lehto, *Int. J. Pharm.* **343**, 141 (2007).
138. C. A. Prestidge, T. J. Barnes, A. Mierczynska-Vasilev, W. Skinner, F. Peddie and C. Barnett, *Phys. Stat. Sol.* **A204**, 3361 (2007).
139. T. A. Desai, D. Hansford and M. Ferrari, *J. Membrane Sci.* **159**, 221 (1999).
140. H. D. Tong, H. V. Jansen, V. J. Gadgil, C. G. Bostan, E. Berenschot, C. J. M. van Rijn and M. Elwenspoek, *Nano Lett.* **4**, 283 (2004).
141. C. C. Striemer, T. R. Gaborski, J. L. McGrath and P. M. Fauchet, *Nature* **445**, 749 (2007).
142. T. A. Desai, *Med. Eng. Phys.* **22**, 595 (2000).
143. L. T. Canham, C. L. Reeves, A. Loni, M. R. Houlton, J. P. Newey, A. J. Simons and T. I. Cox, *Thin Solid Films* **297**, 304 (1997).
144. V. Chin, B. E. Collins, M. J. Sailor and S. N. Bhatia, *Adv. Mater.* **13**, 1877 (2001).
145. S. C. Bayliss, R. Heald, D. I. Fletcher and L. D. Buckberry, *Adv. Mater.* **11**, 318 (1999).
146. A. H. Mayne, S. C. Bayliss, P. Barr, M. Tobin and L. D. Buckberry, *Phys. Stat. Sol.* **A182**, 505 (2000).
147. J. L. Coffer, M. A. Whitehead, D. K. Nagesha, P. Mukherjee, G. Akkaraju, M. Totolici, R. S. Saffie and L. T. Canham, *Phys. Stat. Sol.* **A202**, 1451 (2005).
148. S. P. Low, K. A. Williams, L. T. Canham and N. H. Voelcker, *Biomater.* **27**, 4538 (2006).
149. A. V. Sapelkin, S. C. Bayliss, B. Unal and A. Charalambou, *Biomater.* **27**, 842 (2006).
150. W. Sun, J. E. Puzas, T. J. Sheu, X. Liu and P. M. Fauchet, *Adv. Mater.* **19**, 921 (2007).

INDEX